Effects of the COVID-19 Pandemic on the Use and Perception of Urban Green Space

Effects of the COVID-19 Pandemic on the Use and Perception of Urban Green Space

Editors

Francesca Ugolini
David Pearlmutter

Basel • Beijing • Wuhan • Barcelona • Belgrade • Novi Sad • Cluj • Manchester

Editors
Francesca Ugolini
National Research Council of Italy
Sesto Fiorentino
Italy

David Pearlmutter
Ben-Gurion University of the Negev
Beer-Sheva
Israel

Editorial Office
MDPI
St. Alban-Anlage 66
4052 Basel, Switzerland

This is a reprint of articles from the Special Issue published online in the open access journal *Land* (ISSN 2073-445X) (available at: https://www.mdpi.com/journal/land/special_issues/pandemic_ugs).

For citation purposes, cite each article independently as indicated on the article page online and as indicated below:

Lastname, A.A.; Lastname, B.B. Article Title. *Journal Name* **Year**, *Volume Number*, Page Range.

ISBN 978-3-7258-0095-7 (Hbk)
ISBN 978-3-7258-0096-4 (PDF)
doi.org/10.3390/books978-3-7258-0096-4

Cover image courtesy of David Pearlmutter

Contents

About the Editors

Francesca Ugolini

Francesca Ugolini (Ph.D.) is a researcher at the Institute of BioEconomy of the National Research Council of Italy. She graduated in forestry and environmental sciences and her work focuses on the growth performance of plants and their functions in relation to growth conditions. Her expertise on plant ecophysiology has been applied in the field of agriculture, and also in urban regeneration projects aimed at promoting sustainable urban planning and ecosystem service evaluation. The services and benefits that green spaces provide to citizens are among her major interests, and she has investigated how the planning and management of urban green spaces can be improved via collaboration between different actors. In addition to this, she has contributed to the scientific literature in understanding the effects of drastic movement and accessibility restrictions in green spaces on citizens during the COVID-19 pandemic. Bringing together competencies in agriculture and green space management is her main career objective, as well as recognizing the importance of urban agriculture in terms of environmental, productive and social benefits.

David Pearlmutter

David Pearlmutter (Ph.D.) is an architect and full professor at the Department of Environmental, Geoinformatic and Urban Planning Sciences of Ben-Gurion University of the Negev in Israel. His work focuses on bioclimatic architecture and urban design for improved thermal comfort and energy efficiency in the built environment. This includes observational and modeling studies of microclimatic stress and pedestrian behavior, particularly in urban green spaces, as well as the development and life-cycle assessment of biocomposite building materials, which can contribute to carbon neutrality and environmental health. He has taken a leading role in several international initiatives aimed at promoting urban green infrastructure as a strategic asset for future cities.

 *land*

Article

Attitudes and Behaviors toward the Use of Public and Private Green Space during the COVID-19 Pandemic in Iran

Mohammad Reza Khalilnezhad [1,*,†], **Francesca Ugolini** [2,†] **and Luciano Massetti** [2]

[1] Faculty of Arts, University of Birjand, Birjand 9718854987, Iran
[2] Institute of Bioeconomy–National Research Council, Via Madonna del Piano 10, Sesto Fiorentino, 50019 Firenze, Italy; francesca.ugolini@ibe.cnr.it (F.U.); luciano.massetti@ibe.cnr.it (L.M.)
*   Correspondence: smkhalilnejad@birjand.ac.ir
†   Co-first author.

Abstract: This paper reports the results of an exploratory study carried out in Birjand, Iran, during the first year of the COVID-19 pandemic. The aim of the study was to explore the behavioral change in the use and the motivation to visit a green space (public or private) during the pandemic as compared to the pre-pandemic period, the effect of green spaces (private and public) on users' feelings, the relations between the extent to which the access to green spaces was missed, and characteristics of respondents and the place they live. A survey was carried out through an online questionnaire in winter 2020 and about 400 responses were collected. The results showed a decrease in visitation of public green spaces during the pandemic, and higher visitation of private green spaces such as gardens or courtyards by those with access. In addition, both public and private green spaces enhance positive feelings and decrease the negative ones. Respondents missed access to green spaces, especially when their visitation before the pandemic was high, and women missed them more than men. Therefore, private green spaces might represent an opportunity for psychological respite in time of a pandemic, but also for socialization. The study reports respondents' useful suggestions for urban landscape planning for the city of Birjand that might also be useful for other cities in dry lands; improving the quality of green spaces beyond the quantity may play a role in enhancing the connection to nature in the time of a pandemic, with positive effects on mental health, and this can also can improve recreation opportunities and reduce inequalities.

Keywords: feelings; green areas access; private greenery; public greenery; landscape planning; urban planning

Citation: Khalilnezhad, M.R.; Ugolini, F.; Massetti, L. Attitudes and Behaviors toward the Use of Public and Private Green Space during the COVID-19 Pandemic in Iran. *Land* **2021**, *10*, 1085. https://doi.org/10.3390/land10101085

Academic Editor: Thomas Panagopoulos

Received: 3 September 2021
Accepted: 12 October 2021
Published: 14 October 2021

1. Introduction

1.1. COVID-19 and Urban Green Spaces

The emergence of the COVID-19 pandemic as a global crisis deprived people worldwide of many usual activities, due to governmental restrictions set to limit social gatherings and crowding and to avoid contagion. People experienced to some extent measures of lockdowns, with some governments specifically preventing also possibility of visiting public green spaces for a prolonged period of time [1,2], while in other countries visiting green spaces was allowed. Although many people were afraid to use them [3], in general there was a great demand for urban green space [4]. In the US the public health authorities recommended to design and implement social distancing guidelines specifically for visiting parks and closely monitor people's behaviors and travel patterns to parks [5]. Indeed, the pandemic changed the way many people viewed and interacted with the natural environment and UGS [6]. For example, in Europe going to UGS for reasons that could be considered as non-essential, such as observing nature or meeting people-which could pose a risk for possible contagion, decreased dramatically, while UGS were used for physical activities, passing through them to reach a destination or for taking kids outdoors and by dog walkers [1]. In a study carried out in Tokyo (Japan), the use of UGS changed according to lifestyles and user typologies, who visited specific green spaces depending on their characteristics and their stress-reducing functions [7].

In general, it seems that in the countries where access to UGS was allowed, the use of UGS increased [8], although there were also evidences that UGS use was influenced by the perception of contagion risk. For instance, in New York City, Lopez et al. [9] found that most surveyed users declared to visit UGS less often than before the pandemic because of the crowds and low social distancing. This can be obviously exacerbated by the UGS density and accessibility within a town or neighborhood, and the presence of a private green space [10]. For example, UK and US parks and gardens remained open, although in many big cities this evidenced the inequalities in UGS access between neighborhoods–with communities characterized by socio-economic disadvantages such as lower income, and often with higher ethnic diversity, suffering inadequate quality, functionality and location of UGS [11,12]. Similar results were found in Mexico City [13], where low access to UGS has prevented their use especially in low-income neighborhoods and by women who were also scared by the risk of contagion.

This situation highlighted the significant role that urban nature can play in the health and wellbeing of urban dwellers [14–17]. It is acknowledged how urban green spaces (UGS) in both private and public forms have positive socio-ecological benefits: they provide opportunities for social interactions [18–20], improve numerous health and well-being indicators [14,21–26], and enhance the air quality and thermal comfort in the urban environment [27–29]. Thus, UGS represent a social right that public policy should guarantee, by enhancing equity in access to nature, especially on public lands [30–34]. Also residential yards provide access to nature, and in other studies have been demonstrated as having an important role [35–37] in providing moments of respite, especially during such crises. Ma, Lam, Cheung, & Fok [38] highlighted the importance of accessible natural green areas as they represent an opportunity to withstand the detrimental psychological effects of the pandemic (fear for contagion, anxiety), facilitating the adaptation and resilience of people in term of both mental and physical heath, with consequences in both the short and long run. Maintaining physical activity also has a potential buffering effect against COVID-19 related consequences, and on increasing the immune vigilance [39–42]. In contrast, the restrictions on movement and consequent reduction of physical activity resulting from governmental decisions induced further consequences: in Canada, the majority of children and youth decreased their outdoor activities during the COVID-19 pandemic, increased their screen time on digital devices and other sedentary behaviors [43]; in Brazil, those who reduced their level of physical activity revealed higher levels of mood confusions [44].

The COVID-19 crisis presents an exceptional opportunity for policy makers to take transformative actions regarding appropriate behaviors and adaptative rules to counter the spread of the virus [32] in relation to the use of public spaces, particularly urban green spaces. COVID-19 is also an opportunity to reflect about the spatial distribution of the urban fabric in terms of urban planning [45], urban greening policy and UGS design [46]. Knowing more about the usage and perception of the urban spaces during such a crisis can help define strategies to mitigate the effects of the pandemic [46] and develop urban planning approaches aimed to decrease the inequalities towards poor, marginalized, and vulnerable groups, and it is even an impetus to start a green revolution in terms of green mobility (i.e., cycling and walking) [47], and to rethink the nature of urban space for safer and more sustainable cities [48,49].

1.2. The COVID-19 Pandemic in Iran

Iran was the second country to experience fatalities due to COVID-19, declaring two deaths on 18 February 2020 (less than 50 days after the first fatalities in China), and during the first wave of the pandemic (February 2020–April 2021) it counted the highest number of cases and deaths among the Eastern Mediterranean countries [50]. The situation was exacerbated by the political and economic sanctions imposed on the country [51]. In response to COVID-19, from February to April 2020 the government cancelled public events, closed schools, universities, shopping centers, bazaars, and holy shrines, and banned festival celebrations. Government restrictions were gradually eased from late April 2020, although the number of new cases increased again in May with new peaks reported in July and December 2020 [52]. During the whole year of 2020, theatres, swimming pools, saunas, beauty salons, schools and universities were closed most of the time, and cultural and sports gatherings were banned, though most of the mosques remained open. In contrast, urban parks, gardens and other types of green infrastructure remained open to the public, though access was defined by health protocols, such as social distancing and other rules [51].

1.3. Context of the Study

During the recent decades, urban expansion in Iran has caused a dramatic decrease in UGS [53–55], especially in cities located in a dry climate zones, where urban planning has been occurring without an appropriate consideration of public green space [56]. This has yielded high disparities in the distribution of urban parks and contributed to citizens' unequal access to UGS [57].

In this study, the authors focused on the city of Birjand, located in a highly strategic position in the East of the country, at an elevation of 1491 m. It is characterized by a cold desert climate with extreme drought: an average of 150 mm annual rainfall and temperature extremes of 44 °C in the summer and −15 °C in the winter.

Birjand has a limited number of newly-established contemporary urban parks, but counts many tree-lined streets with pine trees (mostly *Pinus nigra*) which have been planted since the 1940s, forming the green landscape of Birjand. The first city park, called the National Garden, was established in those years while the city entrance park was established in the early 1970s [58].

In the last 30 years of urban development, several new neighborhoods have been built, but only a few small parks have been established in these areas. Moreover, between 1997 and 2020, the population increased from 133,474 inhabitants to about 200,000 people (an annual growth rate of 1.49%), the city surface expanded from 1690 hectares to 3630 hectares (becoming 2.14 times larger) (Figure 1), while in contrast, UGS increased by only 1.6 times (from 160 to 253.54 hectares) [59] (Table S1). The absence of an urban greening policy permitted neighborhoods to grow denser without incorporating green spaces in any meaningful way. Additionally, despite an enhancement of the green space per capita from about 12 m^2 to 13.8 m^2 during the mentioned period, even the access to private green spaces has decreased. People used to have access to nature through the private yards, but the mass construction pattern especially in the last decade has favored housing without private green space, greater building density and the transformation of yards into parking lots [60]. This caused UGS to be insufficiently accessible to urban dwellers, as large urban are parks are usually too far away.

At the same time, many historical gardens such as the landscape heritage gardens have been restored and opened to the public, and they represent the most visited green spaces together with the parks (Figure 2).

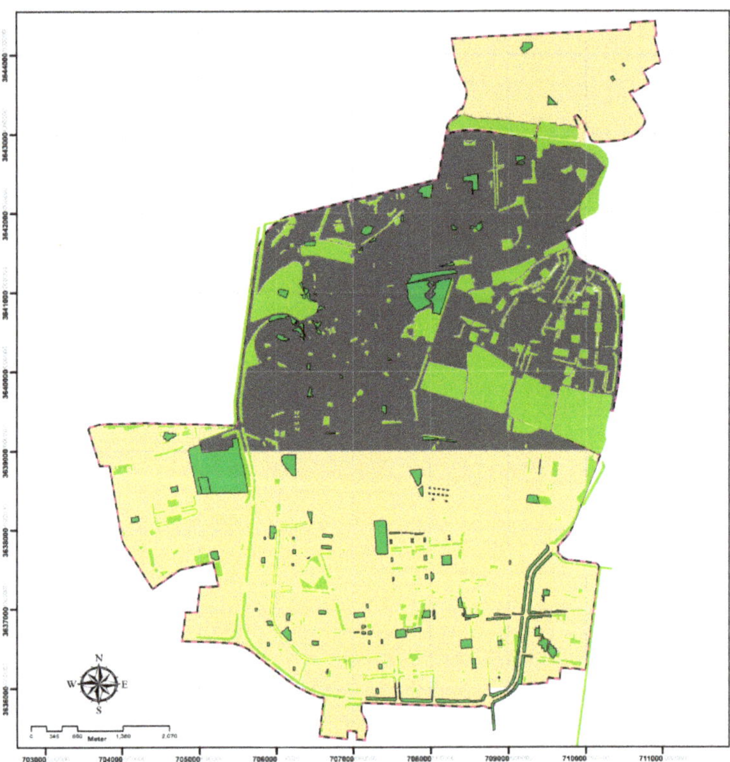

Figure 1. Map of Birjand, old town (grey), urban expansion (light yellow), parks and green spaces (dark green), and planned green spaces (light green) [61].

Today, the city is divided into two districts and counts 57 neighborhoods: 24 located in District 1, and 33 in District 2. District 1 has an area of 1798 hectares, 88,846 inhabitants and 33.02 ha of green spaces while District 2 has an area of 1831 hectares, 111,091 inhabitants and 47.63 ha of green spaces (Table S2). In general, green space per capita, per capita income, property prices and preference for living are higher in District 2 than in District 1 (Table S3) [62].

1.4. Hypotheses and Aims of the Current Study

The benefits of green spaces in a harsh climate are well acknowledged [34,61,63]. However, despite the possibility for Iranian citizens to visit UGS during the pandemic, we hypothesized that the COVID-19 crisis and related government measures may have fundamentally changed the public's use of UGS, with consequent questions about the general public's appreciation of UGS services and benefits [1,64] and the role of public vs. private green spaces. Knowing more about it can aid policy making to design green spaces as a form of "spatial medicine" [65,66] by ensuring safe and unrestricted use of green areas during the next predictable crises [67–69].

The study aimed at exploring the following aspects concerning visitors of public and private green spaces in the town of Birjand: (i) the behavioral changes in the use and in the motivation to visit a green space during the pandemic, as compared to the pre-pandemic period; (ii) the feelings (positive/negative) perceived in a green space; (iii) the extent to which access to a public green space during the pandemic was missed by users; and (iv) the perception of meaningful ecosystem services and benefits connected to green spaces. This was done through an exploratory survey targeting the inhabitants of Birjand.

Figure 2. Photos of cityscape (**a**,**b**), city parks (**c**,**d**), district parks (**e**,**f**), neighborhood parks (**g**,**h**) and private large (**i**) and small (**j**) courtyards in Birjand.

2. Materials and Methods

2.1. The Questionnaire

The exploratory survey was done through an online questionnaire the using Google Modules online platform. Compared to other types of distribution, online forms are generally more efficient and easily allow for complex patterns of question branching and skipping. In this specific case, the choice of an online survey was also made on the basis of the following considerations: affordable and less time-consuming method for both administrators and respondents, the characteristics of the Iranian population, largely urban and young (32 years old is the average age, with 67% between 16 and 64 years old), with a rather high rate of literacy (85% of adult population) and making extensive use of digital devices and the internet. Statistics show an increasing trend of those using the internet (currently 70% of the population) and a 155% rate of mobile connections with respect to the population [70].

Nevertheless, the fact that online surveys do limit the respondents to those who have internet access and use online communication (such as email and social media) raises the probability that the sample is non-representative of the actual population, and thus limits the generalization of the findings.

The minimum sample size (n = 384) was calculated according to the equation:

$$n = N \times X/(X + N - 1)$$

where N is the population size and $X = Z_{\alpha/2}^2 \times p \times (1-p)/MOE^2$, and $Z_{\alpha/2}$ is the critical value of the Normal distribution at $\alpha/2$ (for a confidence level of 95%, α is 0.05 and the critical value is 1.96), MOE is the margin of error (set at 5%), p is the sample proportion (expected number of respondents visiting green spaces = 50%).

The questionnaire was structured in six main sections and contained between 20 and 31 questions (see Supplementary Materials), depending on whether respondents self-identified as those who usually visit a public urban green space or those who do not, and on whether they had access to a private green space or not. The first section asked respondents for information about the characteristics of the place where they live (i.e., type of housing, presence of public green spaces near the residency, and size of the accessible private green space, if any. In order to characterize the neighborhoods of the respondents, respondents were asked to write down the name of their neighborhood in Birjand (as optional) in order to associate it with territorial municipal information. This included the price of urban lands (class rank from 1–most expensive, to 7–cheapest) taken from the Atlas of Spatial Divisions of Birjand [62], the preferred area for residency (class rank from 1–most preferred, to 6–least preferred) taken from the Comprehensive Plan of Birjand [65,71], and the greenery per capita (class rank from 1 to 8 classes of green e.g., class 1 = 61–70 m^2 per capita; . . . ; class 8 = 0–10 m^2 per capita etc.) taken from the official Database of the Organization of Birjand's Urban Parks & Green Spaces [72] (Table S3). This information was used for regression analysis (see Section 2.3).

The second section was about the pre-pandemic usage of public green spaces (A) i.e., frequency of visitation, distance, type of green space and motivation to visit it, and/or of private green spaces (B) i.e., size of the private green space, if it is shared with other people, frequency and motivation of visitation; the third section was about the usage (i.e., frequency and motivation of visitation) and an assessment of the feelings (i.e., happiness, pleasure, excitement, physical energy, anxiety, depression, fear, tension, anger, sadness) in public green spaces (A) and/or in private green spaces (B) during the pandemic (the period which started in February 2020 and it was ongoing at the time of the survey, characterized by diverse types of restrictive measures e.g., lockdowns, social distancing etc., although the access to public green spaces was always allowed). The fourth section was about the level to which respondents missed having access to urban green spaces and "what" was mainly missed in relation to a green space, as an open question. Finally, the fifth section was about personal details (i.e., gender, age, occupation).

The questionnaire included closed questions in order to reduce the probability of errors, some of which allowed multiple selection (up to two) among different options (e.g., reason of visitation of green spaces), and some others included a 4 (or 5)-point Likert scale (e.g., feelings in green spaces, missing level of a green space, and perception of benefits and services of green spaces).

One open question, about what the respondents missed most during the pandemic regarding the visitation of a green space, aimed at leaving respondents the freedom to express their thoughts. For the analysis, the responses were interpreted, simplified and coded according to a common meaning. For instance, we coded as "nature" all texts such as "enjoying nature", "listening to birds and water", "connection to nature"; as "breathing in open air" the sentences related to "breathing outside", "breathing without masks"; or "quiet" those sentences or words related to "calm green areas", "quiet", and "peace". The same coded-answers were further aggregated into a smaller number of categories for the regression analysis (see next chapter). For instance, when respondents expressed concepts like missing "meeting people", "children playing" and "staying together with the family", these were coded "sociality" as they are related to social aspects of green spaces. The new code "Relax and quiet" aggregated answers connected to "relaxation", "quiet green spaces" and "no stress".

The survey was pilot-tested by a small group of people who checked the translation into Persian and identified any words or phrases that were not clear in order to revise them and increase the reliability.

2.2. Administration of the Questionnaire

The questionnaire distribution and data collection took place from the 5 January 2021 to the 14 February 2021, just after the peak of COVID-19 contagions of the previous fall.

Initially, the link to the online survey questionnaire was sent via email to the first author's networks of professional and personal contacts living in Birjand and posted on social media (WhatsApp, Telegram, etc.). The people contacted were kindly invited to forward the link of the survey to their contacts. This created such a snowball effect so that it was difficult to track its distribution and the people invited to participate in the survey.

The introduction to the questionnaire reported a brief explanation about the purpose of the study, the target group (inhabitants of Birjand without any specific characteristics), the time needed to fill it in and information regarding data handling and the responsible institutions and persons. This information had to be checked before accessing the questionnaire.

The questionnaire was anonymous and took approximately 15 min to complete and responses were collected without identifiers via Google Spreadsheets, and downloaded for analysis.

2.3. Statistical Analysis

All the responses were checked if they were fully filled for the analysis.

Among the respondents ($n = 401$), 7 were not considered for the analysis because they did not live in Birjand. Due to the fact that most questions were mandatory–except the one about the neighborhood, we did not have uncompleted questionnaires, since all respondents indicated their neighborhood.

Descriptive statistics (with the frequencies and percentages) were performed for all valid answers. We identified two groups of respondents: those with only access to public green spaces and those with access also to private green spaces.

To assess if the frequency and motivation to visit a public or private green space in the period before the pandemic, were related to the type of green, the Chi Square goodness of fit test was used, assuming as null hypothesis that the visitation does not change with the type of green space.

To assess if the pandemic influenced the frequency and the motivation to visit a green space (public and private), the Wilcoxon signed-rank test was applied for the comparisons of the variables between pre-pandemic and during the pandemic at $p < 0.05$.

The feeling connected to staying outdoors in a green space (private and public) was assessed through a 5-point Likert scale, with the nominal value scale transformed into an ordinal scale (1 = Much less, 2 = Less, 3 = No change, 4 = More, 5 = Much more).

To assess if having access to a private green space changed the frequency and the motivation for visiting a public green space, the Chi Square goodness of fit test was used, assuming as null hypothesis that having access to a private green space would not affect the visitation of public green spaces.

When the frequency in a cell was <5, the Fisher exact test was applied.

In addition, multiple regression analysis with forward stepwise selection method was performed specifying an alpha significance level 0.05, to find any relationship between dependent and independent variables in the two groups, separately: visitors of public green spaces and visitors of private green spaces. Categorical variables such as occupation, type of house, and presence of a private green space were coded as dummy variables (0 not in the category, 1 within the category). The analysis was also performed to check if the visitation of public green spaces was related to the presence of private green spaces and other variables (i.e., type of house, access to private green space–as respondents with private green spaces could visit public ones, frequency of visitation of private green space and distance of a public green space), pre-pandemic and during the pandemic (for the latter, we included also the frequency of visitation of public green spaces before the pandemic).

Regarding gender, male (M) was set to 1, female (F) to 0. Variables with categorial scales such as age, education, frequency of visitation, and the classes related to municipal information of the neighborhood (greenery, urban land price and preference for residency) were used as ordinal value scales. For this reason, age classes ranged from 1 (less than 20 years old) to 8 (more than 80 years old), education ranged from 1 (mandatory) to 4 (post-graduate), frequency of visitation from 1 (Never) to 6 (Every day), distance of the closest public green space from 1 (closest) to 4 (farthest), missing a green space from 1 (not at all) to 4 (a lot) and classes of greenery, urban land price and of preference for residency as indicated in Section 2.1. The following relationships were investigated:

(i)   feelings in a public and private green space (dependent variable) and gender, age, education, occupation, municipal information of the neighborhood, in addition to specific variables for the visitors of public (frequency of visitation during the pandemic) or private green space (size of the private green space and frequency of visitation during the pandemic as independent variables.

(ii) perceived benefits or services and other aspects of public and private green spaces (each item as dependent variable) and the above-mentioned independent variables.

(iii) extent to which a green space was missed (dependent ordinal variables) and gender; age; education; occupation; access to private green space (Yes = 1); frequency of visitation pre and during the pandemic, distance of public green space.

3. Results

3.1. Characteristics of the Respondents

The survey collected 394 valid responses, mostly represented by women, mainly ranging from 20 to 39 years old, while males represented 39% of the sample, mainly ranging from 20 to 49 years old (Table S4). A large majority of respondents (91%) had higher education degree. In addition, most respondents (45%) were employees in public or private companies and self-employed, especially men, while among females, students and housewives were also represented (Table S4).

Regarding the type of house where respondents live, 55% declared to live in a flat in a condominium (n = 218) and 45% in a single house (n = 176) (Table S5). Approximately half of all respondents (n = 185) did not have any private green space available nearby the house. This occurred especially for those living in a flat, while 88% of those living in a single house had access to a private green space. The most frequent size of private green space was between 30 and 100 m^2, especially linked to single houses. Big green spaces (more than 100 m^2) represented 10% of all responses and they were more typical of single houses. Only 5% of respondents living in a condominium declared to have access to a big (more than 100 m^2) private green space.

The large majority of respondents without access to a private green space had a public green space available within 300 m (Table S6), mainly represented by a park (indicated by 40% of those living in flats and 37% living in single houses). Only 8% of respondents declared that they did not have any public green space nearby the residence, the majority of which also did not have a private green space to enjoy.

3.2. Use of Public Green Space before the Pandemic

Before the pandemic, the use of public green spaces was fairly good as 96% of respondents ($n = 379$) declared to visit a public green space. About one third of them (33.8%) used to go to the closest public green space and altogether, 60% visited green spaces no farther than 500 m (Table S7). The most visited green space typology was the urban park (55.1%), followed by green areas out of the town (about 20%) (Table S7). The frequency of visitation was quite high, as 65.2% went many times in a single month (Table S7).

The highest frequency of visitation of public green spaces was related to the lack of a private green space by respondents, although the majority of those with a private green space demonstrated to be habitual visitors of public green spaces, going at least several times a month (Table S8). In addition, the frequency was positively related to the motivations to visit the green space such as *physical exercise* ($p < 0.05$) and *walking* ($p < 0.05$) and to the distance of the public green space (Table S9).

Among the most selected reasons to visit a public green space were: *taking the kids outdoors* and *walking* (Table S7), regardless of access to a private green space (Table S8).

Those respondents who declared to visit a private green space before the pandemic ($n = 209$), did so pretty often as 52% of them declared to visit it many times a week (and much more everyday as compared to visitors of public green spaces) and 22% several times a month (Table S7). Activities performed in the private green space were various, mainly used for *staying outdoors* and *relaxing* but also *gardening* or *passing through the space to go somewhere else*, as compared to the reasons for visiting public green spaces. *Observing nature* and *taking the kids outdoors* we also selected (Table S7).

3.3. During the Pandemic

During the pandemic, the visitation of green spaces in general was much reduced, especially regarding the public green spaces ($p < 0.0001$) with respect to the pre-pandemic time (Table 1), even by respondents who did not have access to a private green space (Table S8). 42% respondents never visited a public green space during the pandemic and 40% did it at most once a month (Table 1). 10% of those with access to private green space never visited it during the pandemic and in general the visitation occurred less frequently (Table 1). The frequency of visitation of public green spaces was positively related to the motivations *physical exercise* (Table S9).

As well as before the pandemic, the main common reasons for visiting a public green space were *walking, observing nature,* and *taking the kids outdoors* (strongly decreased with respect to pre-pandemic though, $p < 0.001$) and to a minor degree *meeting people*, which increased ($p < 0.01$) (Table 1). These were especially selected by respondents without access to a private green space; instead, those with access to private green space significantly reduced the visitation of public green spaces for *taking the kids outdoors* ($p < 0.001$) (Table S8). The motivations to visit a private green space were mostly *taking kids outdoors* and *observing nature* (which increased, $p < 0.001$ and $p < 0.01$, respectively), *but also relaxing* and *staying outdoors* (Tables 1 and S10).

Table 1. Frequency (1) and motivation (2) of visitation of public green spaces (*n* = 379 respondents, upper part in each block) and private green spaces (*n* = 209 respondents, lower part in each block) before and during the pandemic.

		Pre-Pandemic		During the Pandemic		
(1) Frequency of visitation		Count	Percentage	Count	Percentage	*p*-value
Public green space	Every day	33	9%	14	4%	**
	More than once a week	97	26%	25	6%	***
	Several times a month	117	31%	33	8%	***
	Once a month	74	19%	37	9%	***
	Less than once a month	58	15%	122	31%	***
	Never	15	4%	164	42%	***
	Total	394	100%	394	100%	
Private green space	Every day	58	28%	41	20%	*
	More than once a week	50	24%	45	22%	ns
	Several times a month	45	22%	46	22%	ns
	Once a month	18	9%	21	10%	ns
	Less than once a month	31	15%	35	17%	ns
	Never	7	3%	21	10%	**
	Total	209	100%	209	100%	
(2) Motivation of visitation [a]						
Public green space	Gardening	3	0.5%	2	1%	ns
	Meeting people	34	5%	38	10%	**
	Observing nature	110	17%	61	16%	ns
	Passing through it	35	5%	43	11%	***
	Physical exercise	31	5%	23	6%	ns
	Reading	3	0.5%	1	0%	ns
	Relaxing	52	8%	14	4%	**
	Staying outdoors	32	5%	39	10%	***
	Taking the dog out	0	0%	0	0.0%	-
	Taking the kids outdoor	185	28%	56	15%	***
	Walking	175	26%	105	27%	ns
	Total	660	100.0%	382	100%	
Private green space	Gardening	45	13%	30	6%	***
	Meeting people	2	1%	21	4%	***
	Observing nature	46	13%	97	19%	*
	Passing through it else	41	12%	37	7%	*
	Physical exercise	13	4%	24	5%	ns
	Reading	12	3%	7	1%	*
	Relaxing	65	19%	82	16%	ns
	Staying outdoors	75	21%	76	15%	*
	Taking the dog out	0	0%	1	0%	-
	Taking the kids outdoor	39	11%	97	19%	***
	Walking	13	4%	38	8%	*
	Total	351	100%	510	100%	

The Chi Square goodness of fit test was used to compare the items between pre- and during pandemic at $p < 0.05$. Significant differences are showed with * at $p < 0.05$; ** at $p < 0.01$; *** at $p < 0.001$; "ns" for non-significant differences. [a] Respondents could select up to two choices among the given options.

3.4. Green Space Effect on Feelings

Being in a green space (Table 2) resulted in an enhancement of positive feelings, such as *happiness* and *pleasure* and *physical energy* (although rated to a minor extent), without difference between public or private green spaces. In addition, a large majority of respondents indicated that green spaces contribute to reducing a series of negative feelings such as *anxiety* and *tension* and *sadness* and *depression*. Private green spaces contributed more to the reduction of *anxiety and fear* than the public green spaces.

Table 2. Feelings of people who visited a public (upper part) or a private (lower part) green space.

		Happy	Pleased	Excited (Ecstatic)	Physically Energetic	Anxious	Nervous	Tense	Frightened	Irritated	Upset	Depressed	Blue
Public green spaces (n = 231)	More	65%	59%	31%	57%	24% *	11%	10%	13% *	22%	10%	9%	10%
	No change	17%	22%	36%	23%	21%	25%	27%	30%	27%	28%	27%	25%
	Less	19%	19%	33%	20%	54%	63%	63%	57%	50%	62%	65%	65%
	Avg.	3.5	3.4	2.9	3.4	2.5	2.2	2.2	2.3	2.6	2.3	2.1	2.1
Private green spaces (n = 188)	More	65%	62%	28%	55%	14%*	11%	15%	4% *	14%	14%	12%	10%
	No change	17%	19%	43%	25%	19%	23%	17%	30%	26%	17%	21%	19%
	Less	18%	19%	29%	20%	66%	66%	68%	65%	60%	69%	67%	71%
	Avg.	3.6	3.5	2.9	3.4	2.1	2.1	2.1	2.0	2.3	2.1	2.1	2.0

Percentages are reported for each point of the Likert scale and the weighted average (avg) is calculated associating each point a score (1 = Much less; Less = 2; 3 = No change; 4 = More; 5 = Much more). The percentages of the positive scores (much more and more) and the negative scores (less, much less) were added. The Chi Square goodness of fit test was used to compare the feeling in the two types of green spaces at $p < 0.05$. Significant differences are showed with * at $p < 0.05$; ** at $p < 0.01$; *** at $p < 0.001$; ns for non-significant differences.

The regression analysis also showed some interesting results (Table 3): the positive feelings (i.e., happy, pleased) were correlated to the higher price of urban lands ($p < 0.05$) but also to the lower amount of greenery per capita (higher classes), while negative feelings such as anxiety, fear and hanger were more typical of males and greater in less preferred neighborhoods for residency, and tension, hanger and sadness (depression, feeling blue) were greater in neighborhoods with higher greenery per capita.

Table 3. Feelings in a private (left) or public (right) green space in relation to some independent variables.

		Private Green Space (n = 209)			Public Green Space (n = 231)		
		b	Std. Err.	p-Value		b	Std. Err.	p-Value
Happy	Intercept	3.101	0.252	0.000			
	Greenery class [a]	0.087	0.029	0.003			
	Price of urban lands [a]	−0.076	0.037	0.038			
Pleased	Intercept	3.048	0.225	0.000			
	Greenery class [a]	0.078	0.030	0.010			
	Price of urban lands [a]	−0.096	0.037	0.010			
Excited (ecstatic)					Intercept	2.165	0.367	0.000
					Education	0.417	0.102	0.000
					Age	−0.210	0.060	0.001
Physically energetic					Intercept	2.526	0.449	0.000
					Greenery class [a]	0.072	0.035	0.042
Anxious	Intercept	3.116	0.300	0.000	Intercept	2.460	0.247	0.000
	Visit frequency of private green space during pandemic	−0.119	0.048	0.014	Preferred for [a] residency	0.606	0.282	0.033
	Preferred for residency [a]	0.428 [a]	0.212	0.045	Student	0.446	0.182	0.015
Nervous	Intercept	1.970	0.376	0.000	Intercept	2.375	0.394	0.000
	Age	−0.123	0.057	0.031	Preferred for residency [a]	0.665	0.252	0.009
	Male	0.344	0.126	0.007	Greenery class [a]	−0.069	0.034	0.042
	Employee	−0.271	0.135	0.046	Price of urban lands [a]	−0.533	0.240	0.027
Tense	Intercept	1.937	0.395	0.000			
	Male	0.344	0.132	0.010			
	Employee	−0.307	0.142	0.031			
	Price of urban lands [a]	0.081	0.040	0.045			
	Greenery class [a]	−0.063	0.032	0.049			
Upset	Intercept	2.190	0.197	0.000	Intercept	2.536	0.214	0.000
	Student	0.364	0.148	0.015	Preferred for residency [a]	0.120	0.045	0.009
	Greenery class [a]	−0.079	0.031	0.011	Greenery class [a]	−0.086	0.034	0.012
	Preferred for residency [a]	0.100	0.041	0.015			

Table 3. *Cont.*

		Private Green Space (*n* = 209)				Public Green Space (*n* = 231)		
		b	Std. Err.	*p*-Value		b	Std. Err.	*p*-Value
Frightened	Intercept	2.518	0.269	0.000	Intercept	2.834	0.275	0.000
	Preferred for residency [a]	0.097	0.038	0.011	Age	−0.143	0.059	0.017
	Male	0.249	0.119	0.037				
Depressed	Intercept	2.128	0.424	0.000	Intercept	2.406	0.245	0.000
	Male	0.343	0.127	0.007	Preferred for residency [a]	0.122	0.044	0.006
	Price of urban lands [a]	0.090	0.039	0.020				
	Greenery class [a]	−0.085	0.031	0.006				
Blue	Intercept	1.941	0.416	0.000				
	Male	0.265	0.123	0.032				
	Greenery class [a]	−0.087	0.030	0.004				
	Employee	−0.322	0.131	0.015				

[a] for this variable, the higher the class the smaller the value (for instance, class 1 means greater green surface per capita/higher price of the land/most preferred area; class 7 means smaller green surface per capita/cheapest price of the land/least preferred area for residency). Male was the reference (=1), b: non-standardized beta coefficient, Std. Err.: standard error of b; *p*-value: statistical significance. Only statistically significant results are shown.

Regarding the feeling connected to the visitation of public green spaces (Table 3), younger people and with higher education level felt more ecstatic; respondents feeling more anxious and nervous were those living in neighborhoods with greater greenery per capita and those living in less preferred areas.

3.5. Appreciation of Ecosystem Services and Other Aspects Relative to Green Spaces

As Table 4 shows, respondents mostly appreciated the *clean air* and the *comfortable climate* that in general green spaces provide, regardless of the type of green space. Public green spaces were also more appreciated than private ones for *cleanliness* (*p* < 0.01), *natural value* (*p* < 0.01), *biodiversity* (*p* < 0.001) and *socialization* (*p* < 0.001). In contrast, private green spaces were appreciated to a greater extent for the *privacy* (*p* < 0.001). *Safety from contagion* in green spaces was appreciated by about 83% of respondents in both groups, with no differences between public or private green space.

Table 4. To which extent ecosystem services and management aspects of public (top) and private (bottom) green spaces are appreciated by respondents.

		Natural Value	Biodiversity	Clean Air	Comfortable Climate	Quiet/Freedom from Noise	Proximity	Cleanness	Privacy	Socialization	Security from Bad People	Safety from Contagion
Public green space (*n* = 231)	Appreciated	92%	84%	97%	95%	83%	62%	94%	68%	73%	86%	83%
	Avg.	3.5	3.3	3.6	3.6	3.3	2.8	3.6	3.0	3.0	3.4	3.3
Private green space (*n* = 183)	Appreciated	86%	73%	93%	94%	86%	49%	90%	93%	45%	84%	84%
	Avg.	3.3	3.0	3.6	3.6	3.3	2.5	3.4	3.4	2.5	3.3	3.3
	p value	**	***	ns	ns	ns	**	*	***	***	ns	ns

Percentages represent the sum of positive appreciation (very much appreciated and appreciated) of each item related to public (upper rows) and private (lower rows) green spaces, the latter evaluated by those with access to a private green space who responded the question. The weighted average (Avg) was calculated associating each point of the 4-point Likert scale a score (1 = Not at all appreciated; 2 = Little appreciated; 3 = Appreciated; 4 = Very much appreciated). The Chi Square goodness of fit test was used to compare the appreciation for the items in the two types of green spaces at *p* < 0.05. Significant differences are showed with * at *p* < 0.05; ** at *p* < 0.01; *** at *p* < 0.001; "ns" for non-significant differences.

The regression analysis (Table S11) evidenced how the appreciation of *natural value* and *biodiversity* was linked to people living in single houses and in areas with lower price of land ($p < 0.05$) when referred to a private green space, and to females ($p < 0.05$) and respondents living in areas with lower green surface per capita ($p < 0.01$) when referred to a public green space. Housewives ($p < 0.05$) and respondents with higher education level ($p < 0.01$) appreciated the *quiet* of a private green space ($p < 0.05$, $p < 0.01$ respectively) while *socialization* was particularly appreciated in public green spaces, especially when the visitation before the pandemic was greater and the green surface per capita in the neighborhood was lower ($p < 0.05$). In general, women felt less safe than men for the *presence of bad people* in green spaces ($p < 0.01$) and *security* of a private green space was appreciated by respondents living in the least preferred place for residency.

Finally, females and those who did visit private green spaces to greater extent during the pandemic felt *safer from COVID-19 contagion* in private green spaces.

3.6. How Much the Green Area Visitation Is Missed

When respondents were asked to what extent they missed going to a green space, 66% said they missed it very much and 22% rather much. Only 10% missed it only to a small extent. The regression analysis (Table 5) evidenced a positive relationship between the extent to which a green area was missed and the frequency of visitation of public green spaces pre-pandemic ($p < 0.001$) and the distance of public green space ($p < 0.05$), and a negative relationship with male, which means that women ($p < 0.01$) suffered more than men. Having access or not a private green space did not influence the deprivation feeling.

Table 5. Relationship between "Missing of a green space" and independent variables related to private (resulting not significant) and public green spaces.

	b	Std. Err.	*p*-Value
Intercept	3.67	0.13	0
Male	−0.2	0.07	0.006
Frequency of visitation of public green space pre-pandemic	0.11	0.03	0.0002
Distance of public green space	−0.06	0.03	0.032

Multiple regression analysis between the dependent variable "Missing a green space" and independent variables related to users of green spaces before the pandemic ($n = 379$). Only statistically significant results are reported. b: non-standardized beta coefficient, Std. Err.: standard error of b; *p*-value: statistical significance.

What respondents missed (Table S12) was mainly related to *nature* (19%) (staying in connection with nature, observing and listening the sounds of nature), *meeting people* (18%) and *staying with the family* (10%).

Nature was especially missed by women (Table S13) while the "social aspect" (meeting people, family togetherness, children's play) was especially related to males ($p < 0.05$) and younger people ($p < 0.05$) and to the distance of the public green space ($p < 0.01$) or lesser per capita greenery ($p < 0.05$).

Physical activity including walking and doing physical exercise (e.g., running) was especially missed by students ($p < 0.05$) and older people ($p < 0.01$).

Those who declared *not to miss anything*, were mostly respondents who declared low visitation frequency of public green spaces before the pandemic ($p < 0.01$) and those living at shorter distance from a public green space ($p < 0.001$) or near a square with trees ($p < 0.01$).

4. Discussion

The study revealed how green spaces are important elements for the respondents. It should be noted that respondents were mostly young citizens, less than 49 years old, with a large majority of people with higher level education. Despite the inherent bias due to the snowballing approach, which makes it impractical to achieve a truly representative sample of the overall population, the exploratory study reports that the public green spaces are frequently visited by the respondents, although the accessibility of nearby public green spaces (distance set at 300 m) is limited to only one third of all respondents. Also, approximately half of the respondents did not have any private green space available nearby their residence, especially those living in a condominium, and a few did not have any public green space in the surroundings. This depends on the characteristics of the urban fabric, whose urban expansion during the last decades put little attention on the provision of green spaces to urban dwellers. The urban fabric and the harsh climate amplify the importance of green spaces in such a dryland and invited reflection on how the COVID-19 pandemic affected behaviors and perceptions of green spaces. From the analysis of the collected data, we noticed that the respondents were quite habitual users of public green spaces before the pandemic. We identified two groups of respondents, those with no access and those with access to a private green space, in the form of a private garden (mainly in connection to a single house) or yards shared among neighbors, to understand the role of private green space in such peculiar conditions.

The following discussion, in accordance with the objectives and research questions, takes into consideration the type of green space (public vs. private) used by respondents and is presented in three sections: the behavioral change due to the pandemic; the effect of green spaces on feelings, and the perception of services, benefits and other aspects related to green spaces in general, such as the extent to which people missed accessing to them.

1. Behavioral change in the use and in the motivation to visit a green space during the pandemic

The study showed how the visitation of green spaces in Birjand was in general much reduced during the pandemic, especially in public green spaces. This kind of behavioral change was also observed in European contexts where the governments' regulations were more restrictive [1], whereas in many other countries, the visitation of parks and green areas increased [8]. In Asian cities, Lu, Zhao, Wu, & Lo [73] found a 5.3% increase in the odds of people using green spaces regardless of the increase of COVID-19 infections, and Hamidi & Zandiatashbar [5] found that residents of compact counties in the United States were less likely to reduce their trips to parks during the shelter-in-place order. During the same period, Norwegians had and increased need for contact with urban green space, and in particular Venter, Barton, Gundersen, Figari, & Nowell [17] reported a 291% increase in physical activity in green spaces. Also in the UK, people spent more time in nature and visited nature more often during the pandemic [73]. However, in compact counties of the US, the high number of COVID-19 deaths coupled with the small size homes and the lack of private green spaces have demonstrated psychological counteracting effects, with an increase of feelings of anxiety and isolation, while visiting a park represented a real possibility of respite to mitigate physical and psychological challenges [5]. This is why in general, there was an increase in the visitation of green spaces to the extent allowed, at least for essential activities such as for taking the dog out, physical exercise and relaxing in contexts with more restrictive regulations [1].

In Birjand, the reduction in the usage of green spaces declared by respondents was likely due to two main reasons: on one hand, the governmental regulations that posed a limitation in movement for citizens, on the other hand, the fear of exposure to contagion due to gatherings among people visiting urban parks [38]. The decline in the visitation of the urban parks observed was severe, as 42% never visited a public green space (vs. 4% before the pandemic) and 40% did it at most once a month during the time considered. Perhaps such a sharp decline was also exacerbated by the chronic scarce accessibility of public urban green spaces, as urban parks are concentrated in some parts of the city and the long travel distance might have discouraged people to travel for the perceived risk to contagion during travelling [74], as respondents did not perceive private or public green spaces as an unsafe place for contagion risk. The urban expansion characterizing the city of Birjand has not favored a diversification or increase in the number of green spaces over time. Indeed, the per capita green space is about 14 m^2, which is much lower than the optimal amount (50 m^2 per person) [75], and inadequate especially considering the contextual harsh environmental conditions.

Most of the people's interaction with urban nature happens in a few urban parks, historical gardens and private yards. In this regard, this study highlighted the importance of providing accessible green areas or other types of nature-based spaces to facilitate the adaptation of people to disease outbreaks. Private green spaces in the form of yards or gardens constitute a crucial alternative to the lack of public green spaces. The study has demonstrated the great visitation of private green spaces with respect to public ones during the pandemic (66% vs. 18% of respondents, respectively visited at least several times a month), most likely influenced by their proximity. Regarding the reasons to visit a green space, the pandemic did not substantially change those to visit a public green spaces -although the reduction of visitation for *taking the kids outdoors* and its increase in private green spaces, highlighting the possible fear of infection in public spaces [76], where people may not know each other [73]. Consequently, the private green spaces revealed their compensative role for the lack of residential greenery for the family members during the crises [77,78].

Public green spaces were visited for *walking* even by those people with a private green space, while the private ones became places to enjoy nature-centered activities such as *relaxation*, *observing nature* and *staying outdoors*, which were probably elicited by the situation. In addition, there was an increase of people *passing through a green space* to reach a destination, which supports the importance of green corridors made of tree-lined streets and gardens in the urban fabric, not only for the variety of services, such as shade, climate mitigation but also for psychological benefits and wellbeing [79].

The results are in harmony with previous studies which also reported how urban inhabitants feel the residential greenery as calm and safe places to be in contact with nature and to meet other people [80]. Even in Berlin, as green capital in Europe (with much higher per capita greenery than Birjand), public parks located far away during the pandemic have played a very small role in the urban dwellers' everyday life due to the government restrictions on people's movement, while residential greenery was the only accessible green space for many residents [80].

Thus, for future urban and landscape planning it would be crucial to preserve a pattern of public and private greenery-in the form of public pocket green spaces, traditional Iranian yards and gardens, or as collective residential proximity greenery which is more popular in European contexts. Urban regeneration projects aimed to increase the green area surfaces and accessibility will guarantee access in all parts of the town, thus even to most vulnerable users. As Säumel, Hogrefe, Battisti, Wachtel, & Larcher [80] proved, freely accessible residential greenery would play crucial role in the transformation of cities into livable, healthy and ecologically friendly environments.

2. Effect of green spaces on feelings during the pandemic

The results suggest that being in a green space (either public or private) enhances positive feelings such as happiness and pleasure (on average by 65.5% and 61% respondents, respectively declared that they felt happier and more pleased), and also physical energy, although to a lesser extent. However, happiness for those visiting private green spaces seemed to depend also on the neighborhood, especially the ones with lower greenery and lower price of land, enhancing the role of private green pockets for psychological wellbeing. Green spaces in general were effective in the reduction of all negative feelings, and the private ones were more effective in the reduction of anxiety and fear. However, in general, lesser preferred neighborhoods and greater greenery per capita were found to be related to greater anxiety, likely linked to the pandemic. In Poland, due to the COVID-19 epidemic there was observed a considerable decline in physical and psychological health, and the overall quality of life of youth in the period of limited access to public spaces [81]. Under circumstances of lockdown, with no access to public spaces, the importance of private spaces can indeed buffer individuals from depressive and anxiety-related symptoms, and function as a beneficial source of refuge, security, and stability during these stressful times [12,16,73]. While private gardens can compensate for a lack of access to public green spaces, both the public and private green spaces are the essential health resources in times of crisis [16].

The characteristics of a neighborhood play an important role. For instance, private green spaces contribute to the positive feelings (happiness and pleasure) especially if the neighborhood has lower land price. Less preferred areas for residency seemed to be more related to negative feelings such as anxiety and nervousness but also tension and fear. However, the presence of greenery in the neighborhood was found to be related to some negative feelings such as anxiety and nervousness if related to public green spaces. Therefore, beyond the private yards and gardens, strategic planning of neighborhoods that guarantees easy access to UGS and quality natural environments in neighborhoods are crucial to maintain favorable health and wellbeing [73,82], especially in times of pandemic. Neighborhood green spaces would guarantee easy accessibility and less crowding as compared to public urban parks [4,83–87] especially if these are limited in size and the urban fabric is compact [46,88], and also limited is the possibility of visual contact with nature through home windows [89–92]. Thus, landscape design strategies employing a tree-based landscaping neighborhood scale in the arid regions (and not only there) could create more comfortable outdoor spaces to enhance the livability and sustainability of neighborhoods [93]. A study in the UK revealed how a greater land-cover greenness within a 250 m radius around a respondent's postcode was important in predicting higher levels of mental wellbeing during the COVID-19 pandemic.

Although it is hard to add large green spaces in the neighborhoods of Birjand, landscape planners can improve the existing pocket parks and strengthen linkages between nearby green spaces. It was demonstrated that a complete green network can ease walking for recreation and reduce inequality in green space distribution [87]. But in new developed neighborhoods where space is not limited, planners may consider adding new parks, addressing the people's needs that are expressed by the reasons to visit a green space.

3. Perception of services and benefits and other aspects related to the green spaces

We noticed how the respondents were aware about a variety of services provided by green spaces, such as the amelioration of air quality and climate conditions. This seems obvious in a dryland; however, for decades the urban planning of the city has been continuously sealing land, preventing the creation of green spaces. If private greeneries provide immediate access to nature and the privacy component allows creative or reflective moments within one's own private sphere that would be rarely feasible in public green spaces [94], public green spaces were highly appreciated by respondents for the socialization, cleanliness and natural value, highlighting the importance of a quality design and maintenance of public green spaces, beyond mere quantity. Ameliorating the quality of green spaces by targeting specific green functions [95], including natural value [96,97], biodiversity [98–101], but also socialization [102–105], facilities and security will improve the potential benefits of urban nature in wellbeing and health [106]. High-quality green areas also may help improve prosocial behavior in children [107] and the contact with nature in times of pandemic may be a key also for improved well-being post-COVID-19 [108]. In addition, our respondents in general did not consider green spaces as a place with risk of contagion and in this regard, Ma et al. [38] reported how the adaptative benefits of visiting nature may outweigh the risks when proper visitor management measures are implemented. In fact, other studies have shown how the risk of contagion outdoors is actually much less than indoors [109].

4. Extent to which the green space access was missed

Results highlighted that the majority of respondents missed going to a green space very much. The extent to which a green area was missed depended on the usual frequency of visitation before the pandemic, the gender (with women more influenced) and the distance of public green space visited before the pandemic, as if missing a green area was likely related to the actual scarce presence of favorite green spaces nearby. Again, this highlights the importance of equity in urban landscape planning. In simple words, the more greenery available, the more people would not miss anything-and the farther the green space, the more it is missed, as for social activities like those demonstrated by Schipperijn et al. [110]. Respondents confirmed their awareness about the services and benefits provided by green spaces even when we asked what they missed, as most thoughts were related to sensitiveness towards nature, meeting people and staying with the family in green spaces, all aspects that emphasize the importance of the quality green spaces beyond mere quantity.

We have noticed how the "nature" connected to green spaces was missed more by women. Roe et al. [111] showed that there may be a gender difference in perceiving nature as a coping mechanism as female stress levels were remarkably higher in areas of low green space compared to males, and Theodorou et al. [112] revealed that psychopathological distress from the pandemic through green spaces was higher for women and unmarried people. Hostetler & Noiseux [113] found how those spaces where residents can access nature close to home result in a variety of conservation, social, and health benefits. Thus, in the context of pandemic, when many people have lost access to many of their typical sources of socialization, UGS may serve as an important compensatory role in satisfying a variety of psychological needs [114].

5. Conclusions

The results of this study report useful information and suggestions for urban landscape planning of the city of Birjand and other cities in drylands. This is an arid city in the East of Iran that has developed according to an uncontrolled residential planning without enough attention to urban greening challenges. In addition, it also faces environmental challenges such as the lack of water and harsh climate that limit vegetation growth. Therefore, preparing adequate urban green spaces in an arid environment as a coping strategy against the pandemic might be very hard, but the study has demonstrated how important are green spaces in a time of limitation of movement due to restrictions against the spread of the pandemic, as they enhance positive feelings and reduce anxiety and stress, and allow different types of activities, especially linked to social aspects.

Private green spaces such as gardens and courtyards of single houses or condominium shared with neighbors represent the pocket greenery that allow moments of respite in time of pandemic. But, if private green spaces may convey a sense of safety, public green spaces are important for their social value.

Private green spaces mitigate the lack of public green spaces but in a way, they represent an unequal distribution of green areas access, that should be guaranteed to the whole community.

Women demonstrated high sensitiveness and appreciation towards public green spaces, as they likely take kids there-contact with nature is recognized as an important function, and appreciate the socialization, although in general, the perception of insecurity is greater than for men.

Therefore, some specific recommendation for urban and landscape planning in Birjand can be drawn. Firstly, the typology of compact residential development which has become the popular form of the neighborhoods must be integrated with existent and new green spaces. Although in Birjand it is hard to add large green spaces in many of neighborhoods, landscape planners can improve the existing pocket parks and strengthen linkages between nearby green spaces by creating green corridors and preserving the existing green spaces. A complete green network can ease walking for recreation and reduce inequality in green space distribution. In newly planned neighborhoods which are not in the city center and space is not a limiting factor, planners may consider balancing soil sealing with new green spaces. Secondly, promoting the quality of existing urban green spaces (e.g., natural value, biodiversity, security but also socialization) is a greening strategy for the city of Birjand. The findings emphasized that improving the quality of green space beyond quantity may play a role in provision of high-quality nature exposure for urban population in the time of pandemic as well as the key for improved well-being of urbanites post-COVID-19.

Thus, the Coronavirus pandemic has provided an opportunity for urban planners and landscape architects to reflect on the existent greenery and on how to adapt existing greening policies to the challenges of today and the future. Enhancing not only the quantity of green spaces but also their quality by targeting specific green functions which were highlighted by our respondents will improve the wellbeing and livability of the city.

The authors have made clear that this is an exploratory study and it does not claim to reflect the real situation. The obtained sample cannot be representative of the population and UGS users because the survey was not restricted to specific respondents and it was spread to people through a snowball effect. However, we believe that the sample is enough to provide the considerations we made without generalizing the results. In addition, we have provided a justification of the used methods and the limitations of such decision.

Supplementary Materials: The following are available online at https://www.mdpi.com/article/10.3390/land10101085/s1, Table S1. Area of different urban green space typologies in Birjand city, Table S2. Number and area of urban parks by districts, Table S3. Greenery class, price of land and preference for residency for the Birjand neighborhoods where respondents live, Table S4. Descriptive statistics of the sample, Table S5. Users of private green space per size of space and house typology, Table S6. Type of public green space within 300 m from home without/with presence of private green space, Table S7. Characteristics of visitation of public green spaces before the pandemic by users of public green spaces, Table S8. Difference in frequency (1), motivation (2) of visitation and benefits and services perception (3) regarding public green spaces by respondents without (No) and with (Yes) access to a private green space (pre and during the pandemic), Table S9. Frequency of visitation of public green spaces in relation to variables connected to the demographic and the presence of private green spaces, Table S10. Motivation (up to two choices) for visiting a public and private green space during the pandemic, Table S11. Aspects of private (left) and public (right) green space appreciated by respondents, in relationship to independent variables, Table S12. Missed things related to green spaces indicated by respondents, Table S13. What was missed by respondents (n = 394) in relation to demographic information and green space visitation before the pandemic, Survey: Use of green spaces during the pandemic by COVID-19 in Birjand.

Author Contributions: Conceptualization, M.R.K. and F.U.; methodology, M.R.K. and F.U.; formal analysis, F.U. and L.M.; investigation, M.R.K.; data curation, M.R.K. and F.U. and L.M.; writing—original draft preparation, M.R.K. and F.U.; writing—review and editing, M.R.K. and F.U. All authors have read and agreed to the published version of the manuscript.

Funding: This research received no external funding.

Institutional Review Board Statement: The study was conducted according to the guidelines of the 758 Declaration of Helsinki, and approved by the Research Ethics Committee of University of Birjand 759 (Approval ID: IR.BIRJAND.REC.1400.004; Approval Date: 23 August 2021).

Informed Consent Statement: Informed consent was obtained from all subjects involved in the study.

Conflicts of Interest: The authors declare no conflict of interest.

References

1. Ugolini, F.; Massetti, L.; Calaza-Martínez, P.; Cariñanos, P.; Dobbs, C.; Ostoic, S.K.; Marin, A.M.; Pearlmutter, D.; Saaroni, H.; Šaulienė, I.; et al. Effects of the COVID-19 pandemic on the use and perceptions of urban green space: An international exploratory study. *Urban For. Urban Green.* **2020**, *56*, 126888. [CrossRef]
2. Sandford, A. Coronavirus: Half of Humanity now on Lockdown as 90 Countries Call for Confinement. Available online: https://www.euronews.com/2020/04/02/coronavirus-in-europe-spain-s-death-toll-hits-10-000-after-record-950-new-deaths-in-24-hou (accessed on 24 May 2021).
3. Fabisiak, B.; Jankowska, A.; Kłos, R. Attitudes of Polish seniors toward the use of public space during the first wave of the COVID-19 pandemic. *Int. J. Environ. Res. Public Health* **2020**, *17*, 8885. [CrossRef] [PubMed]
4. Zhu, J.; Xu, C. Sina microblog sentiment in Beijing city parks as measure of demand for urban green space during the COVID-19. *Urban For. Urban Green.* **2021**, *58*, 126913. [CrossRef]

5. Hamidi, S.; Zandiatashbar, A. Compact development and adherence to stay-at-home order during the COVID-19 pandemic: A longitudinal investigation in the United States. *Landsc. Urban Plan.* **2021**, *205*, 103952. [CrossRef] [PubMed]

6. Meagher, B.R.; Cheadle, A.D. Distant from others, but close to home: The relationship between home attachment and mental health during COVID-19. *J. Environ. Psychol.* **2020**, *72*, 101516. [CrossRef]

7. Yamazaki, T.; Iida, A.; Hino, K.; Murayama, A.; Hiroi, U.; Terada, T.; Koizumi, H.; Yokohari, M. Use of Urban Green Spaces in the Context of Lifestyle Changes during the COVID-19 Pandemic in Tokyo. *Sustainability* **2021**, *13*, 9817. [CrossRef]

8. Geng, D.C.; Innes, J.; Wu, W.; Wang, G. Impacts of COVID-19 pandemic on urban park visitation: A global analysis. *J. For. Res.* **2021**, *32*, 553–567. [CrossRef]

9. Lopez, B.; Kennedy, C.; Field, C.; Mcphearson, T. Who benefits from urban green spaces during times of crisis? Perception and use of urban green spaces in New York City during the COVID-19 pandemic. *Urban For. Urban Green.* **2021**, *65*, 127354. [CrossRef] [PubMed]

10. da Schio, N.; Phillips, A.; Fransen, K.; Wolff, M.; Haase, D.; Ostoić, S.K.; De Vreese, R. The impact of the COVID-19 pandemic on the use of and attitudes towards urban forests and green spaces: Exploring the instigators of change in Belgium. *Urban For. Urban Green.* **2021**, *65*, 127305. [CrossRef]

11. Mell, I.; Whitten, M. Access to nature in a post covid-19 world: Opportunities for green infrastructure financing, distribution and equitability in urban planning. *Int. J. Environ. Res. Public Health* **2021**, *18*, 1527. [CrossRef]

12. Ye, Y.; Qiu, H. Using urban landscape pattern to understand and evaluate infectious disease risk. *Urban For. Urban Green.* **2021**, *62*, 127126. [CrossRef] [PubMed]

13. Huerta Mayen, C.; Cafagna, G. Snapshot of the Use of Urban Green Spaces in Mexico City during the COVID-19 Pandemic: A Qualitative Study. *Int. J. Environ. Res. Public Health* **2021**, *18*, 4304. [CrossRef]

14. Samuelsson, K.; Barthel, S.; Colding, J.; Macassa, G.; Giusti, M. Urban nature as a source of resilience during social distancing amidst the coronavirus pandemic. *OSF Prepr.* **2020**. [CrossRef]

15. Langemeyer, J.; Madrid-Lopez, C.; Mendoza Beltran, A.; Villalba Mendez, G. Urban agriculture—A necessary pathway towards urban resilience and global sustainability? *Landsc. Urban Plan.* **2021**, *210*, 104055. [CrossRef]

16. Poortinga, W.; Bird, N.; Hallingberg, B.; Phillips, R.; Williams, D. The role of perceived public and private green space in subjective health and wellbeing during and after the first peak of the COVID-19 outbreak. *Landsc. Urban Plan.* **2021**, *211*, 104092. [CrossRef]

17. Venter, Z.S.; Barton, D.N.; Gundersen, V.; Figari, H.; Nowell, M. Urban nature in a time of crisis: Recreational use of green space increases during the COVID-19 outbreak in Oslo, Norway. *Environ. Res. Lett.* **2020**, *15*, 104075. [CrossRef]

18. Kaźmierczak, A. The contribution of local parks to neighbourhood social ties. *Landsc. Urban Plan.* **2013**, *109*, 31–44. [CrossRef]

19. Wright Wendel, H.E.; Zarger, R.K.; Mihelcic, J.R. Accessibility and usability: Green space preferences, perceptions, and barriers in a rapidly urbanizing city in Latin America. *Landsc. Urban Plan.* **2012**, *107*, 272–282. [CrossRef]

20. Rostami, R.; Lamit, H.; Khoshnava, S.M.; Rostami, R. Successful public places: A case study of historical Persian gardens. *Urban For. Urban Green.* **2016**, *15*, 211–224. [CrossRef]

21. Wolf, I.D.; Wohlfart, T. Walking, hiking and running in parks: A multidisciplinary assessment of health and well-being benefits. *Landsc. Urban Plan.* **2014**, *130*, 89–103. [CrossRef]

22. Wolch, J.R.; Byrne, J.; Newell, J.P. Urban green space, public health, and environmental justice: The challenge of making cities 'just green enough'. *Landsc. Urban Plan.* **2014**, *125*, 234–244. [CrossRef]

23. King, D.K.; Litt, J.; Hale, J.; Burniece, K.M.; Ross, C. 'The park a tree built': Evaluating how a park development project impacted where people play. *Urban For. Urban Green.* **2015**, *14*, 293–299. [CrossRef]

24. Lee, A.C.K.; Maheswaran, R. The health benefits of urban green spaces: A review of the evidence. *J. Public Health* **2011**, *33*, 212–222. [CrossRef]

25. Van den Berg, A.E.; Jorgensen, A.; Wilson, E.R. Evaluating restoration in urban green spaces: Does setting type make a difference? *Landsc. Urban Plan.* **2014**, *127*, 173–181. [CrossRef]

26. Jones, R.; Tarter, R.; Ross, A.M. Greenspace interventions, stress and cortisol: A scoping review. *Int. J. Environ. Res. Public Health* **2021**, *18*, 2802. [CrossRef]

27. Papangelis, G.; Tombrou, M.; Dandou, A.; Kontos, T. An urban "green planning" approach utilizing the Weather Research and Forecasting (WRF) modeling system. A case study of Athens, Greece. *Landsc. Urban Plan.* **2012**, *105*, 174–183. [CrossRef]

28. Amani-Beni, M.; Zhang, B.; di Xie, G.; Xu, J. Impact of urban park's tree, grass and waterbody on microclimate in hot summer days: A case study of Olympic Park in Beijing, China. *Urban For. Urban Green.* **2018**, *32*, 1–6. [CrossRef]

29. Kuchcik, M.; Dudek, W.; Błażejczyk, K.; Milewski, P.; Błażejczyk, A. Two faces to the greenery on housing estates–mitigating climate but aggravating allergy. A Warsaw case study. *Urban For. Urban Green.* **2016**, *16*, 170–181. [CrossRef]

30. Shanahan, D.F.; Lin, B.B.; Gaston, K.J.; Bush, R.; Fuller, R.A. Socio-economic inequalities in access to nature on public and private lands: A case study from Brisbane, Australia. *Landsc. Urban Plan.* **2014**, *130*, 14–23. [CrossRef]

31. Gidlow, C.J.; Ellis, N.J.; Bostock, S. Development of the Neighbourhood Green Space Tool (NGST). *Landsc. Urban Plan.* **2012**, *106*, 347–358. [CrossRef]

32. Bahrini, F.; Bell, S.; Mokhtarzadeh, S. Urban Forestry & Urban Greening The relationship between the distribution and use patterns of parks and their spatial accessibility at the city level: A case study from Tehran, Iran. *Urban For. Urban Green.* **2017**, *27*, 332–342. [CrossRef]

33. Wei, F. Greener urbanization? Changing accessibility to parks in China. *Landsc. Urban Plan.* **2017**, *157*, 542–552. [CrossRef]

34. Braubach, M.; Egorov, P.A.; Mudu, A.T.W. The effects of urban green space on environmental health equity and resilience to extreme weather. In *Proceedings of the European Conference "Nature-Based Solutions to Climate Change in Urban Areas and Their Rural Surroundings", Bonn, Germany, 17–19 November 2015*; Kabisch, N., Stadler, J., Duffield, S., Korn, H., Bonn, A., Eds.; Bundesamt für Naturschutz (BfN): Bonn, Germany, 2015; pp. 63–65.
35. Lin, B.B.; Gaston, K.J.; Fuller, R.A.; Wu, D.; Bush, R.; Shanahan, D.F. How green is your garden? Urban form and socio-demographic factors influence yard vegetation, visitation, and ecosystem service benefits. *Landsc. Urban Plan.* **2017**, *157*, 239–246. [CrossRef]
36. Beumer, C. Show me your garden and I will tell you how sustainable you are: Dutch citizens' perspectives on conserving biodiversity and promoting a sustainable urban living environment through domestic gardening. *Urban For. Urban Green.* **2018**, *30*, 260–279. [CrossRef]
37. Taylor, J.R.; Lovell, S.T. Mapping public and private spaces of urban agriculture in Chicago through the analysis of high-resolution aerial images in Google Earth. *Landsc. Urban Plan.* **2012**, *108*, 57–70. [CrossRef]
38. Ma, A.T.H.; Lam, T.W.L.; Cheung, L.T.O.; Fok, L. Protected areas as a space for pandemic disease adaptation: A case of COVID-19 in Hong Kong. *Landsc. Urban Plan.* **2021**, *207*, 103994. [CrossRef]
39. Cunningham, G.B. Physical activity and its relationship with COVID-19 cases and deaths: Analysis of U.S. counties. *J. Sport Health Sci.* **2021**. [CrossRef]
40. Polero, P.; Rebollo-Seco, C.; Adsuar, J.C.; Pérez-Gómez, J.; Rojo-Ramos, J.; Manzano-Redondo, F.; Garcia-Gordillo, M.Á.; Carlos-Vivas, J. Physical Activity Recommendations during COVID-19: Narrative Review. *Int. J. Environ. Res. Public Health* **2020**, *18*, 65. [CrossRef]
41. Filgueira, T.O.; Castoldi, A.; Santos, L.E.R.; de Amorim, G.J.; de Sousa Fernandes, M.S.; Anastácio, W. de L. do N.; Campos, E.Z.; Santos, T.M.; Souto, F.O. The Relevance of a Physical Active Lifestyle and Physical Fitness on Immune Defense: Mitigating Disease Burden, with Focus on COVID-19 Consequences. *Front. Immunol.* **2021**, *12*, 587146. [CrossRef]
42. Sallis, R.; Young, D.R.; Tartof, S.Y.; Sallis, J.F.; Sall, J.; Li, Q.; Smith, G.N.; Cohen, D.A. Physical inactivity is associated with a higher risk for severe COVID-19 outcomes: A study in 48 440 adult patients. *Br. J. Sports Med.* **2021**, *55*, 1099–1105. [CrossRef]
43. Mitra, R.; Moore, S.A.; Gillespie, M.; Faulkner, G.; Vanderloo, L.M.; Chulak-Bozzer, T.; Rhodes, R.E.; Brussoni, M.; Tremblay, M.S. Healthy movement behaviours in children and youth during the COVID-19 pandemic: Exploring the role of the neighbourhood environment. *Heal. Place* **2020**, *65*, 102418. [CrossRef] [PubMed]
44. Puccinelli, P.J.; da Costa, T.S.; Seffrin, A.; de Lira, C.A.B.; Vancini, R.L.; Nikolaidis, P.T.; Knechtle, B.; Rosemann, T.; Hill, L.; Andrade, M.S. Reduced level of physical activity during COVID-19 pandemic is associated with depression and anxiety levels: An internet-based survey. *BMC Public Health* **2021**, *21*, 425. [CrossRef]
45. Bereitschaft, B.; Scheller, D. How Might the COVID-19 Pandemic Affect 21st Century Urban Design, Planning, and Development? *Urban Sci.* **2020**, *4*, 56. [CrossRef]
46. Sharifi, A.; Khavarian-Garmsir, A.R. The COVID-19 pandemic: Impacts on cities and major lessons for urban planning, design, and management. *Sci. Total Environ.* **2020**, *749*, 142391. [CrossRef] [PubMed]
47. Barbarossa, L. The post pandemic city: Challenges and opportunities for a non-motorized urban environment. An overview of Italian cases. *Sustainability* **2020**, *12*, 7172. [CrossRef]
48. Martínez, L.; Short, J.R. The pandemic city: Urban issues in the time of covid-19. *Sustainability* **2021**, *13*, 3295. [CrossRef]
49. Sepe, M. Covid-19 pandemic and public spaces: Improving quality and flexibility for healthier places. *URBAN Des. Int.* **2021**, *26*, 159–173. [CrossRef]
50. Rassouli, M.; Ashrafizadeh, H.; Shirinabadi Farahani, A.; Akbari, M.E. COVID-19 Management in Iran as One of the Most Affected Countries in the World: Advantages and Weaknesses. *Front. Public Health* **2020**, *8*, 2019–2021. [CrossRef]
51. Abdoli, A. Iran, sanctions, and the COVID-19 crisis. *J. Med. Econ.* **2020**, *23*, 1461–1465. [CrossRef]
52. Abdi, M.; Mirzaei, R. Iran without Mandatory Quarantine and with Social Distancing Strategy Against Coronavirus Disease (COVID-19). *Health Secur.* **2020**, *18*, 257–259. [CrossRef]
53. Allahyari, H.; Salehi, E. Presentation of a suitable approach for green programming of urban ways through integrative method CA-Markov: Case study—Azadi Street of Tehran, Iran. *Model. Earth Syst. Environ.* **2020**, *6*, 373–382. [CrossRef]
54. Ali asghar Pilehvar Spatial-geographical analysis of urbanization in Iran. *Humanit. Soc. Sci. Commun.* **2021**, *8*, 63. [CrossRef]
55. Kiani, A.; Javadiyan, M.; Pasban, V. Evaluation of Urban Green Spaces and their Impact on Living Quality of Citizens (Case Study: Nehbandan City, Iran). *J. Civ. Eng. Urban.* **2014**, *4*, 89–95.
56. Bahriny, F.; Bell, S. Patterns of urban park use and their relationship to factors of quality: A case study of tehran, Iran. *Sustain.* **2020**, *12*, 1560. [CrossRef]
57. Fasihi, H. Urban Parks and Their Accessibility in Tehran, Iran. *Environ. Justice* **2019**, *12*, 242–249. [CrossRef]
58. Naseh, M.A. *Birjand Local History at Early Contemporary Period*; Naseh, M.A., Ed.; Institute for Iranian Contemporary Historical Studies: Tehran, Iran, 2016; ISBN 9789642834778.
59. Zista, A. *Master Plan of Birjand City*; U.C.E.: Birjand, Iran, 2020.
60. Shokouhi, M.A.; Eskandari Sani, M.; Mohammadabadi, J. Strategic planning of Birjand city development with SWOT and QSPM models. *Khorasan Socio-Cultural Stud.* **2016**, *11*, 105–125.
61. Aram, F.; Higueras García, E.; Solgi, E.; Mansournia, S. Urban green space cooling effect in cities. *Heliyon* **2019**, *5*, e01339. [CrossRef]

62. Javid, M.A. *Atlas of Birjand Spatial Divisions*, 1st ed.; Fakhimzadeh, H., Ed.; Municipality of Birjand: Birjand, Iran, 2020.
63. Vargas-Hernández, J.G.; Pallagst, K.; Zdunek-Wielgołaska, J. Urban Green Spaces as a Component of an Ecosystem. In *Handbook of Engaged Sustainability*; Dhiman, S., Ed.; Springer International Publishing: Cham, Switzerland, 2018; pp. 1–32, ISBN 978-3-319-53121-2.
64. Carpentieri, G.; Guida, C.; Fevola, O.; Sgambati, S. The Covid-19 pandemic from the elderly perspective in urban areas: An evaluation of urban green areas in ten European capitals. *TeMA-J. Land Use Mobil. Environ.* **2020**, *13*, 389–408.
65. Rice, L. After Covid-19: Urban design as spatial medicine. *URBAN Des. Int.* **2020**. [CrossRef]
66. Roe, M. Editorial: Food and landscape. *Landsc. Res.* **2016**, *41*, 709–713. [CrossRef]
67. Uchiyama, Y.; Kohsaka, R. Access and use of green areas during the covid-19 pandemic: Green infrastructure management in the "new normal". *Sustainability* **2020**, *12*, 9842. [CrossRef]
68. Pamukcu-Albers, P.; Ugolini, F.; La Rosa, D.; Grădinaru, S.R.; Azevedo, J.C.; Wu, J. Building green infrastructure to enhance urban resilience to climate change and pandemics. *Landsc. Ecol.* **2021**, *36*, 665–673. [CrossRef] [PubMed]
69. Cobbinah, P.B.; Erdiaw-Kwasie, M.; Adams, E.A. COVID-19: Can it transform urban planning in Africa? *Cities Health* **2020**, 1–4. [CrossRef]
70. Hootsuite 2021. Available online: https://datareportal.com/reports/digital-2021-iran (accessed on 24 May 2021).
71. Zista, A. *Master Plan of Birjand City*; U.C.E.: Birjand, Iran, 1997.
72. Ebrahimzadeh, A. Birjand Parks and Green Space Organization. 2017. Available online: http://birjandpark.ir/index.php/info/tables (accessed on 10 January 2021).
73. Lu, Y.; Zhao, J.; Wu, X.; Lo, S.M. Escaping to nature during a pandemic: A natural experiment in Asian cities during the COVID-19 pandemic with big social media data. *Sci. Total Environ.* **2021**, *777*, 146092. [CrossRef]
74. Sivak, C.J.; Pearson, A.L.; Hurlburt, P. Effects of vacant lots on human health: A systematic review of the evidence. *Landsc. Urban Plan.* **2021**, *208*, 104020. [CrossRef]
75. WHO. Urban Planning, Environment and Health: From Evidence to Policy Action—Meeting Report. World Health Organization, 2010; p. 119. Available online: https://www.euro.who.int/__data/assets/pdf_file/0004/114448/E93987.pdf (accessed on 24 May 2021).
76. Dong, N.; Chen, J.; Zhang, S. Safety Research of Children's Recreational Space in Shanghai Urban Parks. *Procedia Eng.* **2017**, *198*, 612–621. [CrossRef]
77. Groenewegen, P.P.; van den Berg, A.E.; Maas, J.; Verheij, R.A.; de Vries, S. Is a Green Residential Environment Better for Health? If So, Why? *Ann. Assoc. Am. Geogr.* **2012**, *102*, 996–1003. [CrossRef]
78. Carver, A.; Lorenzon, A.; Veitch, J.; Macleod, A.; Sugiyama, T. Is greenery associated with mental health among residents of aged care facilities? A systematic search and narrative review. *Aging Ment. Health* **2020**, *24*, 1–7. [CrossRef]
79. Massetti, L.; Petralli, M.; Napoli, M.; Brandani, G.; Orlandini, S.; Pearlmutter, D. Effects of deciduous shade trees on surface temperature and pedestrian thermal stress during summer and autumn. *Int. J. Biometeorol.* **2019**, *63*, 467–479. [CrossRef]
80. Säumel, I.; Hogrefe, J.; Battisti, L.; Wachtel, T.; Larcher, F. The healthy green living room at one's doorstep? Use and perception of residential greenery in Berlin, Germany. *Urban For. Urban Green.* **2021**, *58*, 126949. [CrossRef]
81. Szczepańska, A.; Pietrzyka, K. The COVID-19 epidemic in Poland and its influence on the quality of life of university students (young adults) in the context of restricted access to public spaces. *J. Public Health* **2021**, *7*, 1–11. [CrossRef]
82. Flouri, E.; Midouhas, E.; Joshi, H. The role of urban neighbourhood green space in children's emotional and behavioural resilience. *J. Environ. Psychol.* **2014**, *40*, 179–186. [CrossRef]
83. Xie, J.; Luo, S.; Furuya, K.; Sun, D. Urban parks as green buffers during the COVID-19 pandemic. *Sustainability* **2020**, *12*, 6751. [CrossRef]
84. Chen, J.; Chang, Z. Rethinking urban green space accessibility: Evaluating and optimizing public transportation system through social network analysis in megacities. *Landsc. Urban Plan.* **2015**, *143*, 150–159. [CrossRef]
85. Zhang, H.; Chen, B.; Sun, Z.; Bao, Z. Landscape perception and recreation needs in urban green space in Fuyang, Hangzhou, China. *Urban For. Urban Green.* **2013**, *12*, 44–52. [CrossRef]
86. Gong, L.; Mao, B.; Qi, Y.; Xu, C. A satisfaction analysis of the infrastructure of country parks in Beijing. *Urban For. Urban Green.* **2015**, *14*, 480–489. [CrossRef]
87. Wen, C.; Albert, C.; Von Haaren, C. Equality in access to urban green spaces: A case study in Hannover, Germany, with a focus on the elderly population. *Urban For. Urban Green.* **2020**, *55*, 126820. [CrossRef]
88. Lennon, M. Green space and the compact city: Planning issues for a 'new normal'. *Cities Health* **2020**, 1–4. [CrossRef]
89. Grinde, B.; Patil, G.G. Biophilia: Does visual contact with nature impact on health and well-being? *Int. J. Environ. Res. Public Health* **2009**, *6*, 2332–2343. [CrossRef] [PubMed]
90. Ugolini, F.; Massetti, L.; Pearlmutter, D.; Sanesi, G. Usage of urban green space and related feelings of deprivation during the COVID-19 lockdown: Lessons learned from an Italian case study. *Land Use Policy* **2021**, *105*, 105437. [CrossRef]
91. Ulrich, R. View through a window may influence recovery from surgery. *Science* **1984**, *224*, 420–421. [CrossRef] [PubMed]
92. Gilchrist, K.; Brown, C.; Montarzino, A. Workplace settings and wellbeing: Greenspace use and views contribute to employee wellbeing at peri-urban business sites. *Landsc. Urban Plan.* **2015**, *138*, 32–40. [CrossRef]
93. Birge, D.; Mandhan, S.; Qiu, W.; Berger, A.M. Potential for sustainable use of trees in hot arid regions: A case study of Emirati neighborhoods in Abu Dhabi. *Landsc. Urban Plan.* **2019**, *190*, 103577. [CrossRef]

94. Chalmin-Pui, L.S.; Roe, J.; Griffiths, A.; Smyth, N.; Heaton, T.; Clayden, A.; Cameron, R. "It made me feel brighter in myself"—The health and well-being impacts of a residential front garden horticultural intervention. *Landsc. Urban Plan.* **2021**, *205*, 103958. [CrossRef]
95. Seidl, M.; Saifane, M. A green intensity index to better assess the multiple functions of urban vegetation with an application to Paris metropolitan area. *Environ. Dev. Sustain.* **2021**, *23*, 15204–15224. [CrossRef]
96. Irvine, K.N.; Fuller, R.A.; Devine-Wright, P.; Tratalos, J.; Payne, S.R.; Warren, P.H.; Lomas, K.J.; Gaston, K.J. Ecological and Psychological Value of Urban Green Space. In *Dimensions of the Sustainable City. Future City*; Jenks, M., Jones, C., Eds.; Springer: Dordrecht, The Netherlands, 2010; Volume 2, pp. 215–237. [CrossRef]
97. Chang, C.-R.; Chen, M.-C.; Su, M.-H. Natural versus human drivers of plant diversity in urban parks and the anthropogenic species-area hypotheses. *Landsc. Urban Plan.* **2021**, *208*, 104023. [CrossRef]
98. Carrus, G.; Scopelliti, M.; Lafortezza, R.; Colangelo, G.; Ferrini, F.; Salbitano, F.; Agrimi, M.; Portoghesi, L.; Semenzato, P.; Sanesi, G. Go greener, feel better? The positive effects of biodiversity on the well-being of individuals visiting urban and peri-urban green areas. *Landsc. Urban Plan.* **2015**, *134*, 221–228. [CrossRef]
99. Marselle, M.R.; Irvine, K.N.; Lorenzo-Arribas, A.; Warber, S.L. Does perceived restorativeness mediate the effects of perceived biodiversity and perceived naturalness on emotional well-being following group walks in nature? *J. Environ. Psychol.* **2016**, *46*, 217–232. [CrossRef]
100. Henwood, K.; Pidgeon, N. Talk about woods and trees: Threat of urbanization, stability, and biodiversity. *J. Environ. Psychol.* **2001**, *21*, 125–147. [CrossRef]
101. Harvey, D.J.; Montgomery, L.N.; Harvey, H.; Hall, F.; Gange, A.C.; Watling, D. Psychological benefits of a biodiversity-focussed outdoor learning program for primary school children. *J. Environ. Psychol.* **2020**, *67*, 101381. [CrossRef]
102. Hadavi, S.; Kaplan, R.; Hunter, M.C.R. Environmental affordances: A practical approach for design of nearby outdoor settings in urban residential areas. *Landsc. Urban Plan.* **2015**, *134*, 19–32. [CrossRef]
103. Mouratidis, K.; Poortinga, W. Built environment, urban vitality and social cohesion: Do vibrant neighborhoods foster strong communities? *Landsc. Urban Plan.* **2020**, *204*, 103951. [CrossRef]
104. Pham, T.T.H.; Labbé, D.; Lachapelle, U.; Pelletier, É. Perception of park access and park use amongst youth in Hanoi: How cultural and local context matters. *Landsc. Urban Plan.* **2019**, *189*, 156–165. [CrossRef]
105. Cameron-Faulkner, T.; Melville, J.; Gattis, M. Responding to nature: Natural environments improve parent-child communication. *J. Environ. Psychol.* **2018**, *59*, 9–15. [CrossRef]
106. Zhang, L.; Tan, P.Y.; Diehl, J.A. A conceptual framework for studying urban green spaces effects on health. *J. Urban Ecol.* **2017**, *3*, 1–13. [CrossRef]
107. Putra, I.G.N.E.; Astell-Burt, T.; Cliff, D.P.; Vella, S.A.; Feng, X. Association between caregiver perceived green space quality and the development of prosocial behaviour from childhood to adolescence: Latent class trajectory and multilevel longitudinal analyses of Australian children over 10 years. *J. Environ. Psychol.* **2021**, *74*, 101579. [CrossRef]
108. Olszewska-Guizzo, A.; Fogel, A.; Escoffier, N.; Ho, R. Effects of COVID-19-related stay-at-home order on neuropsychophysiological response to urban spaces: Beneficial role of exposure to nature? *J. Environ. Psychol.* **2021**, *75*, 101590. [CrossRef]
109. Bulfone, T.C.; Malekinejad, M.; Rutherford, G.W.; Razani, N. Outdoor Transmission of SARS-CoV-2 and Other Respiratory Viruses: A Systematic Review. *J. Infect. Dis.* **2021**, *223*, 550–561. [CrossRef]
110. Schipperijn, J.; Ekholm, O.; Stigsdotter, U.K.; Toftager, M.; Bentsen, P.; Kamper-Jørgensen, F.; Randrup, T.B. Factors influencing the use of green space: Results from a Danish national representative survey. *Landsc. Urban Plan.* **2010**, *95*, 130–137. [CrossRef]
111. Roe, J.J.; Ward Thompson, C.; Aspinall, P.A.; Brewer, M.J.; Duff, E.I.; Miller, D.; Mitchell, R.; Clow, A. Green space and stress: Evidence from cortisol measures in deprived urban communities. *Int. J. Environ. Res. Public Health* **2013**, *10*, 4086–4103. [CrossRef] [PubMed]
112. Theodorou, A.; Panno, A.; Carrus, G.; Carbone, G.A.; Massullo, C.; Imperatori, C. Stay home, stay safe, stay green: The role of gardening activities on mental health during the Covid-19 home confinement. *Urban For. Urban Green.* **2021**, *61*, 127091. [CrossRef]
113. Hostetler, M.; Noiseux, K. Are green residential developments attracting environmentally savvy homeowners? *Landsc. Urban Plan.* **2010**, *94*, 234–243. [CrossRef]
114. Keefer, L.A.; Landau, M.J.; Rothschild, Z.K.; Sullivan, D. Attachment to objects as compensation for close others' perceived unreliability. *J. Exp. Soc. Psychol.* **2012**, *48*, 912–917. [CrossRef]

Article

How Has the COVID-19 Pandemic Affected the Perceptions of Public Space Employees?

Soyoung Han [1], **Cermetrius Lynell Bohannon** [1] and **Yoonku Kwon** [2,*]

[1] College of Architecture and Urban Studies, Virginia Tech, Blacksburg, VA 24060, USA; syhan@vt.edu (S.H.); cbohanno@vt.edu (C.L.B.)

[2] Department of Landscape Architecture, Chonnam National University, Gwangju 61186, Korea

* Correspondence: ykkwon@chonnam.ac.kr

Abstract: The purpose of this study is to derive the subjective perception about COVID-19 of public space employees and to identify the characteristics of COVID-19 related issues. By using the Q-method, 24 workers in four public spaces located in Blacksburg, Virginia, USA were selected as P-Sample and Q-Sorting was conducted. Three types of perceptions were identified; Type 1 (Expansion of Non-Face-To-Face Service), Type 2 (Expansion of Professional Labor), and Type 3 (Expansion of Welfare Service Type). All three types recognized that when a confirmed case occurs in a public space, the right and safety of users or communities to know is important, and accurate information must be provided, because it is necessary to prevent the spread of infection. Above all, these results show another side of the COVID-19 situation, as the participants in this study are currently in charge of various tasks such as quarantine and service provision in public spaces. This study can be used as basic data for policy response and system improvement of public spaces in the event of an infectious disease such as COVID-19 in the future.

Keywords: COVID-19; public space; perceptions; employees; Q-methodology

Citation: Han, S.; Bohannon, C.L.; Kwon, Y. How Has the COVID-19 Pandemic Affected the Perceptions of Public Space Employees?. *Land* **2021**, *10*, 1332. https://doi.org/10.3390/land10121332

Academic Editors: Francesca Ugolini and David Pearlmutter

Received: 24 October 2021
Accepted: 2 December 2021
Published: 3 December 2021

Publisher's Note: MDPI stays neutral with regard to jurisdictional claims in published maps and institutional affiliations.

1. Introduction

Public spaces including urban parks and plazas are universal leisure facilities where people spend a lot of time, and their functions are very diverse. Public spaces play a role in managing the health of people, such as operating various leisure programs, volunteering, and exercising. As the scenery of the public life, these places are of prime significance in various fields of study in urbanism and the history of city planning [1,2]. Various types of public spaces are significant elements of any habitable, sustainable urban development [3–5]. The concept of a public space raises different functions of outdoor public assembly spots [6,7] and neighboring spaces between buildings [8,9]. Key examples of public spaces in urban areas are cafes, retail bazaars, theme parks, streets, and pedestrian walkways [10–12]. Due to these various functions, high-quality public space is positioned as a crucial system in our society in improving the quality of life of people.

However, the COVID-19 pandemic, which has been going on since early 2020, is having a huge impact on our lives. Preventive measures against infectious diseases that have never been experienced, such as social distancing and wearing masks, had to be implemented in daily life. As this situation continues, members of society of all ages are having a hard time enduring various types of difficulties [13]. These difficulties are particularly evident in public spaces. Restrictions on the use of public spaces are one of the key policy measures to reduce the spread of COVID-19 and protect public health [14,15]. In other words, in a pandemic environment, public spaces often limit service provision and are repeatedly closed for a certain period of time. As the media and many scholars have already pointed out, there has been widespread recognition that the COVID-19 pandemic would demand a New Normal and will completely change the flow of world civilization [16].

Most of the preceding research on COVID-19 has been limited to establishing an infectious disease response system in the health sector. These studies mainly related to the quarantine method, monitoring system and information sharing, epidemiological investigation, diagnostic testing, crisis situation determination and response process, preparation and response for infectious disease management in medical institutions, and local responses [17–19]. Moreover, in the medical field, research on infectious disease mechanisms, treatment technologies, vaccines, diagnostic technologies, surveillance epidemiology, and infrastructure construction are being carried out [20,21]. When it comes to infectious diseases, policies aimed at reducing social contact and limiting mobility have been used for centuries [22–24]. Movement has been shown in numerous studies to influence disease transmission and incidence [25,26]. Therefore, most of the studies on the interrelation between the pandemic and spatial aspects aim to measure the level of social interaction and mobility for a large population in a large geographic area and then quantify its transmission rate and impact. To date, various studies worldwide have researched COVID-19 using geographic information systems (GIS), which focused on spatiotemporal analysis, data mining, environmental variables, and health and social geography [27–29].

As social distancing measures are vital part of mitigating pandemics [30], planners, designers, architects, landscape designers, and journalists are already predicting and studying how this crisis will change the interactions between people in public spaces [15,31]. In a pandemic situation, employees of public space are members of organizations who play an important role in providing services in accordance with the government's quarantine guidelines. So far, there have been a few studies regrading employees' physical health, but no concrete studies on the perception of their workspace and the workload for which they are responsible. They may experience various social, psychological, health difficulties, and conflicts as they have to take on the role of supporters in managing visitors on the front line from the risk of an infectious disease they have never experienced before. However, as mentioned above, most of the studies related to infectious diseases have been conducted as large-scale and quantitative studies. Measuring and quantifying the level of social interaction and mobility for large populations over large geographic areas is often not feasible. Thus, the purpose of this study is to derive the subjective perception about COVID-19 of public space employees and to identify the characteristics of COVID-19 related issues. To this end, Q-Methodology, which is a useful analysis method for measuring the subjective perception on a specific topic [32,33], has been used in this study. In particular, we posit the preparation in relation to policy development for public space management by examining the perception of workers in public places about COVID-19 in the process of coping with the pandemic.

2. Public Space and Psychological Distance during the Pandemic

Recent studies have shown that access to urban spaces has positive effects during the situation of social distancing [34,35]. Therefore, in order to accurately predict the user pattern of public spaces in the post-pandemic era, it is necessary to understand the factors affecting outdoor activities due to the pandemic at the level of individuals who are consumers and providers of outdoor activities.

Historically, public spaces have been regarded as timeless, transformative, and elusive entities for a city's urbanism [36]. A public space that reflects a wider society [37] has been perceived as providing different functions to users in different ways; this became a precursor in questioning the public space beyond a concept of singularity [38]. It is critical to first take into account the conscious structuring of places in an attempt to minimize unsettling encounters with people who could threaten the accepted narratives of a specific place [39]. In addition, the ways that public spaces are inclusive and exclusive at the same time presents a paradoxical situation. For example, public spaces can be relatively more accessible while being more closed, depending upon the individual and the associated public.

Before COVID-19, the discussion around the term *public space* led to definitions that captured various characteristics. Definitions of public space evolved around considerations of ownership [39], human behavior [40], the democraticness and responsiveness of places [41], and accessibility [42]. Seemingly shared and accessible public open spaces have been replaced by more orderly places exposed to control, power, exclusion, and inaccessible narratives [39,43–46]. Uncertainty of diversity, urban spontaneity, and captivation of urban flavor have been replaced by expectations and knowledge of the quality of the urban environment. This is a cumulative result of a planning and governance structure that responds (or fails to respond) to deeper structural changes occurring in society [47].

However, COVID-19 has significantly highlighted the lack of accessible and usable public space. Continuous development trends have deprived people of adequate local public spaces or suitable alternatives in their homes, including semi-private or semi-public spaces [48].

The low proximity between social actors in an urban environment is a natural side effect of increasing urban density, and in return, it appears to be associated with the spread of infectious diseases [49]. In this regard, urban morphology has been shown to influence the spread of COVID-19 in a variety of ways, for example, access to open spaces [50]. Obviously, the transmission of COVID-19 has a strong proximal dimension as it requires close personal contact [51]. Therefore, effective policies for social distancing come from changing social norms or limiting activities in public spaces where reduction in personal distance is inevitable.

During the COVID-19 pandemic, most people isolated themselves from physical contact while working from home, using digital connections, and using digital public spaces to maintain social distancing [52]. Most people have shifted from the traditional way of working to a "do it yourself in the living room" approach, where the home is now a production space where workshops are organized to perform work tasks and duties. Salama [30] maintains that currently we are in a transition period called the 'new normal', which will eventually become a stable condition of the actual normal. However, from a psychological point of view, social distance is classified as a type of psychological distance [53,54]. Psychological distance refers to how close an object is psychologically to the now, here, or self. Types of psychological distance include temporal distance, spatial (or spatial) distance, social distance, and virtual (or probabilistic) distance [53]. In this psychological distance, differences lead to differences in responses to objects. For example, people's perception of global warming was different when they heard that a natural disaster related to global warming occurred in their area. [55–59]. Therefore, examining public space employees' perceptions of COVID-19 could be a way of seeing the psychological distance to COVID-19. In addition, this may provide a clue for the direction of the space operation program in the post-COVID-19 era.

3. Methodology

3.1. Q-Methodology

This study used the Q-methodology to identify and categorize the subjective perceptions of COVID-19 workers in public spaces. The Q-methodology is one of the factor analysis methods in which the unit of analysis becomes a human and groups people with similar response patterns. In particular, this method is suitable for this study in that it is a method for objectively measuring values, beliefs, attitudes, etc., which can be regarded as the subjective domain of human beings, in that a more objective approach to the domain of human subjectivity is possible. In other words, in objectifying subjectivity, it is a method of gathering people who have similar reactions to a specific object or issue and confirming the contents of these people's reactions.

3.2. Q-Statements

The Q-Statement is a statement that is representative of the research topic and functions similarly to the scale of quantitative research. Q-Statements refer to the totality of feelings and opinions shared within a culture as the sum of all subjective statements that each individual can communicate about a research topic [60,61]. Q-Statements collection can be accomplished by mobilizing various methods, for example, the researcher contacts people related to the research topic and conducts an in-depth interview or reviews various types of literature related to the research topic [60].

In this study, Q-Statements were selected by conducting in-depth interviews and literature research on public space workers. In-depth interviews were conducted with five urban park managements using unstructured open-ended questions. Based on the results of these in-depth interviews and literature review, a total of 98 questions directly and indirectly related to the response of public spaces workers to COVID-19 were derived as the first statement. Among the first statements, similar or duplicate items were deleted and merged, and 55 items were extracted secondly, and after verification by two professors of landscape architecture, 32 items were finally selected as a Q-Statements. The 32 Q-Statements are shown in Table 1 below.

Table 1. Finalized Q-Statements.

No.	Q-Statements
1	I am very concerned about the possibility of infection among employees and users in public spaces.
2	Users are reluctant to use the service due to concerns about the COVID-19 infection.
3	When a confirmed case occurs in a public space, there is concern about it being reported in the media.
4	Users' physical function was greatly reduced due to the closure of public spaces and restrictions on outdoor activities.
5	Psychological anxiety, depression, and feelings of isolation increased due to restrictions on the use of visitors who used public spaces on a regular basis.
6	The closure of public spaces will affect the mental health of users who frequently use periodic activity programs.
7	The most serious problem is that the vulnerable are at risk by closing public spaces or providing non-face-to-face services.
8	Because vulnerable groups have many limitations in using non-face-to-face services, the problem of closure of public spaces cannot be solved with non-face-to-face services alone.
9	In order to be able to provide online services, it is necessary to educate public space employees on the production of related contents.
10	Budget support and infrastructure (Wi-Fi, video production, and editing facilities, etc.) for providing online services in public spaces should be provided.
11	The provision of online services has limitations due to publicity, gaps in users' ability to use, and decreased access to devices.
12	Due to COVID-19, office work such as recording and reporting is overloaded and difficult.
13	Due to COVID-19, disinfection and quarantine work is overloaded.
14	Wearing a mask all the time is difficult.
15	In response to COVID-19, staff in public spaces are well coordinated and controlled.
16	Public spaces are rather safe zones because they are thoroughly controlled and quarantined as much as possible.
17	Currently, local governments are responding relatively well to COVID-19 in consideration of the overall situation of public spaces.
18	It is important to actively cooperate with government policies on quarantine.
19	The operation of public spaces is actively responding well to the prevention of COVID-19.
20	Because public spaces are closed, it's hard because people assume that employees are resting.
21	Education on health and infectious diseases, such as quarantine and disinfection, is sufficiently provided.
22	In public spaced, efforts are being made to develop programs suitable for non-face-to-face situations or services for high-risk groups.
23	In the future, measures should be taken to examine and prepare for various aftereffects on dogs caused by COVID-19.

Table 1. *Cont.*

No.	Q-Statements
24	The minimum face-to-face programs or services that must be operated should be maintained.
25	A response system should be prepared for crises and emergency cases.
26	Support for quarantine items and quarantine should be given priority (for example, daily disinfection support by a professional quarantine company, portable thermal imaging camera, etc.).
27	It is necessary to diversify the working patterns of public space employees (such as telecommuting, flexible, selective work, etc.).
28	When a confirmed case occurs, it is necessary to have a press guideline that reports only the type and area of the welfare center without disclosing the name of the public space.
29	It is also important for each public space to make its own efforts to respond to COVID-19.
30	Replacement labor should be hired for the exclusion of self-quarantining employees.
31	There is also a need for medical personnel who can reside in public spaces.
32	During the COVID-19 period, any program that can maintain the previous daily life should be strengthened.

3.3. P-Sample

The P-Sample means selecting the research subjects who will respond to the Q-Sample. Since the Q-Methodology aims to grasp the subjective perception of a specific topic rather than the generalization of the research, the subjects related to the research purpose are generally composed of the P-Sample [62,63].

In the Q-methodology, the P-sample is based more on the depth of experience and information on the research question than on the representativeness of the population [61,64]. Therefore, in this study, it was considered that the workers related to public space programs were appropriate for analysis in that they had a relatively comprehensive understanding of public space management. The sampling method was snow-balling. This is a strategy mainly used by researchers to increase social access to upper echelon research subjects. On the other hand, the disadvantage of this method is that members in the human network based on friendship can maintain homogeneous values and perspectives. In order to control this, the researcher explicitly revealed these issues to the research subjects during the sampling process and asked them to direct individuals who are connected based on formal-functional relationships rather than acquaintances. The sampling results are presented in Table 2 below. In consideration of the social sensitivity of the research subject, the study subjects were anonymized, and only the minimum identification information meaningful to the study was disclosed. Therefore, this study consisted of 24 workers in four public spaces located in Blacksburg, Virginia in USA as a P-Sample. Although the number of P sets was small, major actors from the public spaces were generally included.

Table 2. Composition of P sets.

General Category	Specific Category	Number
Public Park	Program Director	3
	Recreation Coordinator	5
	Recreation Assistant	6
Trail	Program Director	2
	Activities Assistant	4
Park and Natural Area	Conservation Manager	1
	Program Assistant	3
Total		24

All the data for analysis were collected from 13 October to 10 December 2020. The data collection was conducted by visiting each public space directly by the researcher and was conducted in compliance with the quarantine guidelines for the prevention of COVID-19. In order to comply with research ethics, the purpose of the research was fully explained before the investigation, consent to participate in the research was obtained, confidentiality of personal information, free participation in research, and guidance on Q-Methodology were individually implemented.

When examining the characteristics of the P-Sample participating in this study, focusing on gender, age, and work experience, the gender was found to be 61.6% female and 38.4% male. By age group, 21.7% were in their 20s, 43.5% in their 30s, and 30.4% in their 40s.

3.4. Q-Sorting

Q-Sorting is the process of classifying the subjects according to the degree to which they agree or disagree with each Q-statement. This process focuses on how the entire item is distributed to each individual rather than for or against a specific item. One can arrange the samples according to the relative importance of the Q-Statements. There are two types of Q-Statements classification methods: the free distribution method, which is arbitrarily classified by the research subjects, and the forced distribution method, which usually presents the classification framework in the form of a normal distribution [33,65]. This study was classified according to the forced distribution method. First, the subjects of the P sample read each statement card and grasped its contents, and then, in a broad framework, were first classified into three groups: disagree, neutral, and agree. Finally, the Q-Statements were arranged to have a distribution as shown in Figure 1 below, and after classification, the reasons for choosing each of the two statements with the most agreement or the most disagreement were specifically stated.

Card	−4	−3	−2	−1	0	1	2	3	4
(Score)	(2)	(3)	(4)	(4)	(6)	(4)	(4)	(3)	(2)

Figure 1. Distributions of Q-Sorting.

3.5. Q-Analysis

The stage of analysis in the Q-method refers to the stage of categorizing individuals according to the degree of cognitive similarity through the scores surveyed based on the responses of the questionnaire discussed above, and capturing features or finding correlations based on the typed results [61,62,66]. The response values given to the questions were integer values ranging from −4 (most negative) to 4 (most positive). In previous studies, values ranging from −5 to +5 were often used, but since the number of questions (N = 32) used in this study was relatively small, the range of response values was reduced to control outliers in the results.

The PQMethod analysis program (ver 2.35) was used for typing and correlation analysis, and correlations and factors of items were analyzed by varimax rotation. The researcher gave appropriate labels to the clusters of the types that appeared as a result of the analysis to intuitively indicate the classification and characteristics between the groups. Response values corresponding to each type were standardized and replaced with a 'Z-Score', and their intensity was measured as an indicator of the relative distances in which the responses were distributed. Because Z-Scores have a standardized distribution, they provide a standardized framework for comparing the strengths of each response value in comparisons between types. In the interpretation of the results, it was necessary to accompany the process of supplementing the interpretation based on the fact that response results with factor weights of 1.00 or more usually contain singularities. Thus, each comment (follow-up interview) was received from the participants for the statements placed at both ends, and the contents of representative comments were used as quotations if necessary to help the reader understand.

4. Results

4.1. Recognition Types and Structures

As a result of the analysis, the perception of workers in public places about COVID-19 can be divided into three types. Each type was derived based on the significance of uniqueness and correlation between types. The eigen values of each type were 6.6418 for Type 1, 3.5115 for Type 2, and 2.1238 for Type 3. The explanatory variables of each type were 23% for Type 1, 14% for Type 2, and 14% for Type 3, and the total explanatory variance for the three types was 51%. The number of workers belonging to the three types was twelve in Type 1, six in Type 2, and five in Type 3, a total of twenty-three people, and the other one was not included in any of the types (Table 3).

Table 3. Eigenvalues and explanatory variables of types.

	Type 1	Type 2	Type 3
Eigen value	6.6418	3.5115	2.1238
Explanatory variable	23%	14%	14%

As presented in Table 4, the correlation between Type 1 and Type 3 was r = 0.462, Type 1 and Type 2 was r = 0.062, and the correlation between Type 2 and Type 3 was r = 0.221, indicating a relatively low correlation. A low correlation indicates that each type has high independent explanatory power, and it can be interpreted that the subjective perceptions of respondents belonging to each type are different from each other.

Table 4. Correlation between types.

	Type 1	Type 2	Type 3
Type 1	1		
Type 2	0.062	1	
Type 3	0.462	0.221	1

4.2. Characteristics by Type

The analysis of the characteristics of each type is as follows, focusing on the reasons for selecting the statement for each type and the statement with the most agreement or disagreement.

4.2.1. Type 1: Expansion of Non-Face-To-Face Service

The statements with which Type 1 agrees or disagrees at a standard score of +1 or higher are shown in Table 5 below. Participants belonging to Type 1 agreed that making efforts to develop programs suitable for non-face-to-face situations or services for high-risk groups in public spaces is significant. They were also actively responding well to the prevention of COVID-19, and recognized that it was important to actively cooperate with the government's quarantine policy. In addition, they anticipated that regular visitors felt that depression, anxiety, and feelings of isolation increased due to restrictions on the use of public spaces, and they perceived limitations in providing online services due to the gap in visitors' ability to use and access. On the other hand, they did not have any concerns about a confirmed case among public space users being reported to the media, and did not agree with the need for reporting guidelines such as not disclosing the specific name of a public space when a confirmed case occurs. In addition, they disagreed that disinfection and office work were overloaded due to the COVID-19, and that visitors were reluctant to use public spaces due to concerns about COVID-19 infection.

Table 5. Representative Q Statements and Standard Scores of Type 1.

No.	Q-Statement	Z-Score
22	In public space, efforts are being made to develop programs suitable for non-face-to-face situations or services for high-risk groups.	1.724
11	The provision of online services has limitations due to publicity, gaps in users' ability to use, and decreased access to devices.	1.303
19	The operation of public spaces is actively responding well to the prevention of COVID-19.	1.275
18	It is important to actively cooperate with government policies on quarantine.	1.181
5	Psychological anxiety, depression, and feelings of isolation increased due to restrictions on the use of visitors who used public spaces on a regular basis.	1.178
3	When a confirmed case occurs in a public space, there is about it being reported in the media.	−1.806
13	Due to COVID-19, disinfection and quarantine work is overloaded.	−1.788
28	When a confirmed case occurs, it is necessary to have a press guideline that reports only the type and area of the welfare center without disclosing the name of the public space.	−1.780
2	Users are reluctant to use the service due to concerns about COVID-19 infection.	−1.728
12	Due to COVID-19, office work such as recording and reporting is overloaded and difficult.	−1.691

Looking more specifically at the reasons for choosing the statements that the type 1 participants most agreed or disagreed with, first, they said that public space actively copes with the situation; "We are developing and providing a variety of face-to-face and non-face-to-face programs that can be conducted flexibly in accordance with the quarantine guidelines. This has resulted in a positive response from visitors (Respondents 2, 12, 19, and 21)". On the other hand, they also revealed the limitations of providing non-face-to-face or online services; "it is regrettable because we know that online access is very low for the vulnerable group (Respondent 14), most seniors are vulnerable to digital capabilities, and some programs of public spaces are limited in non-face-to-face application to them because social-interaction is essential (Respondents 6 and 21), and that the provision of services using smartphones is limited by region and by class (digital underprivileged) is a problem that is already being discussed in the field (Respondent 22)".

Meanwhile, "COVID-19 is contagious, so it is important to cooperate with the government guidelines prepared by experts to prevent the spread of infection quickly, and the safety of users with weak immunity is more important than worrying about the news and publicity. It is necessary to be careful (Respondents 11 and 14); if there is a confirmed case, the information should be disclosed that the area is at risk (Respondent 2 and 13)". In addition, "there are many inquiries about the use of public spaces, and users have a lot of opinions that it is better to proceed with the service, and they think that the desire to use the service has rather increased as it has been prolonged (Respondents 2, 5, and 11). They recognized that information about the occurrence of an infected person in the relevant place should be clearly communicated.

4.2.2. Type 2: Expansion of Professional Manpower

Statements in which Type 2 agreed or disagreed at a standard score of +1 or higher are presented in Table 6 below. Workers belonging to Type 2 thought that medical experts who can reside in public spaces are required, and they also recognized that they are making efforts to develop programs suitable for non-face-to-face situations or services for high-risk users. In addition, although the employees are cooperating and controlling the COVID-19 situation well, they also felt difficulties as the disinfection, quarantine, and office work were overloaded due to COVID-19.

On the other hand, they did not agree on the necessity of reporting guidelines, such as not disclosing the name of a public place when a confirmed case occurred. They also felt that local governments were not handling the situation in public spaces well. They disagreed on the necessity of supporting living services for visitors or the importance of the efforts of public places themselves in responding to COVID-19. They recognized that wearing a mask does not feel burdensome to them.

Table 6. Representative Q Statements and Standard Scores of Type 2.

No.	Q-Statement	Z-Score
31	There is also a need for medical personnel who can reside in public spaces.	1.724
22	In public space, efforts are being made to develop programs suitable for non-face-to-face situations or services for high-risk groups.	1.303
13	Due to COVID-19, disinfection and quarantine work is overloaded.	1.275
12	Due to COVID-19, office work such as recording and reporting is overloaded and difficult.	1.181
15	In response to COVID-19, staff in public spaces are well coordinated and controlled.	1.178
28	When a confirmed case occurs, it is necessary to have a press guideline that reports only the type and area of the welfare center without disclosing the name of the public space.	−1.806
17	Currently, local governments are responding relatively well to COVID-19 in consideration of the overall situation of public spaces.	−1.788
32	During the COVID-19 period, a program that can maintain the previous daily life should rather be strengthened.	−1.780
29	It is also important for each public space to make its own efforts to respond to COVID-19.	−1.728
14	Wearing a mask all the time is difficult.	−1.691

Specifically, looking at the reasons why public space workers belonging to Type 2 chose the statements that they most agreed or disagreed with, first of all, regarding the need for medical experts as employees in public spaces, "because many visitors need them (Respondent 1), all employees can learn medical knowledge through education, but their expertise is low, and the current workforce is too heavy (Respondents 3 and 7)". In addition, about the overload of work due to the COVID-19, "I am embarrassed to do things I have never experienced before. The burden and fear of the process are great (Respondent 1), and although it is natural process for disinfection and quarantine to be carried out for visitors, there are situations in which the quarantine task is more important than the actual program (Respondent 17)". In addition, when it comes to reporting guidelines when a confirmed patient occurs, "Basically, in public facilities, users have the right to know, so it is necessary to provide accurate information about the confirmed patient (Respondent 18), the burden is high and each public space has its own characteristics, so guidelines that take into account the situation are needed (Respondent 1), although individual efforts are important at the level of public space, a joint response is needed more (Respondent 1)".

Combining the above analysis, Type 2 was labeled 'Expansion of Professional Labor'. The most prominent characteristic of public space workers belonging to Type 2 is that they recognize that medical experts as employees in public spaces are needed above all else. In addition, this is a type that feels that the work such as disinfection, quarantine, and office work is overloaded due to COVID-19.

4.2.3. Type 3: Expansion of Welfare Service Type

The statements with which Type 3 agree or disagree at the standard score of +1 or higher are shown in Table 7 below. Participants belonging to Type 3 recognized that the most serious problem is that regular visitors who visit for health care are at risk due to the closure of public spaces, which increases psychological anxiety and depression among regular users. For this reason, they believed that non-face-to-face service alone cannot solve the problem of closure, and that online service provision has limitations due to gaps in users' ability to use and reduced accessibility. In addition, they were concerned that an infected person may come out among public space employees and users, but they recognized that public space management is responding well to COVID-19 prevention. They also considered that budget support and infrastructure should be prepared for providing online services within public institutions.

Table 7. Representative Q Statements and Standard Scores of Type 3.

No.	Q-Statement	Z-Score
7	The most serious problem is that the vulnerable are at risk by closing public spaces or providing non-face-to-face services.	1.794
5	Psychological anxiety, depression, and feelings of isolation increased due to restrictions on the use of visitors who used public spaces on a regular basis.	1.566
8	Because the vulnerable group has many limitations in using non-face-to-face services, the problem of closure of public spaces cannot be solved with non-face-to-face services alone.	1.350
1	I am very concerned about the possibility of infection among employees and users in public spaces.	1.293
19	The operation of public spaces is actively responding well to the prevention of COVID-19.	1.183
11	The provision of online services has limitations due to publicity, gaps in users' ability to use, and decreased access to devices.	1.119
10	Budget support and infrastructure (Wi-Fi, video production, and editing facilities, etc.) for providing online services in public spaces should be provided.	1.047
31	There is also a need for medical personnel who can reside in public spaces.	−1.808
26	Support for quarantine items and quarantine should be given priority (for example, daily disinfection support by a professional quarantine company, portable thermal imaging camera, etc.).	−1.324
30	Replacement labor should be hired for the exclusion of self-quarantining employees.	−1.297
28	When a confirmed case occurs, it is necessary to have a press guideline that reports only the type and area of the welfare center without disclosing the name of the public space.	−1.253
32	During the COVID-19 period, a program that can maintain the previous daily life should rather be strengthened.	−1.216
13	Due to COVID-19, disinfection and quarantine work is overloaded.	−1.159
27	It is necessary to diversify the working patterns of public space employees (such as telecommuting, flexible, selective work, etc.).	−1.145
14	Wearing a mask all the time is difficult.	−1.074

On the other hand, they do not agree with the necessity of medical personnel residing in public spaces, priority support for quarantine, and replacement of staff for self-quarantine. In addition, they do not agree with the strengthening of daily life services, the diversification of work patterns for public space employees in the future, the confidentiality of the public space name in case of a confirmed case. Wearing a mask while working is not as uncomfortable for them as Type 2.

Specifically, looking at the reasons why workers belonging to Type 3 chose the statement that they most agreed or disagreed with, first, regarding the vulnerable group in the context of the suspension of public space operation, "Since the meaning of public space is very significant in the daily life of the elderly of the vulnerable class, it is appropriate to say that daily life has collapsed, and proper services are not provided to them at present. (Respondents 16 and 23), I am feeling and witnessing it (respondents 4, 16, and 24), in the era of the COVID-19 pandemic, face-to-face and non-face-to-face services must be conducted together, so prepared infrastructure for environment is essential (Respondent 24)". Also, regarding the need for medical experts in public spaces, "currently, medical personnel are resident (Respondent 15); employees are well-trained, and it is more burdensome for more medical personnel to come (Respondent 4)". On the other hand, there were opinions about the quarantine tasks such as "as it is a necessary and natural task, I do not feel that the quarantine tasks are excessive, it must be done unconditionally to the extent possible and all employees share the task (Respondents 16 and 24)".

Combining the above analysis, the Type 3 were named 'Expansion of Welfare Service Type'. Public space workers belonging to Type 3 perceived that the health of visitors deteriorates due to closure of public spaces and the increase in vulnerable groups at risk are the most serious problems. While they felt that welfare services should be expanded in preparation for this, they also had a negative perception of the need for medical personnel or substitute personnel in public spaces.

4.3. Comparison between Types

First of all, Type 1 and Type 2 had different perceptions of work overload due to COVID-19. Type 2 thought that work was overloaded due to COVID-19, whereas Type 1 disagreed (see Table 8).

Table 8. Comparison between Type 1 and Type 2.

	Q-Statement	Z-Score		
		Type 1	Type 2	Difference
13	Due to COVID-19, disinfection and quarantine work is overloaded.	−1.788	1.608	−3.396
12	Due to COVID-19, office work such as recording and reporting is overloaded and difficult.	−1.691	1.484	−3.175

There was no significant difference between Type 1 and Type 3. On the other hand, Type 2 and Type 3 had conflicting perceptions about the need for resident labor in public spaces, as shown in Table 9. In other words, Type 2 recognized the need for medical personnel to reside in public spaces, whereas Type 3 did not.

Table 9. Comparison between Type 2 and Type 3.

	Q-Statement	Z-Score		
		Type 1	Type 2	Difference
31	There is also a need for medical personnel who can reside in public spaces.	1.987	−1.808	−3.795

Meanwhile, the three types of common perception are shown in Table 10. All three types actively supported the disclosure of the names of public spaces in the event of a confirmed case in common. This suggests that all three types comply with the local residents' right to know by providing accurate information and agree with the community's safety and infection prevention policy.

Table 10. Common items between the 3 types.

	Q-Statement	Z-Score
28	There is also a need for medical personnel who can reside in public spaces.	−1.654

5. Discussion

All three types recognized that when a confirmed case occurs in public space, the right and safety of users or communities to know is crucial, and accurate information must be provided as it is necessary to prevent the spread of infection. Above all, these results show other sides of the COVID-19 situation as the participants of this study are currently in charge of various tasks such as quarantine and service provision in public spaces. In seeking specific countermeasures based on these types of perceptions of workers, it is necessary to first consider the common perceptions of each type. On the other hand, for the part that shows contradictory perceptions, it is necessary to examine the regulations and circumstances surrounding public space more precisely. Based on this, specific policy recommendations for the operation of public spaces when an infectious disease such as COVID-19 spreads in the future are presented as follows.

First of all, due to the closure of public spaces and the limitations of service provision due to COVID-19, users who regularly visit public spaces for the purpose of exercise face various difficulties such as physical, psychological, and daily life. In particular, since this deteriorates the health and quality of life of the vulnerable, such as the elderly [67,68], it is necessary to establish a safe and stable service provision system even in the case of an infectious disease. In other words, in the event of an infectious disease outbreak, specific measures should be prepared so that the service can be safely and stably provided, rather than a total service cessation. In particular, since the elderly gain vitality from meeting and talking in public spaces and checking each other's safety, face-to-face services as small as possible should be provided while complying with infectious disease-related regulations. For this, it is necessary to consider how to organize operating costs, subdivide labor, or restructure staffing as needed.

Secondly, the current operating guidelines for public spaces do not presuppose the situation of infectious diseases. Therefore, taking into account the opinions of the medical community who are concerned about the continued occurrence of infectious diseases [69,70], it is necessary to prepare operational guidelines and evaluation guidelines suitable for the situation and conditions of public spaces that can respond to infectious diseases in the future. In an existing study, an Australian case presented a manual on procedures, education and worker training, and a monitoring system to provide care to clients at the time of MERS [71], and the elderly at home in response to this COVID-19 As in the case of the publication of rules for providing meal services for children [72], specific operational guidelines and appropriate public space evaluation guidelines should be prepared so that public spaces can provide services even in an infectious disease situation. In addition, as the provision of non-face-to-face services such as phone safety checks increases, work standards for telecommuting, etc., should be revised together.

Thirdly, it is necessary to prepare an integrated face-to-face and non-face-to-face welfare platform to respond more actively to COVID-19. One of the methods of providing services in public spaces in response to infectious diseases is the provision of non-face-to-face services using IT technology [72–74]. To this end, access to media should be strengthened for users and workers in public spaces, and not only equipment and facilities for online education should be reinforced, but also online utilization capabilities should be improved. In particular, since digital access is weak in the case of the elderly, it is necessary to prepare an education program to bridge the gap. In addition, it is necessary to strengthen the capacity of workers to produce and edit online videos or operate equipment, so that video production is possible in the public space itself. Therefore, it is necessary to secure the budget for these changes and to flexibly reorganize the institution.

Fourthly, programs that should be operated as face-to-face services and tasks to discover and manage target audiences must be combined. The occurrence of infectious diseases is more dangerous in vulnerable groups [75,76], and it is necessary to demonstrate expertise in case management for the vulnerable to minimize the negative impact of infectious diseases on the vulnerable [75,77,78]. In addition, as a result of this study, some cognitive types (Type 2) suggested the need to expand physical and mental health professionals such as nurses and mental health experts from the existing social worker-centered human resources structure in public spaces. This provides a clue that it may be necessary to provide services in public spaces to reorganize the workforce structure to constantly respond to infectious diseases in the future.

Fifthly, according to the research of Northwestern University in the USA, in the correlation between vitamin D and the death rate of COVID-19, patients in countries with high mortality rates such as Italy had lower vitamin D levels compared to countries that did not. These results show the importance of outdoor programs where people can be in sunlight. Therefore, in preparation for the post-coronavirus era, it is necessary to prepare various public space programs that are not limited to indoor public space programs, but expand outdoor programs such as walking on outdoor trails taking social distancing into consideration.

Finally, to prepare for the post-COVID-19 era, a big data management system to overcome the crisis should be established. In order to properly intervene by time, target, high-risk person, and project while considering the infectious disease warning stage and the distancing stage [79], some data on public space users must be established. This should provide individual customized services by building data on users, such as whether emergency measures are necessary according to the status of general visitors of public spaces.

6. Conclusions

This study attempted to investigate the situation of public spaces due to COVID-19 in more detail by understanding the types of perceptions of COVID-19 workers in public spaces. Based on this, we attempted to explore appropriate countermeasures for public spaces in the outbreak of a large-scale infectious disease such as COVID-19. By using the Q-method, 24 workers in four public spaces located in Blacksburg, Virginia, USA were selected as P-Sample and Q-Sorting was conducted. To summarize the study results, there were three types of subjective perception types of public space workers regarding COVID-19. Type 1 was named as 'Expansion of Non-Face-To-Face Service', and employees in this type perceived that they were actively and flexibly responding to the situations arousing in public space due to COVID-19. In addition, they were concerned about the worsening situation of particularly vulnerable users due to restrictions on the use of the public space. They also recognized that information about the occurrence of an infected person in the public space should be clearly communicated. Type 2 was named as 'Expansion of Professional Labor', and it was found that workers belonging to Type 2 recognized that medical personnel as employees in public spaces were needed above all else. As a type of person who felt that disinfection, quarantine, and office work were overloaded due to the COVID-19, they considered that a government response was necessary in consideration of the overall situation of public spaces. Type 3 was labeled as 'Expansion of Welfare Service Type', and they perceived that the health of visitors deteriorates due to closure of public spaces and the increase in vulnerable groups at risk are the most serious problems. While it is necessary to expand welfare services, they had a negative perception about the need for resident medical personnel or substitutes in public spaces.

This study attempted to make policy recommendations on the perception of infectious diseases for public space workers in the context of infectious diseases and the direction of service provision in public spaces based on this. By doing so, this study can be used as basic data for policy response and system improvement of public spaces in the event of an infectious disease such as COVID-19 in the future. This study also has an implication that the results provide a starting point for public spaces to provide continuous services and programs even in the case of future infectious diseases. The spread of the disease has caused a state of being classified as a new normal due to social distancing measures, and is characterized by separation, isolation, and involvement in the virtual world, which focuses on telecommuting through the use of information and communication technology. As a result of attempts to limit the spread of the disease, the acceptability of the new normal seems to be a stable state in the future, that is, a catalyst for the actual normal. While dealing with health in the post-epidemic virtual world, negative consequences appear for many people around the world to work in public spaces.

However, although this study systematically classifies workers' perceptions of public space services in the context of infectious diseases, there is a limitation in not providing concrete and specific reasons why participants perceive them as they are. Moreover, given the novelty of COVID-19 and other factors that may impede predictions of behavioral changes (age, politics, and necessity), it is difficult to predict whether COVID-19 will lead to long-term changes in human behavior in public places [80]. Therefore, in-depth studies that complement these limitations should be continued in the future. Another strand of qualifying studies examining the influence of infectious diseases on social interaction in our daily life should be conducted to understand the mechanism between social interaction and mobility [81].

Author Contributions: Conceptualization, S.H.; methodology, S.H.; writing—original draft preparation, S.H.; writing—review and editing, Y.K.; revisions, C.L.B. All authors have read and agreed to the published version of the manuscript.

Funding: This research received no external funding.

Institutional Review Board Statement: Ethical review and approval were waived for this study, due to involving no more than minimal risk.

Informed Consent Statement: Informed consent was obtained from all subjects involved in the study.

Data Availability Statement: Not applicable.

Conflicts of Interest: The authors declare no conflict of interest.

References

1. Bada, Y.; Farhi, A. Experiencing Urban Spaces: Isovists Properties and Spatial Use of Plazas. *Courr. Du Savoir* **2009**, *9*, 101–112.
2. Bada, Y.; Guney, Y.I. Visibility and Spatial Use in Urban Plazas: A Case Study from Biskam Algeria. In Proceedings of the 7th International Space Syntax Symposium, Stockholm, Sweden, 8–11 June 2009.
3. Ancell, S.; Thompson-Fawcett, M. The Social Sustainability of Medium Density Housing: A Conceptual Model and Christchurch Case Study. *Hous. Stud.* **2008**, *23*, 423–442. [CrossRef]
4. Hajjari, M. Improving Urban Life through Urban Public Spaces: A Comparison between Iranian and Australian Cases. *Universitas* **2009**, *21*, 15–20.
5. Németh, J. Defining a Public: The Management of Privately Owned Public Space. *Urban. Stud.* **2009**, *46*, 2463–2490. [CrossRef]
6. Law, L. Defying Disappearance: Cosmopolitan Public Spaces in Hong Kong. *Urban. Stud.* **2002**, *39*, 1625–1645. [CrossRef]
7. Yeoh, B.S.; Huang, S. Negotiating Public Space: Strategies and Styles of Migrant Female Domestic Workers in Singapore. *Urban. Stud.* **1998**, *35*, 583–602. [CrossRef]
8. AbuGhazzeh, T.M. Reclaiming Public Spaces in Space: The Ecology of Neighborhood Open the Town of Abu4Nuseir. *Jordan Landsc. Urban. Plan.* **1996**, *36*, 16.
9. Ford, L.; Trachte, J.A. *The Spaces between Buildings*; JHU Press: Baltimore, MD, ISA, 2000.
10. Chua, B.H.; Edwards, N. *Public Space: Design, Use and Management*; NUS Press: Singapore, 1992.
11. Cybriwsky, R. Changing Patterns of Urban Public Space: Observations and Assessments from the Tokyo and New York Metropolitan Areas. *Cities* **1999**, *16*, 223–231. [CrossRef]
12. Kayden, J.S. *Privately Owned Public Space: The New York City Experience*; John Wiley & Sons: Hoboken, NJ, USA, 2000.
13. Low, S.; Smart, A. Thoughts about Public Space during COVID-19 Pandemic. *City Soc.* **2020**, *32*. [CrossRef]
14. Cheshmehzangi, A. 10 Adaptive Measures for Public Places to Face the COVID 19 Pandemic Outbreak. *City Soc.* **2020**, *32*. [CrossRef] [PubMed]
15. Honey-Rosés, J.; Anguelovski, I.; Chireh, V.K.; Daher, C.; Konijnendijk van den Bosch, C.; Litt, J.S.; Mawani, V.; McCall, M.K.; Orellana, A.; Oscilowicz, E.; et al. The Impact of COVID-19 on Public Space: An Early Review of the Emerging Questions–Design, Perceptions and Inequities. *Cities Health* **2020**, 1–17. [CrossRef]
16. Han, S.; Sim, J.; Kwon, Y. Recognition Changes of the Concept of Urban Resilience: Moderating Effects of COVID-19 Pandemic. *Land* **2021**, *10*, 1099. [CrossRef]
17. Reshetnikov, V.; Mitrokhin, O.; Shepetovskaya, N.; Belova, E.; Jakovljevic, M. Organizational Measures Aiming to Combat COVID-19 in the Russian Federation: The First Experience. *Expert Rev. Pharm. Outcomes Res.* **2020**, *20*, 571–576. [CrossRef] [PubMed]
18. Krstic, K.; Westerman, R.; Chattu, V.K.V.; Ekkert, N.; Jakovljevic, M. Corona-Triggered Global Macroeconomic Crisis of the Early 2020s. *Int. J. Environ. Res. Public Health* **2020**, *17*, 9404. [CrossRef] [PubMed]
19. Lai, K.Y.; Webster, C.; Kumari, S.; Sarkar, C. The Nature of Cities and the COVID-19 Pandemic. *Curr. Opin. Environ. Sustain.* **2020**, *46*, 27–31. [CrossRef]
20. Ranabhat, C.L.; Jakovljevic, M.M.; Kim, C.-B. COVID-19 Pandemic: An Opportunity for Universal Health Coverage? *Front. Public Health* **2021**, *9*, 1057. [CrossRef]
21. Carvalho, K.; Vicente, J.P.; Jakovljevic, M.; Teixeira, J.P.R. Analysis and Forecasting Incidence, Intensive Care Unit Admissions, and Projected Mortality Attributable to COVID-19 in Portugal, the UK, Germany, Italy, and France: Predictions for 4 Weeks Ahead. *Bioengineering* **2021**, *8*, 84. [CrossRef]
22. Cetron, M.; Simone, P. Battling 21st-Century Scourges with a 14th-Century Toolbox. *Emerg. Infect. Dis.* **2004**, *10*, 2053. [CrossRef]
23. Gensini, G.F.; Yacoub, M.H.; Conti, A.A. The Concept of Quarantine in History: From Plague to SARS. *J. Infect.* **2004**, *49*, 257–261. [CrossRef] [PubMed]
24. Tognotti, E. Lessons from the History of Quarantine, from Plague to Influenza A. *Emerg. Infect. Dis.* **2013**, *19*, 254. [CrossRef]
25. Mari, L.; Bertuzzo, E.; Righetto, L.; Casagrandi, R.; Gatto, M.; Rodriguez-Iturbe, I.; Rinaldo, A. Modelling Cholera Epidemics: The Role of Waterways, Human Mobility and Sanitation. *J. R. Soc. Interface* **2012**, *9*, 376–388. [CrossRef] [PubMed]
26. Jang, W.M.; Jang, D.H.; Lee, J.Y. Social Distancing and Transmission-Reducing Practices during the 2019 Coronavirus Disease and 2015 Middle East Respiratory Syndrome Coronavirus Outbreaks in Korea. *J. Korean Med. Sci.* **2020**, *35*, e220. [CrossRef]
27. Andersen, L.M.; Harden, S.R.; Sugg, M.M.; Runkle, J.D.; Lundquist, T.E. Analyzing the Spatial Determinants of Local COVID-19 Transmission in the United States. *Sci. Total. Environ.* **2021**, *754*, 142396. [CrossRef] [PubMed]
28. Franch-Pardo, I.; Napoletano, B.M.; Rosete-Verges, F.; Billa, L. Spatial Analysis and GIS in the Study of COVID-19. A Review. *Sci. Total. Environ.* **2020**, *739*, 140033. [CrossRef] [PubMed]
29. Hu, M.; Roberts, J.D.; Azevedo, G.P.; Milner, D. The Role of Built and Social Environmental Factors in COVID-19 Transmission: A Look at America's Capital City. *Sustain. Cities Soc.* **2021**, *65*, 102580. [CrossRef]

30. Salama, A.M. Coronavirus Questions That Will Not Go Away: Interrogating Urban and Socio-Spatial Implications of COVID-19 Measures. *Emerald Open Res.* **2020**, *2*. [CrossRef]
31. Bereitschaft, B.; Scheller, D. How Might the COVID-19 Pandemic Affect 21st Century Urban Design, Planning, and Development? *Urban. Sci.* **2020**, *4*, 56. [CrossRef]
32. Watts, S.; Stenner, P. Doing Q Methodological Research. In *Theory, Method & Interpretation*; SAGE Publications Ltd.: Thousand Oaks, NY, USA, 2012.
33. Zabala, A.; Sandbrook, C.; Mukherjee, N. When and How to Use Q Methodology to Understand Perspectives in Conservation Research. *Conserv. Biol.* **2018**, *32*, 1185–1194. [CrossRef]
34. Aktay, A.; Bavadekar, S.; Cossoul, G.; Davis, J.; Desfontaines, D.; Fabrikant, A.; Gabrilovich, E.; Gadepalli, K.; Gipson, B.; Guevara, M.; et al. Google COVID-19 Community Mobility Reports: Anonymization Process Description (Version 1.1). *arXiv* **2004**, arXiv:2004.04145.
35. Samuelsson, K.; Barthel, S.; Colding, J.; Macassa, G.; Giusti, M. Urban Nature as a Source of Resilience during Social Distancing amidst the Coronavirus Pandemic. *DiVA* **2020**. [CrossRef]
36. Sengupta, U. Ruptured Space and Spatial Estrangement: (Un)Making of Public Space in Kathmandu. *Urban. Stud.* **2017**, *55*, 2780–2800. [CrossRef]
37. Madanipour, A. *Whose Public Space. Whose Public Space?: International Case Studies in Urban Design and Development*; Routledge: London, UK, 2010; p. 237.
38. Webster, C. Property Rights, Public Space and Urban Design. *Town Plan. Rev.* **2007**, *78*, 81–101. [CrossRef]
39. Kohn, M. *Brave New Neighborhoods: The Privatization of Public Space*; Routledge: London, UK, 2004.
40. Gehl, J.; Gemzoe, L. *Public Spaces Public Life*; The Royal Danish Academy of Fine: Copenhagen, Denmark, 1996.
41. Carr, S.; Francis, M.; Rivlin, L.G.; Stone, A.M. *Public Space*; Cambridge University Press: Cambridge, UK, 1992.
42. Carmona, M.; De Magalhaes, C.; Hammond, L. *Public Space: The Management Dimension*; Routledge: London, UK, 2008.
43. Davis, M. Fortress Los Angeles: The Militarization of Urban Space. In *Variations on a Theme Park: The New American City and the End of Public Space*; Sorkin, M., Ed.; Hill & Wang: New York, NY, USA, 1992; pp. 154–180.
44. Davis, M. Fortress Los Angeles: The Militarization of Urban Space. In *Cultural Criminology*; Routledge: London, UK, 2017; pp. 287–314.
45. Mitchell, D. Introduction: Public Space and the City. *Urban. Geogr.* **1996**, *17*, 127–131. [CrossRef]
46. Mitchell, D. *The Right to the City: Social Justice and the Fight for Public Space*; Guilford Press: New York, NY, USA, 2003.
47. Henry, C.; Foss, L.; Fayolle, A.; Walker, E.; Duffy, S. Entrepreneurial Leadership and Gender: Exploring Theory and Practice in Global Contexts. *J. Small Bus. Manag.* **2015**, *53*, 581–586. [CrossRef]
48. Kordshakeri, P.; Fazeli, E. How the COVID-19 Pandemic Highlights the Lack of Accessible Public Spaces in Tehran. *Cities Health* **2020**, 1–3. [CrossRef]
49. Stier, A.J.; Berman, M.G.; Bettencourt, L. COVID-19 Attack Rate Increases with City Size. *arXiv* **2020**, arXiv:2003.10376.
50. Liu, L. Emerging Study on the Transmission of the Novel Coronavirus (COVID-19) from Urban Perspective: Evidence from China. *Cities* **2020**, *103*, 102759. [CrossRef] [PubMed]
51. Iranmanesh, A.; Alpar Atun, R. Reading the Changing Dynamic of Urban Social Distances during the COVID-19 Pandemic via Twitter. *Eur. Soc.* **2021**, *23*, S872–S886. [CrossRef]
52. Ülkeryıldız, E. Transformation of Public and Private Spaces: Instrumentality of Restrictions on the Use of Public Space during COVID 19 Pandemic. In Proceedings of the International Conference of Contemporary Affairs in Architecture and Urbanism (ICCAUA-2020), Alanya, Turkey, 6–8 May 2020; Volume 6, p. 8.
53. Trope, Y.; Liberman, N. Construal-Level Theory of Psychological Distance. *Psychol. Rev.* **2010**, *117*, 440. [CrossRef]
54. Boothby, E.J.; Smith, L.K.; Clark, M.S.; Bargh, J.A. Psychological Distance Moderates the Amplification of Shared Experience. *Personal. Soc. Psychol. Bull.* **2016**, *42*, 1431–1444. [CrossRef]
55. Brügger, A.; Dessai, S.; Devine-Wright, P.; Morton, T.A.; Pidgeon, N.F. Psychological Responses to the Proximity of Climate Change. *Nat. Clim. Chang.* **2015**, *5*, 1031–1037. [CrossRef]
56. Chu, H.; Yang, J.Z. Emotion and the Psychological Distance of Climate Change. *Sci. Commun.* **2019**, *41*, 761–789. [CrossRef]
57. McDonald, R.I.; Chai, H.Y.; Newell, B.R. Personal Experience and the 'Psychological Distance' of Climate Change: An Integrative Review. *J. Environ. Psychol.* **2015**, *44*, 109–118. [CrossRef]
58. Richards, G. The European Cultural Capital Event: Strategic Weapon in the Cultural Arms Race? *Int. J. Cult. Policy* **2000**, *6*, 159–181. [CrossRef]
59. Schuldt, J.P.; Rickard, L.N.; Yang, Z.J. Does Reduced Psychological Distance Increase Climate Engagement? On the Limits of Localizing Climate Change. *J. Environ. Psychol.* **2018**, *55*, 147–153. [CrossRef]
60. Armatas, C.A.; Venn, T.J.; Watson, A.E. Applying Q-Methodology to Select and Define Attributes for Non-Market Valuation: A Case Study from Northwest Wyoming, United States. *Ecol. Econ.* **2014**, *107*, 447–456. [CrossRef]
61. Chapman, R.; Tonts, M.; Plummer, P. Exploring Perceptions of the Impacts of Resource Development: A Q-Methodology Study. *Extr. Ind. Soc.* **2015**, *2*, 540–551. [CrossRef]
62. Bartlett, J.E.; DeWeese, B. Using the Q Methodology Approach in Human Resource Development Research. *Adv. Dev. Hum. Resour.* **2015**, *17*, 72–87. [CrossRef]
63. Brown, S.R. Q Methodology and Qualitative Research. *Qual. Health Res.* **1980**, *6*, 561–567. [CrossRef]

64. Stephenson, W. Q-Methodology, Interbehavioral Psychology, and Quantum Theory. *Psychol. Rec.* **1982**, *32*, 235.
65. Cross, R.M. Exploring Attitudes: The Case for Q Methodology. *Health Educ. Res.* **2005**, *20*, 206–213. [CrossRef] [PubMed]
66. Addams, H.; Proops, J.L. *Social Discourse and Environmental Policy: An Application of Q Methodology*; Edward Elgar Publishing: Cheltenham, UK, 2000.
67. Greenberg, N.; Docherty, M.; Gnanapragasam, S.; Wessely, S. Managing Mental Health Challenges Faced by Healthcare Workers during COVID-19 Pandemic. *BMJ* **2020**, *368*. [CrossRef] [PubMed]
68. Koh, D. Migrant Workers and COVID-19. *Occup. Environ. Med.* **2020**, *77*, 634–636. [CrossRef] [PubMed]
69. Chan, K.H.; Poon, L.L.; Cheng, V.; Guan, Y.; Hung, I.; Kong, J.; Yam, L.Y.; Seto, W.H.; Yuen, K.Y.; Peiris, J.S.M. Detection of SARS Coronavirus in Patients with Suspected SARS. *Emerg. Infect. Dis.* **2004**, *10*, 294. [CrossRef] [PubMed]
70. He, F.; Deng, Y.; Li, W. Coronavirus Disease 2019: What We Know? *J. Med Virol.* **2020**, *92*, 719–725. [CrossRef]
71. Felemban, O.; John, W.S.; Shaban, R.Z. Infection Prevention and Control in Home Nursing: Case Study of Four Organisations in Australia. *Br. J. Community Nurs.* **2015**, *20*, 451–457. [CrossRef]
72. Middleton, A.; Simpson, K.N.; Bettger, J.P.; Bowden, M.G. COVID-19 Pandemic and beyond: Considerations and Costs of Telehealth Exercise Programs for Older Adults with Functional Impairments Living at Home—Lessons Learned from a Pilot Case Study. *Phys. Ther.* **2020**, *100*, 1278–1288. [CrossRef]
73. Godwin, K.M.; Mills, W.L.; Anderson, J.A.; Kunik, M.E. Technology-Driven Interventions for Caregivers of Persons with Dementia: A Systematic Review. *Am. J. Alzheimers Dis. Other Dement.* **2013**, *28*, 216–222. [CrossRef] [PubMed]
74. Somerville, C.; Coyle, C.; Mutchler, J. *Responding to COVID-19: How Massachusetts Senior Centers Are Adapting*; UMASS Boston: Boston, MA, USA, 2020.
75. Arango, C. Lessons Learned from the Coronavirus Health Crisis in Madrid, Spain: How COVID-19 Has Changed Our Lives in the Last 2 Weeks. *Biol. Psychiatry* **2020**, *88*, e33. [CrossRef]
76. Nikolaidis, A.; Paksarian, D.; Alexander, L.; Derosa, J.; Dunn, J.; Nielson, D.M.; Droney, I.; Kang, M.; Douka, I.; Bromet, E.; et al. The Coronavirus Health and Impact Survey (CRISIS) Reveals Reproducible Correlates of Pandemic-Related Mood States across the Atlantic. *Sci. Rep.* **2021**, *11*, 8139. [CrossRef]
77. Hunter, P. The Spread of the COVID-19 Coronavirus: Health Agencies Worldwide Prepare for the Seemingly Inevitability of the COVID-19 Coronavirus Becoming Endemic. *EMBO Rep.* **2020**, *21*, e50334. [CrossRef] [PubMed]
78. Walter-McCabe, H.A. Coronavirus Health Inequities in the United States Highlight Need for Continued Community Development Efforts. *Int. J. Community Soc. Dev.* **2020**, *2*, 211–233. [CrossRef]
79. Lai, J.; Ma, S.; Wang, Y.; Cai, Z.; Hu, J.; Wei, N.; Wu, J.; Du, H.; Chen, T.; Li, R.; et al. Factors Associated with Mental Health Outcomes among Health Care Workers Exposed to Coronavirus Disease 2019. *JAMA Netw. Open* **2020**, *3*, e203976. [CrossRef] [PubMed]
80. James, A.C. Don't Stand so Close to Me: Public Spaces, Behavioral Geography, and COVID-19. *Dialogues Hum. Geogr.* **2020**, *10*, 187–190. [CrossRef]
81. Gholami, A.; Shekari, T.; Aminifar, F.; Shahidehpour, M. Microgrid Scheduling with Uncertainty: The Quest for Resilience. *IEEE Trans. Smart Grid* **2016**, *7*, 2849–2858. [CrossRef]

land

MDPI

Article

Who Started, Stopped, and Continued Participating in Outdoor Recreation during the COVID-19 Pandemic in the United States? Results from a National Panel Study

B. Derrick Taff [1], William L. Rice [2,*], Ben Lawhon [3] and Peter Newman [1]

[1] Department of Recreation, Park and Tourism Management, Pennsylvania State University, University Park, State College, PA 16802, USA; bdt3@psu.edu (B.D.T.); pbn3@psu.edu (P.N.)
[2] Parks, Tourism, and Recreation Management Program, University of Montana, Missoula, MT 59812, USA
[3] The Leave No Trace Center for Outdoor Ethics, Boulder, CO 80306, USA; ben@LNT.org
* Correspondence: william.rice@umontana.edu; Tel.: +1-406-243-5477

Citation: Taff, B.D.; Rice, W.L.; Lawhon, B.; Newman, P. Who Started, Stopped, and Continued Participating in Outdoor Recreation during the COVID-19 Pandemic in the United States? Results from a National Panel Study. *Land* **2021**, *10*, 1396. https://doi.org/10.3390/land10121396

Academic Editors: Francesca Ugolini and David Pearlmutter

Received: 4 November 2021
Accepted: 14 December 2021
Published: 17 December 2021

Publisher's Note: MDPI stays neutral with regard to jurisdictional claims in published maps and institutional affiliations.

Abstract: The COVID-19 pandemic has been proposed as a catalyst for many U.S. residents to re-engage in outdoor recreation or engage in outdoor recreation for the first time. This manuscript describes the results of a representative U.S. national panel study aimed at better understanding the socio-demographic profile (gender, ethnicity, community type, income, and age) of those participants new to outdoor recreation since the start of the COVID-19 pandemic. In doing so, we address how these new outdoor recreationists differ from (1) those who frequently participated in outdoor recreation prior to the pandemic and continue to participate in outdoor recreation, (2) those who did not frequently participate in outdoor recreation prior to the pandemic and remain un-engaged, and (3) those who frequently participated in outdoor recreation prior to the pandemic but stopped their frequent participation following the onset of the pandemic. Results from this U.S. national study suggest that 35.8% of respondents indicated that they did not participate regularly in outdoor recreation prior to the pandemic or during the pandemic, 30.4% indicated that they did participate regularly in outdoor recreation prior to the pandemic and continued to do so regularly during the pandemic, and 13.5% indicated that they did participate regularly in outdoor recreation prior to the pandemic, but did not continue to do so during the pandemic. More than 20% of the sample indicated that they were new outdoor recreationists. The majority of respondents in all categories, including those that were new to outdoor recreation amidst the pandemic, identified as being white, however these new outdoor recreationists were also the least ethnically diverse. The previously but no longer outdoor recreationist respondents were significantly more ethnically diverse than the other three groups, and they tended to live in more urbanized settings. Discussion of these results includes implications for outdoor recreation managers, and researchers who seek to better understand who the COVID-19 pandemic has influenced with regard to outdoor recreation participation. Implications regarding social justice, access and equity to public places that facilitate outdoor recreation, and health-related policies are discussed.

Keywords: COVID-19; pandemic; outdoor recreation; health; participation

1. Introduction

Outdoor recreation participation in the U.S. has been steadily increasing since the 1960s [1,2]. However, the COVID-19 pandemic, declared by the World Health Organization in March of 2020, has been touted as significantly accelerating this broader increase in outdoor recreation [3], although impacting different settings and demographics disproportionately [4,5]. This has led to concerns and challenges for public land area managers and tourism operators as they attempted to manage outdoor recreation as it fluctuated (i.e., non-existent during closures; in some settings, all-time highs during openings) as the pandemic and associated health measures evolved [6].

Research regarding outdoor recreation trends associated with the pandemic suggest that despite some pandemic-related closures of recreation areas, outdoor recreation use spiked significantly in some settings [7–9]. For example, in large western U.S. national parks such as Yellowstone, Grand Canyon, and Yosemite, visitation increased significantly immediately following reduced restrictions on access [10]. Another study suggested that the number of hikers in the U.S. increased by approximately 135% from 2019 to 2020 [11]. However, the above findings reflect certain locations and demographics, as other research has noted decreases in use in locations such as urban parks, and discrepancies in visitation frequency with specific demographics [4,5]. Similarly, recreation behaviors of avid outdoor recreationists living in urban areas in the U.S. were more severely impacted by the pandemic than those residing in rural areas [12], and decisions to pursue outdoor recreation were influenced heavily by perceptions of risk [13,14]. Motivations for pursuing outdoor recreation during this period have been largely driven by needs associated with personal health and well-being [15], such as relieving stress and supporting mental health, and facilitating physical health through outdoor exercise [16]. And, research in the U.S. indicates that outdoor recreation during the pandemic has in fact, led to improved mental health outcomes [17,18]. Similar health-related trends have been discovered internationally [5,19,20].

Despite the limited but growing body of research regarding outdoor recreation as it relates to the pandemic, very little is known about U.S. outdoor recreation participants during this time. For example, are there new outdoor recreationists, driven to participation by the pandemic? The Outdoor Industry Association and Naxion Research Consulting [21] began exploring this question through a national panel of $n = 613$ participants that had either "engaged in outdoor activities for the first time or for the first time in more than a year since the onset of the COVID-19 pandemic" or "participated in outdoor activities before the pandemic" (p. 21). This study indicated that outdoor recreationists during the pandemic tend to consist of more females, are generally younger, more likely to live in urban areas, and represent a slightly lower socioeconomic bracket than pre-pandemic outdoor recreation participants [21]. While these findings represent an important initial step in improving understanding of outdoor recreationists during the pandemic, the criteria for being considered as a new or existing outdoor recreation participant was as little as one occurrence [21]. Furthermore, the study was purposefully weighted to match the U.S. population for gender, ethnicity, age, household income, and region [21]. This robust census-matching approach is common for U.S. panel studies, but by purposefully defining the demographic criteria, it may inaccurately represent actual new or existing outdoor recreation participants. Additionally, the criteria for being considered as a "new outdoor participant" was fairly minimal (i.e., as little as one occurrence) and not mutually exclusive (i.e., first time in more than a year since the onset of the COVID-19 pandemic" or "participated in outdoor activities before the pandemic") [21]. Finally, other pandemic and recreation-related research has highlighted inequities and decreases in recreational opportunities for some demographics amidst the pandemic [4,5,12], further driving the need for more understanding regarding recreation behaviors and demographics during this unique period of time.

The purpose of this study was to further advance this initial research by examining U.S. outdoor recreation demographic trends amidst the pandemic, by applying slightly different methodological approaches than those used by the Outdoor Industry Association and Naxion Research Consulting [21]. Additionally, given that there are currently no national studies examining the demographics of new, current, and non-outdoor recreationist during this unique and challenging time (at the time of this publication nearly two years since the World Health Organization's pandemic declaration), this study advances understanding of these trends as well. Specifically, the purpose of this study is to examine how new outdoor recreationists differ from those who frequently participated in outdoor recreation prior to the pandemic and continue to participate in outdoor recreation, those who did not frequently participate in outdoor recreation prior to the pandemic and remain un-engaged, and finally, those who frequently participated in outdoor recreation prior to the pandemic but stopped their frequent participation following the onset of the pandemic.

2. Materials and Methods

This study used panel data, collected through Qualtrics International Inc. following Institutional Review Board approval through the lead author's University. Data were collected through a paid panel of participants purposefully targeting adults age 18 or older, residing within one of four geographically representative regions of the U.S., including the Northeast, South, Midwest, and West [22]. While one main purpose was to further define new outdoor recreation participants amidst the pandemic, we first screened participants to determine who is, who is not, and who was but is no longer an outdoor recreationist. Our goal was to collect a geographically representative sample of $n = 1000$ new outdoor recreationists in the U.S. This sample estimate was informed by empirical political science studies regarding sample size and inference, and recent nationwide COVID-19 vaccination compliance research using similar sample sizes e.g., [23,24].

These data were not purposefully controlled to match the U.S. census, but rather were allowed to be collected in a manner that represents the U.S. geographically, based on outdoor recreation participation prior to and during the pandemic. This allowed us to determine who these individuals are, rather than place census parameters in which demographics may not match these outdoor recreationists groups. Panel respondents were screened by inquiring whether or not they were "new or returning to frequent outdoor recreation since the COVID-19 pandemic began" (Table 1). This was defined for respondents as "someone who has started pursuing activities undertaken for recreation in nature-based environments that generally involve some level of intentional physical exertion, at least ONCE PER MONTH since March 2020, when the COVID-19 pandemic began."

Table 1. Example of screening questions and subsequent groupings.

	Participant Group			
	Non-Outdoor Recreationist	Existing Outdoor Recreationist	Previous, but No Longer Outdoor Recreationist	New Outdoor Recreationist
Participated in outdoor recreation regularly *prior to* pandemic?	No	Yes	Yes	No
Participated in outdoor recreation regularly *during* pandemic?	No	Yes	No	Yes

The definition of outdoor recreation was derived from the U.S. Bureau of Economic Analysis's definition: "All recreational activities undertaken for pleasure that generally involve some level of intentional physical exertion and occur in nature-based environments outdoors" [25] (p. 4). This more stringent definition and criteria for evaluating outdoor recreation participation was purposefully chosen to provide information for land managers and the outdoor recreation industry regarding consistent and fairly frequent outdoor recreationists. Outdoor recreation activities such as hiking, running outdoors, fishing, gardening, camping or RVing, boating or sailing, and surfing for example, listed in the survey instrument, mimicked previous research on national outdoor recreation participation [21]. Respondents were categorized into the four groups listed in Table 1 and responses in all categories propagated until the quota of $n = 1000$ new outdoor recreationists in the U.S. was met. Thus, data collection began on the 10th of May 2021, and concluded on the 13th of June 2021, allowing for more than one-year of pandemic conditions to be considered as context for the respondents.

Qualtrics conducted ongoing data cleaning throughout this period to ensure the sample was collecting valid responses. Two binary questions asking about the same behaviors were placed at different points in the survey to ensure identical responses.

When identical responses were not recorded, the survey was removed from the dataset. Further cleaning measures ensured that respondents were not "straight-lining" (answering the same response to each question) [26]. Statistical comparisons between the four groups were analyzed using analyses of variance and Kruskal-Wallis H tests through social science statistics software (SPSS)—depending on the nature of the variables measured (nominal vs. continuous).

3. Results

3.1. Non, Existing, Previous but No Longer, and New Outdoor Recreationists Groups

Based on the initial screening questions, $n = 1763$; 35.8% of respondents indicated that they did not participate regularly in outdoor recreation prior to the pandemic or during the pandemic (Table 2). These individuals are noted as *"non-outdoor recreationists."* A total $n = 1501$; 30.4% indicated that they did participate regularly in outdoor recreation prior to the pandemic, and continued to do so regularly during the pandemic. These individuals are noted as *"existing outdoor recreationists."* The smallest subset of respondents ($n = 665$; 13.5%) indicated that they did participate regularly in outdoor recreation prior to the pandemic, but did not continue to do so during the pandemic. We note these respondents as *"previous, but no longer outdoor recreationists"*. Finally, the sample quota of $n = 1000$ (20.3% of the overall sample) *"new outdoor recreationists"* were collected for comparison.

Table 2. Breakdown of non, existing, no longer, and new outdoor recreationists.

	Participant Group			
	Non-Outdoor Recreationist	**Existing Outdoor Recreationist**	**Previous, but No Longer Outdoor Recreationist**	**New Outdoor Recreationist**
Participated in outdoor recreation regularly *prior to* pandemic?	No	Yes	Yes	No
Participated in outdoor recreation regularly *during* pandemic?	No	Yes	No	Yes
	($n = 1763$)	($n = 1501$)	($n = 665$)	($n = 1000$) [a]
Percent of total	35.8%	30.4%	13.5%	20.3%

[a] Sampling ceased once 1000 valid responses were collected for the No → Yes group.

3.2. Demographics and Respondent Categories

Demographic comparisons were examined to determine statistical differences between the groups based on gender, ethnicity, community type, income, and age (Table 3). There were significantly more respondents identifying as women in the *non-outdoor recreationist* group compared to the other categories (existing, those that stopped, and those that are new). While the overall sample had more female respondents than male, there were not statistically significant differences in gender between the *existing, previous but no longer*, and *new outdoor recreationists groups*.

With regard to ethnicity, the majority of respondents in all four categories identified as being white. Within the *non, existing*, and *new outdoor recreationist* groups, the percentage of white respondents ranged between 73.4% to 76.6%. Ethnicity was not statistically different between these groups, and the least ethnically diverse group resulted from the *new outdoor recreationists* category. The *previously but no longer outdoor recreationists* respondents were significantly different that the other three categories, as they were more ethnically diverse compared to the other groups.

Table 3. Breakdown of non, existing, no longer, and new outdoor recreationists by demographics.

	Participant Group			
	Non-Outdoor Recreationist (*n* = 1763); 35.8%	**Existing Outdoor Recreationist** (*n* = 1501); 30.4%	**Previous, but No Longer Outdoor Recreationist** (*n* = 665); 13.5%	**New Outdoor Recreationist** (*n* = 1000); 20.3%
Gender [1]				
Woman	71.0% (1252)	61.9% (929)	62.1% (413)	61.7% (617)
Man	27.2% (479)	36.0% (541)	35.5% (236)	36.2% (362)
Non-binary	0.9% (15)	1.5% (23)	1.4% (9)	1.7% (17)
Prefer not to disclose	0.6% (11)	0.5% (7)	0.9% (6)	0.2% (2)
Prefer to self-describe	0.3% (6)	0.1% (1)	0.2% (1)	0.2% (2)
Differences between groups [6]	A	B	B	B
Ethnicity [2]				
White	74.8% (1318)	73.4% (1101)	61.8% (411)	76.6% (766)
Hispanic or Latina/Latino/Latinx	7.1% (126)	7.4% (111)	11.9% (79)	7.5% (75)
Black or African American	9.2% (163)	11.3% (170)	14.7% (98)	8.0% (80)
Native American, American Indian, or Alaska Native	1.2% (21)	1.2% (18)	2.0% (13)	1.0% (10)
Asian or Pacific Islander	5.0% (88)	4.6% (69)	6.3% (42)	5.5% (55)
Other	1.4% (24)	1.3% (19)	1.7% (11)	1.1% (11)
Prefer not to say	1.3% (23)	0.9% (13)	1.7% (11)	0.3% (3)
Differences between groups [6]	A	A	B	A
Community Type [3],*				
Rural (population < 5000)	30.9% (545)	31.2% (469)	25.9% (172)	29.7% (297)
Urban cluster (population between 5000 and 50,000)	25.0% (440)	26.6% (400)	22.4% (149)	24.6% (246)
Urbanized area (population > 50,000)	44.1% (778)	42.1% (632)	51.7% (334)	45.7% (457)
Differences between groups [6]	A	A	B	A
Income [4]				
$0–$40,000	53.8% (948)	40.7% (611)	47.2% (314)	39.3% (393)
$41,000–$60,000	17.8% (314)	22.6% (339)	20.2% (134)	21.1% (221)
$61,000–$80,000	10.6% (187)	14.5% (217)	13.1% (87)	14.5% (145)
>$80,000	17.8% (314)	22.3% (334)	19.5% (130)	25.1% (251)
Differences between groups [6]	A	B	C	B
Age [5]				
Mean	54	46	44	42
Differences between groups [7]	A	B	B,C	C

[1] A Kruskal–Wallis H test showed that a statistically significant difference exists in gender between the participant groups at a 99.9% confidence interval. [2] A Kruskal–Wallis H test showed that a statistically significant difference exists in ethnicity between the participant groups at a 99.9% confidence interval. [3] A Kruskal–Wallis H test showed that a statistically significant difference exists in community types between the participant groups at a 99.9% confidence interval. [4] A Kruskal–Wallis H test showed that a statistically significant difference exists in income between the participant groups at a 99.9% confidence interval. [5] Welch's ANOVA test showed a statistically significant difference exists in age between the participant groups at a 99.9% confidence interval. [6] Different letters indicate significantly different (*p* < 0.05) group compositions using a Dunn's post-hoc test. [7] Different letters indicate significantly different (*p* < 0.05) means using a Tamhane's T2 post-hoc test. * New outdoor recreationists indicated residing in the following geographic areas of the U.S.: Northeast *n* = 210; South *n* = 190; Midwest *n* = 370; West *n* = 230. Note: Tests used to analyze differences between participant groups were determined based on guidance from [26,27].

The results of residential community type mimicked the ethnicity findings. More than half of the *previous but no longer outdoor recreationists* group indicated residing in urbanized areas with a population of more than 50,000. The *previous but no longer recreationists* group also represented the smallest percentage of individuals living in rural communities under 5000. The *non, existing,* and *new* did not differ significantly by residential area type, but the largest proportions for these three groups lived in urbanized areas, ranging from approximately 42.1% to 45.7%.

With regard to income only the *new* and *existing outdoor recreationists* were similar. These two groups differed significantly (i.e., generally larger annual income) from the *non* and *previous but no longer outdoor recreationists*, and the *non* and *previous but no longer* resulted in statistically significant differences as well. The *new outdoor recreationists* had the largest percentage of individuals reporting annual incomes of greater than $80,000 (25.1%) as well as the smallest percentage of individuals making $0–$40,000 annually. The majority of *non outdoor recreationists* indicated earing $0–$40,000 annually (53.8%), which was the largest percentage of respondents compared to the other groups, followed by *previous but no longer* (47.2%).

The *new outdoor recreationists* were significantly younger than *non outdoor recreationists*, as this category, on average was the youngest group ($M = 42$). The *non outdoor recreationists* were also significantly older ($M = 54$) than the other groups (*existing* $M = 46$; *previous but no longer* $M = 44$).

4. Discussion

This is one of the first studies to examine *new outdoor recreationists* during this unique and challenging time in history, while also exploring who is, who is not, and who was but is no longer an outdoor recreationist. These results highlight the substantive amount of U.S. residents that now consider themselves as outdoor recreationists. Based on this geographically-representative sample, this could be as much as half of the U.S. population.

It is concerning from the public health standpoint that 13.5% of respondents stopped outdoor recreation during the pandemic. This points to inequities associated with access to public lands as health resources [28], based on demographic differences. For example, *previous but no longer* respondents were significantly more ethnically diverse, and tended to reside in more urban environments and earn less annually than the other groups. This aligns with recent findings from an exploration of park use in New York City during the pandemic, where the gap between a sense of belonging in greenspace grew across racial and economic lines [29], and aforementioned research as well [4,5]. This is alarming for land managers, land use planners, and policy makers, and points to the need for continuing efforts that increase equity and access with regard to outdoor recreation opportunities and public park and protected areas.

On the other hand, it is also encouraging from the public health perspective that amidst one of the more challenging periods in recent history, that a new group of U.S. outdoor recreationists (20%) have emerged during the pandemic, likely benefiting from the health outcomes they receive from these activities [16–20,28]. This creates a tremendous opportunity to engage this group of new recreationists in life-long healthy recreation and conservation behaviors that promote the preservation of the resources, sustaining outdoor recreation.

However, it must be highlighted again, that with regard to equity, these *new outdoor recreationists* tend to be predominantly white, and in general are less diverse, and have higher annual incomes than the other groups analyzed in this study. Furthermore, the *new outdoor recreationists* are largely not very different demographically, from *existing outdoor recreationists*. They are predominantly white, on average in their early 40s, with perhaps more disposable income. These two groups do however, differ from *non-outdoor recreationists* in the U.S. For example, on average *new outdoor recreationists* are twelve years younger than *non-outdoor recreationists*, on average. Finally, there were significantly more women, proportionately, in the *non-outdoor recreationists* group, pointing to the large body of empirical research that has highlighted gender inequities and associated constraints with regard to outdoor recreation participation e.g., [30,31].

While this study purposefully did not recruit or weight data based on the U.S. census (beyond region), these results in some circumstances align with other new outdoor recreation studies e.g., [21] suggesting that these individuals are slightly younger, and living in more urban environments. However, our results did not align with previous research [21] with regard to ethnicity or income, as the new outdoor recreationists in this study are predominantly white with higher reported annual incomes. These results also align with other national research regarding outdoor recreation trends [32], not associated with the pandemic, suggesting that approximately half of the U.S. population now participates in outdoor recreation. However, the national outdoor recreation participation frequency criteria established in this study (i.e., once per month) is more stringent than previous national studies, suggesting a substantial increase in participation since the pandemic began.

There were several notable limitations to this research, which should be addressed by future studies. We used a paid Qualtrics panel, which may have influenced responses and representativeness. The classification of outdoor recreation activities [21,25] may have missed some important behaviors, particularly more passive activities such as simply relaxing outdoors. Future research should consider broadening outdoor recreation activities to include both active and passive behaviors. This study purposefully did not weight the sample to match the U.S. census, which could be done with future samples for comparison. Data collection took place in May and June of 2021, over a year into the pandemic, which could have influenced responses in a variety of ways. It would be informative if future research could ask the same questions of these respondents at multiple points over the next few years. The grouped samples as a whole tended to consist of more female respondents. While national trends indicate that an increasing proportion of women are participating in outdoor recreation, our overarching results are perhaps skewed slightly based on the gender distribution we received from our sampling approach.

This study used Qualtrics to facilitate the paid panel, but comparing these results with other paid panel services (e.g., MTurk) would be useful methodologically. While the *new outdoor recreationists* may not differ substantially from *existing* with regard to demographics, it is unknown how their behaviors might differ. Additional analyses examining how these new outdoor recreationists behave (i.e., activities, locations, frequency, anticipated future behaviors) are needed to further inform long-term public land management and associated policies. We did not ask *previous but no longer outdoor recreationists* why they no longer participate, and future research must address why the *previous but no longer outdoor recreationists* are not able to pursue these activities any longer. Is it access, fiscal constraints, fear of virus transmission, or other reasons that have led to a halt in these activities? Finally, it is important to recognize that annual income and leisure time may have been abnormal for respondents during this unique period in time, and may have influenced results. Future research exploring outdoor recreation behaviors should consider asking about specific fluctuations in income and spare time as it pertains to the pandemic.

5. Conclusions

This study examined who is new to outdoor recreation amidst the COVID-19 pandemic. This research also explored who is not, those who were but are no longer, and finally, those who have been and still remain an outdoor recreationist in the U.S. To our knowledge, this is one of the first national studies to examine these important groups and differences between them demographically. This research suggests that amidst the pandemic, approximately half of the U.S. population now participates in outdoor recreation, at least once per month (at the time of this study/manuscript).

Several important demographic differences exist between the groups discovered through this study, as *new* recreationists tend to be younger and wealthier than the other three groups. *Non outdoor recreationists* are older, earn less annually, and tend to consists of more females than the other groups. *Non, existing,* and *new outdoor recreationists* are predominantly white. *Previous but no longer outdoor recreationists* are the most ethnically diverse, and this group tends to live in more urban environments and make less annually than the other groups. Demographically, *new outdoor recreationists* are not vastly different from *existing outdoor recreationists*.

This study highlights demographic inequity issues with regard to outdoor recreation pursuits, particularly during the pandemic, meriting greater focus on this important social justice issue. Access to outdoor recreation, the parks and protected areas that facilitate recreational health benefits remains a concern in the U.S. Future research must expand upon why these inequities continue to exist and inform strategies and policies that promote equal opportunities and access to outdoor recreation, for all.

Policy makers must work harder to ensure that access and availability to outdoor recreation is equitable. Finally, land managers and public health officials may consider employing additional and new ways to encourage diverse demographics to enjoy outdoor recreation. A vital strategy that merits more attention is the promotion of outdoor recreation activities and the places they occur, as health resources [28].

Author Contributions: Conceptualization, P.N., B.L., W.L.R. and B.D.T.; methodology, W.L.R., P.N., B.L. and B.D.T.; formal analysis, W.L.R.; original draft preparation, B.D.T.; writing—review and editing, P.N., B.L. and W.L.R. All authors have read and agreed to the published version of the manuscript.

Funding: This research was funded through the Recreation, Park and Tourism Management Department at Pennsylvania State University and the Leave No Trace Center for Outdoor Ethics.

Institutional Review Board Statement: The study was conducted according to the guidelines of the Declaration of Helsinki, and approved by the Institutional Review Board (or Ethics Committee) of Pennsylvania State University (STUDY00017511; date of approval 4-23-2021).

Informed Consent Statement: Informed consent was obtained from all subjects involved in the study.

Conflicts of Interest: The authors declare no conflict of interest.

References

1. Clawson, M. Outdoor recreation: Twenty-five years of history, twenty-five years of projection. *Leis. Sci.* **1985**, *7*, 73–99. [CrossRef]
2. Cordell, H.K.; McDonald, B.L.; Teasley, R.J.; Bergstrom, J.C.; Martin, J.; Bason, J.; Leeworthy, V.R. Outdoor recreation participation trends. In *Outdoor Recreation in American Life: A National Assessment of Demand and Supply Trends*; Cordell, H.K., Betz, C., Bowker, J.M., English, D.B.K., Mou, S.H., Bergstrom, J.C., Teasley, R.J., Tarrant, M.A., Loomis, J., Eds.; Sagamore Publishing: Champaign, IL, USA, 1999; pp. 219–321.
3. Chronback, U. National parks are logging record crowds. Here's how to visit safely. *Pop. Sci.* 2020. Available online: https://www.popsci.com/story/environment/national-parks-record-crowds-covid-pandemic/ (accessed on 15 October 2020).
4. Larson, L.R.; Zhang, Z.; Oh, J.I.; Beam, W.; Ogletree, S.S.; Bocarro, J.N.; Lee, K.J.; Casper, J.; Stevenson, K.T.; Hipp, J.A.; et al. Urban park use during the COVID-19 pandemic: Are socially vulnerable communities disproportionately impacted? *Front. Sustain. Cities* **2021**, *3*, 710243. [CrossRef]
5. Astell-Burt, T.; Feng, X. Time for 'Green' during COVID-19? Inequities in Green and Blue Space Access, Visitation and Felt Benefits. *Int. J. Environ. Res. Public Health* **2021**, *18*, 2757. [CrossRef]
6. Spenceley, A.; McCool, S.; Newsome, D.; Baez, A.; Barborak, J.R.; Blye, C.-J.; Bricker, K.; Cahyadi, H.S.; Corrigan, K.; Halpenny, E.; et al. Tourism in protected and conserved areas amid the COVID-19 pandemic. *Parks* **2021**, *27*, 103–118. Available online: https://researchrepository.murdoch.edu.au/id/eprint/60318/1/conserved%20areas.pdf (accessed on 4 August 2021). [CrossRef]
7. Oftedal, A.O. *Trail Use Has Surged in Response to the COVID-19 Pandemic*; Parks & Trails Council of Minnesota: St. Paul, MN, USA, 2020; Available online: https://www.parksandtrails.org/2020/04/27/trail-use-covid19/ (accessed on 4 August 2021).
8. Fernandez, P.; Bron, G.; Kache, P.; Tsao, J.; Bartholomay, L.; Hayden, M.; Ernst, K.; Berry, K.; Diuk-Wasser, M. Shifts in outdoor activity patterns in the time of COVID-19 and its implications for exposure to vector-borne diseases in the United States. *Res. Sq.* **2021**. [CrossRef]
9. Venter, Z.; Barton, D.; Gunderson, V.; Figari, H.; Nowell, M. Urban nature in a time of crisis: Recreational use of green space increases during the COVID-19 outbreak in Oslo. *Natl. Environ. Res. Lett.* **2020**, *15*, 104075. [CrossRef]
10. Kupfer, J.A.; Li, Z.; Ning, H.; Huang, X. Using mobile device data to track the effects of the COVID-19 Pandemic on spatiotemporal patterns of national park visitation. *Sustainability* **2021**, *13*, 9366. [CrossRef]
11. Ronto, P. Hiking in the U.S. Has Never Been More Popular. *Run Repeat.* 2021. Available online: https://runrepeat.com/hiking-never-more-popular (accessed on 1 November 2021).
12. Rice, W.L.; Mateer, T.J.; Reigner, N.; Newman, P.; Lawhon, B.; Taff, B.D. Changes in recreational behaviors of outdoor enthusiasts during the COVID-19 pandemic: Analysis across urban and rural communities. *J. Urban Ecol.* **2020**, *6*, juaa020. [CrossRef]
13. Landry, C.E.; Bergstrom, J.; Salazar, J.; Turner, D. How has the COVID-19 pandemic affected outdoor recreation in the U.S.? A revealed preference approach. *Appl. Econ. Perspect. Policy* **2020**, *43*, 443–457. [CrossRef]
14. Rice, W.L.; Mateer, T.J.; Newman, P.; Lawhon, B.; Reigner, N.; Taff, B.D. Outdoor recreationists' perceptions of risk, agency trust, and visitor capacities during the COVID-19 pandemic. *J. Park Recreat. Adm.* **2021**. [CrossRef]
15. Harris, B.; Rigolon, A.; Fernandez, M. Hiking during the COVID-19 pandemic: Demographic and visitor group factors associated with public health compliance. *J. Leis. Res.* **2021**, 1–9. [CrossRef]
16. Mateer, T.J.; Rice, W.L.; Taff, B.D.; Lawhon, B.; Reigner, N.; Newman, P. Psychosocial factors influencing outdoor recreation during the COVID-19 pandemic. *Front. Sustain. Cities* **2021**, *3*, 621029. [CrossRef]
17. Jackson, B.S.; Stevenson, K.T.; Larson, L.R.; Peterson, N.M.; Seekamp, E. Outdoor activity participation improves adolescents' mental health and well-being during the COVID-19 pandemic. *Int. J. Environ. Res. Public Health* **2021**, *18*, 2506. [CrossRef] [PubMed]

18. Larson, L.R.; Mullenbach, L.E.; Browning, M.H.; Rigolon, A.; Thomsen, W.J.; Covelli, M.W.E.; Labib, S.M. Greenspace and park use associated with less emotional distress among college students in the United States during the COVID-19 pandemic. *Environ. Res.* **2021**, *204*, 112367. [CrossRef]

19. Labib, S.M.; Browning, M.H.E.M.; Rigolon, A.; Helbich, M.; James, P. Nature's Contributions in Coping with a Pandemic in the 21st Century: A Narrative Review of Evidence during COVID-19. *EcoEvo* 2021. Available online: https://ecoevorxiv.org/j2pa8/ (accessed on 1 November 2021).

20. Ribeiro, A.I.; Triguero-Mas, M.; Santos, C.J.; Gomez-Nieto, A.; Cole, H.; Anguelovski, I.; Silva, F.M.; Baro, F. Exposure to nature and mental health outcomes during COVID-19 lockdown. A comparison between Portugal and Spain. *Environ. Int.* **2021**, *154*, 106664. Available online: https://pubmed.ncbi.nlm.nih.gov/34082237/ (accessed on 1 August 2021). [CrossRef]

21. Outdoor Industry Association; Naxion Research Consulting. 2021 Special Report: The New Outdoor Participant (COVID and Beyond). *Outdoor Industry Association*. 2021. Available online: https://outdoorindustry.org/resource/2021-special-report-new-outdoor-participant-covid-beyond/ (accessed on 1 August 2021).

22. Qualtrics ESOMAR's 28 Questions to Help Buyers of Online Samples. 2012. Available online: https://www.esomar.org/uploads/public/knowledge-and-standards/documents/ESOMAR-28-Questions-to-Help-Buyers-of-Online-Samples-September-2012 .pdf (accessed on 1 August 2021).

23. Mullinix, K.J.; Leeper, T.J.; Druckman, J.N.; Freese, J. The generalizability of survey experiments. *J. Exp. Political Sci.* **2015**, *2*, 109–138. [CrossRef]

24. Khubchandani, J.; Sharma, S.; Price, J.H.; Wiblishauser, M.J.; Sharma, M.; Webb, F.J. COVID-19 vaccination hesitancy in the United States: A rapid national assessment. *J. Community Health* **2021**, *46*, 270–277. [CrossRef]

25. Highfill, T.; Franks, C.; Georgi, P.S.; Howells, T.F. Introducing the outdoor recreation satellite account. *Surv. Curr. Bus.* **2018**, *98*, 1–9. Available online: https://apps.bea.gov/scb/2018/03-march/0318-prototype-statistics-for-the-outdoor-recreation-satellite-account.htm (accessed on 1 August 2021).

26. Vaske, J. *Survey Research and Analysis: Applications in Parks, Recreation and Human Dimensions*; Venture Publishing: State College, PA, USA, 2008.

27. Calver, M.; Fletcher, D. When ANOVA isn't ideal: Analyzing ordinal data from practical work in biology. *Am. Biol. Teach.* **2020**, *82*, 289–294. [CrossRef]

28. Razani, N. Nature and health in practice. *Park Steward. Forum* **2021**, *37*, 17–22. Available online: https://escholarship.org/uc/item/1rq7m979 (accessed on 1 November 2021). [CrossRef]

29. Pipitone, J.M.; Jovic, S. Urban green equity and COVID-19: Effects on park use and sense of belonging in New York City. *Urban For. Urban Green.* **2021**, *65*, 127338. [CrossRef]

30. Lee, J.-H.; Scott, D.; Floyd, M.F. Structural inequalities in outdoor recreation participation: A multiple hierarchy stratification perspective. *J. Leis. Res.* **2001**, *3*, 4. [CrossRef]

31. Johnson, C.Y.; Bowker, J.M.; Cordell, K.H. Outdoor recreation constraints: An examination of race, gender, and rural dwelling. *J. Rural. Soc. Sci.* **2001**, *17*, 6. Available online: https://egrove.olemiss.edu/jrss/vol17/iss1/6/ (accessed on 1 August 2021).

32. Outdoor Industry Association, 2021 Outdoor Participation Trends Report. 2021. Available online: https://ip0o6y1ji424m064 1msgjlfy-wpengine.netdna-ssl.com/wp-content/uploads/2015/03/2021-Outdoor-Participation-Trends-Report.pdf (accessed on 1 November 2021).

 *land*

Article

Green Space Visits and Barriers to Visiting during the COVID-19 Pandemic: A Three-Wave Nationally Representative Cross-Sectional Study of UK Adults

Hannah Burnett *, Jonathan R. Olsen and Richard Mitchell

MRC/CSO Social and Public Health Sciences Unit, University of Glasgow, Glasgow G12 8QQ, UK;
jonathan.olsen@glasgow.ac.uk (J.R.O.); richard.mitchell@glasgow.ac.uk (R.M.)
* Correspondence: h.burnett.1@research.gla.ac.uk

Abstract: Green spaces have been found to promote physical activity, social contact, and mental wellbeing, however, there are inequalities in the use and experience of green spaces. The United Kingdom's (UK) response to the COVID-19 pandemic imposed very substantial changes on its citizens' lives which could plausibly affect their willingness to visit green spaces. These sudden lifestyle changes severely affected the population's mental health, leading to a greater dependency on the positive influence of nature in reducing stress and improving mood. Whilst early cross-sectional evidence suggested an increased orientation to nature and visits to green spaces as a response to COVID-19 'lockdowns', there is little longitudinal evidence about how sustained and equal these changes may have been. This study explored green space visits, barriers to visiting, and the inequalities of both of those over an entire year of the pandemic in the UK. Three waves of nationally representative cross-sectional surveys were administered by YouGov in April 2020, November 2020, and April 2021 (N = 6713). Data included reported visits to green spaces and, for those with no or infrequent visiting, perceived barriers including those plausibly related to the risk of COVID-19. Green space visits increased over the year as lockdown restrictions were relaxed; 68% of respondents reported green space visits in April 2021, compared with 49% in April 2020. However, the socio-economic inequalities in use were sustained and increased. COVID-19 related barriers fell over time, but there were indications of increased interest in green spaces among younger people. Further action is required to ensure that the positive impacts of green spaces are experienced equally, and that good quality green space is accessible to all.

Keywords: green space; COVID-19; inequalities; barriers; health

Citation: Burnett, H.; Olsen, J.R.;
Mitchell, R. Green Space Visits and
Barriers to Visiting during the
COVID-19 Pandemic: A Three-Wave
Nationally Representative
Cross-Sectional Study of UK Adults.
Land **2022**, *11*, 503. https://doi.org/
10.3390/land11040503

Academic Editors: Francesca Ugolini
and David Pearlmutter

Received: 23 February 2022
Accepted: 29 March 2022
Published: 31 March 2022

Publisher's Note: MDPI stays neutral
with regard to jurisdictional claims in
published maps and institutional affiliations.

1. Introduction

In late 2019, a virulent novel coronavirus was identified. Commonly known as COVID-19, the disease is spread from an infected person's mouth or nose in small liquid particles [1]. To reduce the rate of infection, restrictions on people's interaction and movement have been implemented within and between many countries around the world. At their most extreme, these restrictions included 'lockdowns' requiring most citizens to stay at, or close to, home. The precise timing, duration, and severity of restrictions varied according to the stage of the epidemic in each country.

The first wave of COVID-19 infections took hold in the UK in early 2020. The first lockdown started on 23 March 2020. Most people were only permitted to leave their homes for 'essential' reasons including access to health care, food, and to undertake one form of outdoor exercise [2]. In the subsequent two years, restrictions on movement and social interaction waxed and waned depending on the number of COVID-19 cases and the emergence of new variants. The next substantial set of UK restrictions was imposed between January and April 2021 when another 'stay-at-home' order was implemented and individuals were prohibited from social mixing. In contrast to March 2020 however, restrictions were not placed on the form and frequency of outdoor exercise.

The pandemic and its lockdown restrictions have had a substantial impact on individual behaviours, routines, and habits worldwide [3,4]. Research has highlighted profound effects on population health. In addition to the morbidity and mortality from COVID-19 itself, both reduced access to health care and the experience of the pandemic has resulted in considerable morbidity. Studies in the UK, for example, found that the mental health of adults followed an 'up and down' cycle coinciding with periods of national lockdown and high COVID-19 case rates [5,6]. This negative impact on mental health has been felt internationally as well [7–10]. For example, in a study undertaken in Germany, perceived stress increased across all age groups over the six weeks following the mid-March 2020 COVID-19 outbreak [7]. The exact mental health burden of the pandemic is still too early to measure, with many countries experiencing new waves of virus cases [8]. Furthermore, the impacts of both COVID-19 and its wider consequences have not been even across society. The pandemic appears to have exacerbated pre-existing socio-demographic and geographic health inequalities. In the UK, for example, case rates grew faster and had higher peaks within more deprived areas [11,12].

Visiting green spaces has been shown to benefit mental and physical health outcomes. In the UK, considerable media attention was lavished on how green spaces could be a helpful resource during the pandemic and might mitigate the negative impacts on mental health [13,14]. Cross-sectional research suggested that during the early phases of the COVID-19 pandemic, the frequency of visits to green spaces increased and that this may have had a positive impact on mental health [15–18]. However, as restrictions were lifted and then reimposed, and the population adjusted to the 'new normal', it is probable that visit levels continued to change. Moreover, increased visits to green spaces and their perceptions of mental health benefits were not experienced equally across socio-demographic groups [17].

There are considerable social, economic, and gender inequalities in visits to green spaces; less advantaged populations are far less likely to do so [19–21]. Barriers to the use of green spaces are socially and spatially unequal. Not everyone has good access to a suitable space, or the time and inclination to visit it. This paper is concerned with measuring green space use from an environmental justice and inequalities perspective by exploring the differences according to socio-economic characteristics at a neighbourhood and individual level [22,23]. This paper aimed to provide insights that may add to the existing environmental justice framework as applied to green spaces, with a focus on the use and barriers rather than access and quality [24]. Whereas much environmental justice research explores inequalities in exposure to pathogenic environments, our work focuses on justice in both access to, and use of, a salutogenic environment. We explore how the COVID-19 pandemic may have exacerbated environmental injustice, particularly for those of differing social grades/occupations. It is important to explore environmental injustice alongside the "equigenic hypothesis" [25] which suggests that some environments (such as natural environments) can support the health of the less advantaged as much as, or even more than, the more advantaged [26,27]. By exploring green space visits and barriers during the pandemic, we can strengthen our understanding of who is not using green spaces, why, and how the pandemic impacted existing inequalities.

COVID-19 added several other pandemic-specific barriers including the desire and/or urgent need to socially distance, a sense of severe vulnerability to the virus, and sharp divides between who is, and is not, able or required to work at home or has access to outdoor space there [28–30]. These barriers are of particular interest in this paper. As vaccination proceeded, the population 'got used' to COVID-19, and as the public health response phased up and down in reaction to case-numbers and variants, the significance of COVID-19-specific barriers could plausibly change. A recent study in Korea found that individuals with decreased visits to green spaces between September–December 2020 had 2.06 higher odds of probable major depression at the time of the survey compared to those whose visits had increased or stayed the same [31]. This emphasises the importance of researching the barriers to green space use during the pandemic by sociodemographic characteristics and exploring how the barriers changed over its first year.

Further research is required to describe whether the increased visit numbers were sustained as the COVID-19 pandemic continued, how barriers to visiting changed following subsequent easing and reintroduction of COVID-19 measures, and the inequalities in these. There is a lack of literature in this area, with no studies to date focusing on the change in green space use and barriers in the UK over the first year of the pandemic. It is important to explore this gap, both in order to understand how the experience of the UK relates to that in other countries, and also to promote resilient green spaces and wider societies in the face of future pandemics [32]. The research objectives of this paper were, therefore:

1. To describe variation in green space visiting during the COVID-19 pandemic.
2. To describe reported barriers to green space visits during the COVID-19 pandemic.
3. To explore variation in green space visits and barriers over time and by sex, age, and socio-economic position.

2. Materials and Methods

2.1. Survey Sample

Three waves of a repeat cross-sectional survey were administered by YouGov between April 2020 and April 2021 [33]. Each wave was drawn from YouGov's UK/GB Omnibus of 800,000 panellists. Respondents were selected at random from this panel by YouGov and then sent a survey link to complete. All three waves were nationally representative of the UK population when weightings were applied. The survey waves are described below.

Wave 1: The first wave was administered between 30 April and 1 May 2020, with a sample of 2252 UK adults aged 18+. At that point, the UK population was in the first 'lockdown', also known as the 'stay at home phase'. From 23 March 2020, people were only permitted to leave home for limited purposes, including collecting medicines, doing essential shopping, and doing one form of exercise per day [2]. When wave 1 was administered, the same lockdown restrictions were implemented across constituent nations of the UK.

Wave 2: The second survey wave was administered on 26 November 2020, with a sample of 2246 UK adults. When this wave was undertaken, COVID-19 policies and restrictions differed among the constituent nations of the UK. England was in a winter lockdown, with the population asked to stay at home and only leave for limited reasons such as education, essential shopping, exercise, health care, or to care for vulnerable people [34]. Wales was just out of a strict lockdown, with gyms, schools, and restaurants being reopened [35]. Scotland was operating localized lockdowns with almost half of its population under strict restrictions, including a ban on indoor household socialising and only essential shops being open [36].

Wave 3: The third wave was administered from 29–30 April 2021, exactly one year after wave 1, with a sample of 2215 UK adults. At this time, lockdown restrictions had been eased across the UK. Non-essential shops had reopened, outdoor gatherings for up to six people were allowed, and the population was able to travel outside of their local area [37].

2.2. Survey Content

Wave 1 was initially designed and implemented as a one-off cross-sectional survey but, as the COVID-19 pandemic persisted and evolved, and the UK introduced further restrictions, the two subsequent surveys were commissioned. Some question wording, therefore, differed very slightly between wave 1, and waves 2 and 3 (Table 1).

At every wave, respondents were asked about their green space visiting frequency. Green spaces were defined as 'places where you can see and experience plants, trees, and nature outside of the household (e.g., public parks, sports fields, agricultural land, woodlands, coastal paths, and nature reserves)'. Those that had not visited green spaces or had visited infrequently (once every 2 weeks or once in the last 4 weeks in waves 2 and 3) were asked about the barriers to visiting green spaces and reasons for their non or low-frequency visiting (Table 1). The surveys covered a range of reported barriers to the use of green spaces, with 12 barriers included in wave 1 and 15 barriers in waves 2 and 3. In this analysis, we focused largely on those particularly relevant to COVID-19 and the lockdown restrictions. These were reported as: 'worried about social distancing in green space', 'green spaces are too busy', 'fear for health when outdoors (i.e., contracting COVID-19)', 'member of household/individual at risk of being severely affected by COVID-19', 'using an outdoor space at home instead', and 'not interested'.

Table 1. The survey themes, question wording (by wave), and response categories.

Themes	Wave 1 Questions	Waves 2 and 3 Questions	Response Categories
Visitation frequency	Have you visited a green space since the movement restrictions have been enforced in the UK? (i.e., since 23 March 2020).	Have you visited ANY green spaces in the last 4 weeks	'Yes, I have' 'No, I have not'
Barriers to visits	[If the respondent had not visited green space since lockdown was implemented] Which, if any, of the following are reasons for you not visiting green spaces since the restrictions were introduced (i.e., 23 March 2020)? (Please select all that apply) 1. I am worried that I will not be able to socially distance from others in these spaces (i.e., remain 2 metres away) 2. Green spaces are much busier now 3. I fear for my health when I go outdoors (i.e., contracting Coronavirus (COVID-19) 4. I/a member of my household is at higher risk of being severely affected by Coronavirus (COVID-19) 5. I am using an outside space at my home (e.g., garden) instead 6. I am not interested in visiting green spaces	[If the respondent had not visited green space or had visited infrequently in the last 4 weeks] You previously said you have not regularly visited a green space in the last 4 weeks … Which, if any, of the following are your reasons for this? (Please select all that apply) 1. I am worried that I will not be able to socially distance from others in these spaces (i.e., remain 2 metres away) 2. Green spaces are too busy for me (e.g., I can't enjoy them when they are crowded, they aren't peaceful enough, I feel uncomfortable surrounded by that many people etc.) 3. I fear for my health when I go outdoors (i.e., contracting Coronavirus (COVID-19)) 4. I/a member of my household is at higher risk of being severely affected by Coronavirus (COVID-19) 5. I am using an outside space at my home (e.g., garden etc.) instead 6. I am not interested in visiting green spaces	'Yes' 'No'

Individual demographic and socio-economic (henceforth socio-demographic) characteristics, known to be associated with green space visiting, were also collected. These were: sex (male, female); age group (18–24 years, 25–64 years, 65+ years); and social grade (higher social grade, lower social grade) categorised by YouGov using combined occupational social grade categories (Table S1). Higher social grades included non-manual workers, such as senior managers, whilst lower social grades included all manual workers, such as labourers [38]. The demographic and socio-economic variables were consistent across all survey waves.

2.3. Statistical Analyses

Multiple statistical analyses were conducted to cover each of the research objectives. Firstly, to explore general patterns of green space use and barriers, descriptive statistics were run in R (version 3.5.1) [39]. This included calculating the weighted count and proportion of respondents who had visited green spaces and reported each barrier. These were explored by sex, age, social grade, and survey wave. Cross-tabulations with Pearson's X2 were also used to test for significant differences between the groups.

Next, multiple binary logistic regression analyses were conducted to assess the associations between visiting green space, survey wave, the socio-demographic variables, and the reporting of each barrier. Separate models were run for each barrier.

Finally, interaction terms were added in order to investigate change over time (i.e., between waves) in relationships between the socio-demographic variables, survey wave, and green space visits, and between the socio-demographic variables, survey wave, and each reported barrier. The significance of each interaction was assessed via Wald tests and only those which reached a threshold of $p < 0.05$ were examined in detail. Predicted probabilities were derived to ease the interpretation of the significant interaction terms. Weightings were applied during all analyses to ensure the sample was representative of the UK adult population. Sample weights were calculated and provided by YouGov [40].

3. Results

3.1. Descriptive Statistics

In wave 1 (April 2020), 49% of respondents reported that they had visited green spaces in the 4 weeks prior to the survey. This increased to 65% of respondents in wave 2 (November 2020), and 68% in wave 3 (April 2021) (Table 2, Figure 1). The differences in socio-demographic characteristics can be found in Table 2 and Figure 2.

In all survey waves, the most common reason for not visiting green space was "I am using an outside space at my home (e.g., garden) instead" (wave 1: 47%, wave 2: 26%, and wave 3: 32% (Table 2)). In wave 1, the second most commonly reported barrier was, "I fear for my health when I go outdoors (i.e., contracting Coronavirus (COVID-19)" (27%). In wave 2, the second most commonly reported barrier was "I/a member of my household is at higher risk of being severely affected by Coronavirus (COVID-19)" (15%). By wave 3, the second most common reason was "Green spaces are too busy for me (e.g., I can't enjoy them when they are crowded, they aren't peaceful enough, I feel uncomfortable surrounded by that many people, etc.)" (18%).

Table 2. Proportions visiting green spaces or reporting barriers to doing so.

		Those Who Visited Green Space in the Previous 4 Weeks	Those Who Either Did Not Visit Green Space in the Previous 4 Weeks or Did So Infrequently ˆ						
		% (n)	% (n)	Worried about Social Distancing in Green Spaces	Green Spaces Are Too Busy	Fear for Health When Outdoors (i.e., Contracting COVID-19)	Member of House-hold/Individual at Risk of Being Severely Affected by COVID-19	Using an Outdoor Space at Home Instead	Not Interested in Visiting Green Spaces
Wave									
1 April 20	%	49% (1086)	50% (1123)	25%	9%	27%	26%	47%	8%
2 Nov 20	%	65% (1421)	45% (1020)	14%	9%	14%	15%	26%	9%
3 April 21	%	68% (1479)	45% (987)	14%	18%	10%	8%	32%	10%
Sex *									
Male	%	61% (1934)	45% (1476)	17%	10%	16%	16%	30%	12%
Female	%	61% (2052)	48% (1654)	19%	13%	19%	18%	39%	7%
Age group *									
18–24	%	60% (411)	46% (340)	19%	14%	15%	12%	22%	14%
25–64	%	62% (2685)	46% (2031)	18%	12%	16%	14%	30%	8%
65+	%	58% (890)	49% (758)	18%	10%	23%	27%	54%	9%
Social grade *									
Higher social grade	%	66% (2492)	41% (1578)	20%	12%	18%	16%	36%	9%
Lower social grade	%	53% (1494)	54% (1552)	16%	11%	17%	18%	33%	10%

* Responses by demographic variables combined for all three waves, Chi^2 p-values < 0.05. All N's were weighted to account for survey response bias. ˆ Infrequently defined as once every 2 weeks or once in the previous 4 weeks.

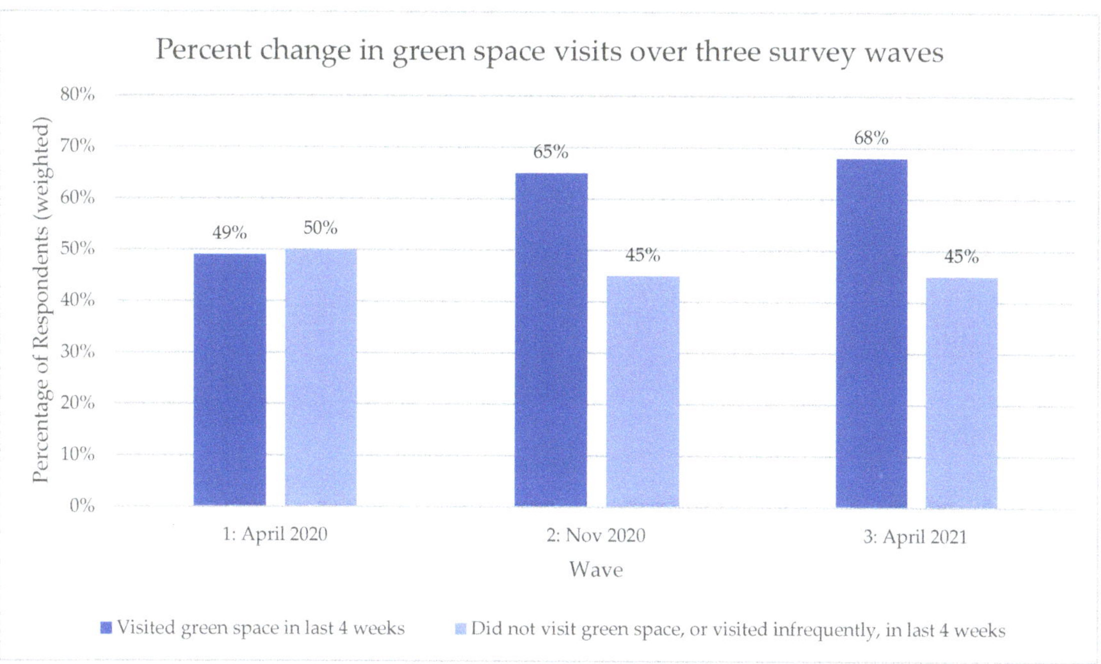

Figure 1. Change in the percentage of respondents stating visits to green space over the three survey waves (all significant Chi^2, *p*-values < 0.05).

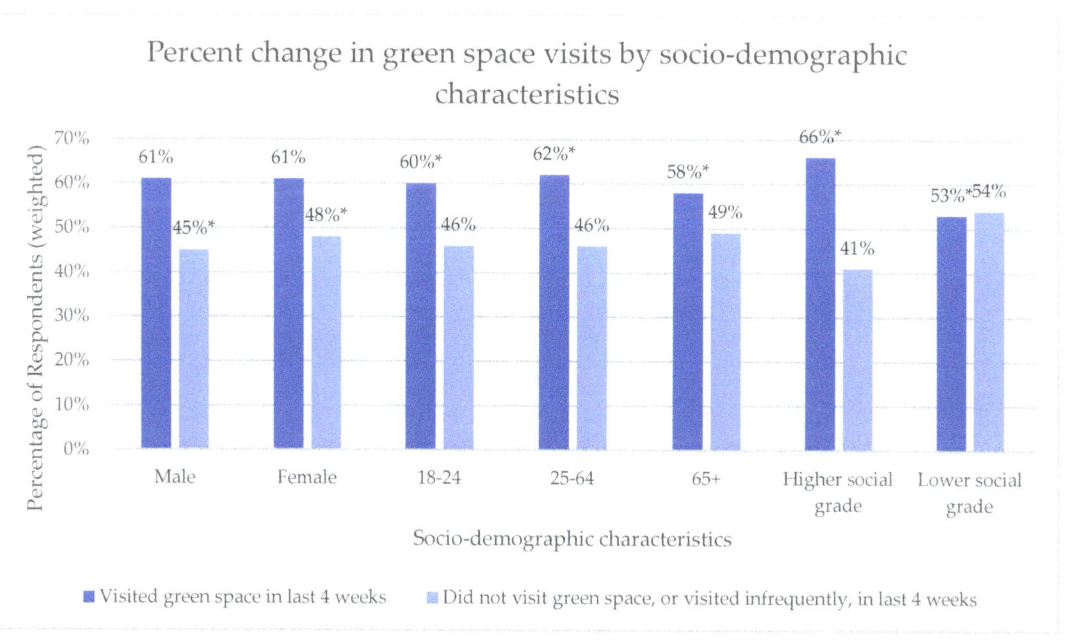

Figure 2. Change in the percentage of respondents stating visits to green space by socio-demographic characteristics (* Chi^2 *p*-values < 0.05).

3.2. Variation in Green Space Visits and Barriers to Visits by Wave

After adjustment for socio-demographic characteristics, the odds of respondents reporting visiting green spaces were significantly higher in wave 2 (OR: 1.28 (95% CI: 1.15–1.44)) than in wave 1 (Table 3). By wave 3, they were significantly higher than both waves 1 and 2 (OR: 2.25 (95% CI: 1.99–2.55)).

Table 3. Odds ratios (95% CI) from logistic regression models predicting either visiting green space or reported barriers to visiting.

	Visited Green Space in Previous 4 Weeks	Barriers Reported by Those Who Either Did Not Visit Green Spaces in the Previous 4 Weeks or Did So Infrequently					
		Worried about Social Distancing in Green Spaces	Green Spaces Are Too Busy	Fear for Health When Outdoors	Member of Household/Individual at Risk of Being Severely Affected by COVID-19	Using an Outdoor Space at Home Instead	Not Interested in Visiting Green Spaces
Wave							
1 (ref) (April 20)	1.00	1.00	1.00	1.00	1.00	1.00	1.00
2 (November 20)	1.28 (1.15–1.44) ***	0.51 (0.41–0.64) ***	1.06 (0.79–1.44)	0.43 (0.34–0.53) ***	0.49 (0.40–0.62) ***	0.39 (0.33–0.47) ***	1.20 (0.88–1.64)
3 (April 21)	2.25 (1.99–2.55) ***	0.50 (0.40–0.62) ***	2.40 (1.84–3.13) ***	0.31 (0.25–0.40) ***	0.25 (0.19–0.32) ***	0.49 (0.40–0.59) ***	1.37 (1.01–1.85) *
Sex							
Male (ref)	1.00	1.00	1.00	1.00	1.00	1.00	1.00
Female	1.03 (0.93–1.14)	1.26 (1.05–1.52) *	1.31 (1.04–1.63) *	1.18 (0.97–1.43)	1.13 (0.93–1.38)	1.56 (1.34–1.83) ***	0.53 (0.41–0.68) ***
Age group							
18–24	0.90 (0.76–1.06)	1.06 (0.79–1.42)	1.11 (0.78–1.56)	0.95 (0.69–1.31)	0.90 (0.63–1.28)	0.63 (0.48–0.83) ***	1.76 (1.24–2.51) **
25–64 (ref)	1.00	1.00	1.00	1.00	1.00	1.00	1.00
65+	0.82 (0.73–0.93) **	0.94 (0.76–1.18)	0.77 (0.58–1.02)	1.48 (1.20–1.84) ***	2.25 (1.83–2.78) ***	2.70 (2.27–3.22) ***	1.12 (0.83–1.51)
Social grade							
Higher social grade (ref)	1.00	1.00	1.00	1.00	1.00	1.00	1.00
Lower social grade	0.57 (0.52–0.63) ***	0.72 (0.60–0.87) ***	0.91 (0.73–1.14)	0.92 (0.76–1.12)	1.14 (0.94–1.38)	0.83 (0.71–0.97) *	1.22 (0.95–1.56)

*** = $p < 0.001$, ** = $p < 0.01$, * $p < 0.05$.

In general, by waves 2 and 3, respondents were less likely than in wave 1 to report worrying about social distancing in green spaces, fear for health when outdoors, a member of their household (or themselves) being at risk of severe consequences of COVID-19, or using an outdoor space at home as barriers to visiting green spaces (Table 3). These were quite substantially reduced odds, typically 0.40 or 0.50. In contrast, by wave 3 respondents were more likely than in wave 1 to report busy green spaces and/or a lack of interest in visiting green spaces as barriers to visits (Table 3). The odds of reporting busy green spaces as a barrier were substantially increased (OR: 2.40 (95% CI: 1.84–3.13)).

3.3. Variation in Green Space Visits and Barriers to Visits by Sex

There were no significant differences between male and female respondents in the likelihood of reporting visits to green spaces. However, female respondents were more likely than males to report three barriers to green space visits: being worried about social distancing in green space, green spaces being too busy, and using an outdoor space at home (Table 3). Odds for females reporting these, relative to males, were typically around 1.30. Female respondents were less likely than males to report a lack of interest as a barrier to visiting (OR: 0.53 (95% CI: 0.41–0.68)).

3.4. Variation in Green Space Visitation and Barriers by Age

Respondents aged 65+ were somewhat less likely to have visited green spaces in the last 4 weeks than those aged 25–64 (Table 3). This older group was also more likely to report fear for their health when outdoors, that they or a member of their household were at risk of severe consequences of COVID-19, and that they were using an outdoor space at home instead as barriers to visiting. These were the strongest associations seen in the models—the OR for using green spaces at home relative to the mid-age group was 2.70 (95% CI: 2.27–3.22), for example. In contrast, younger respondents were less likely to report using an outdoor space at home as a reason for not visiting green spaces. This age group was also more likely to report not being interested in visiting green spaces, with a relatively large odds ratio of 1.76 (95% CI: 1.24–2.51).

3.5. Variation in Green Space Visitation and Barriers by Social Grade

Respondents in the lower social grade group were less likely than those in the higher grade group to have visited green spaces in the last 4 weeks (Table 3). The lower grade group were also less likely to report being worried about social distancing in green spaces and/or using an outdoor space at home instead as a barrier to visiting green spaces. These associations were relatively modest.

3.6. Change over Time in Associations

The addition of interaction terms to the models suggested several significant shifts over time (i.e., between waves) in the association between the socio-demographic variables and both visits and reporting barriers. Details of models with significant interactions are provided in Tables S2 and S3.

The association between visiting green spaces and social grade differed significantly between waves. This was the only socio-demographic variable to show a significant between-wave change in association with visits. The predicted probability plot (Figure 3—panel A) shows that, whilst the likelihood of visiting increased over time for both social grades, the increase was much sharper between waves 1 (April 2020) and 2 (November 2020) for higher social grades, followed by a more modest increase between waves 2 and 3 (April 2021). In contrast, the increase was relatively constant, wave to wave, for those in lower social grades. The result of these differences was an increased socio-economic inequality in visits in wave 3 compared to wave 1. The association between sex and reporting green spaces as being too busy to visit also differed significantly between waves. Figure 3—panel B suggests a reduction in the difference between men and women such that, by wave 3, the difference is lost.

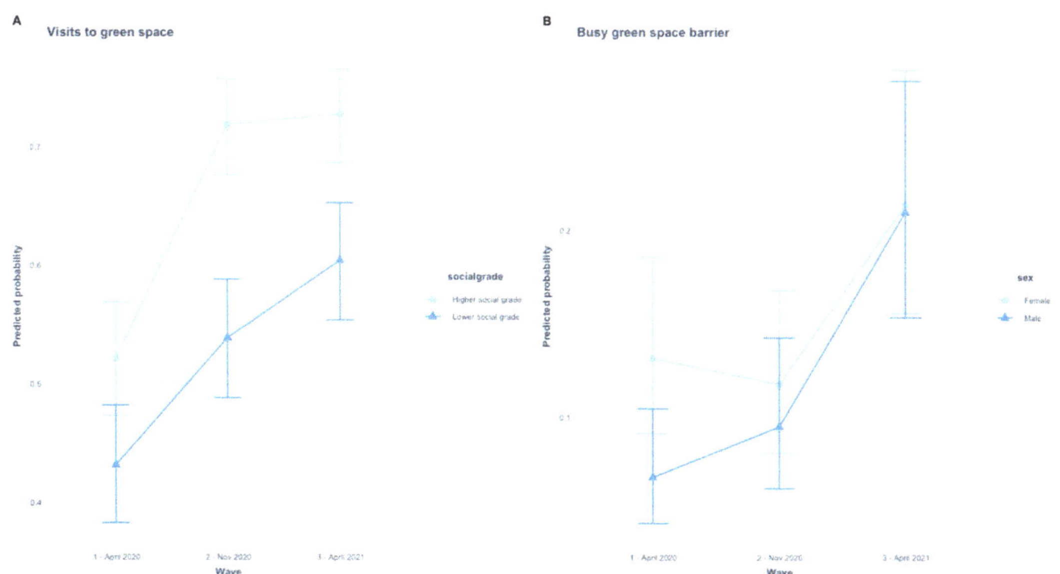

Figure 3. Predicted probability from logistic regression models with significant interaction for (**A**) visits to green space by wave and social grade and (**B**) the barrier 'green spaces are too busy' by wave and sex.

Whilst those were the only significant interactions involving social grade or sex, there were five significant interactions with age group (Figure 4). Figure 4—panel A suggests that whilst both 18–24 and 25–64 years old became less likely to report difficulty social distancing in green spaces as a barrier to use, respondents aged 65+ reporting that barrier remained relatively constant. A dip at wave 2 in reporting busy green spaces as a barrier to use by the youngest age group, and an overall steeper rise through time by the oldest age group probably explains the significant interaction (Figure 4—panel B). There were substantial falls for all age groups in reporting 'fear for my health' as a barrier to green space visits, but the fall was furthest and sharpest for the youngest age group, whilst it reduced between waves 2 and 3 for the older age groups (Figure 4—panel C).

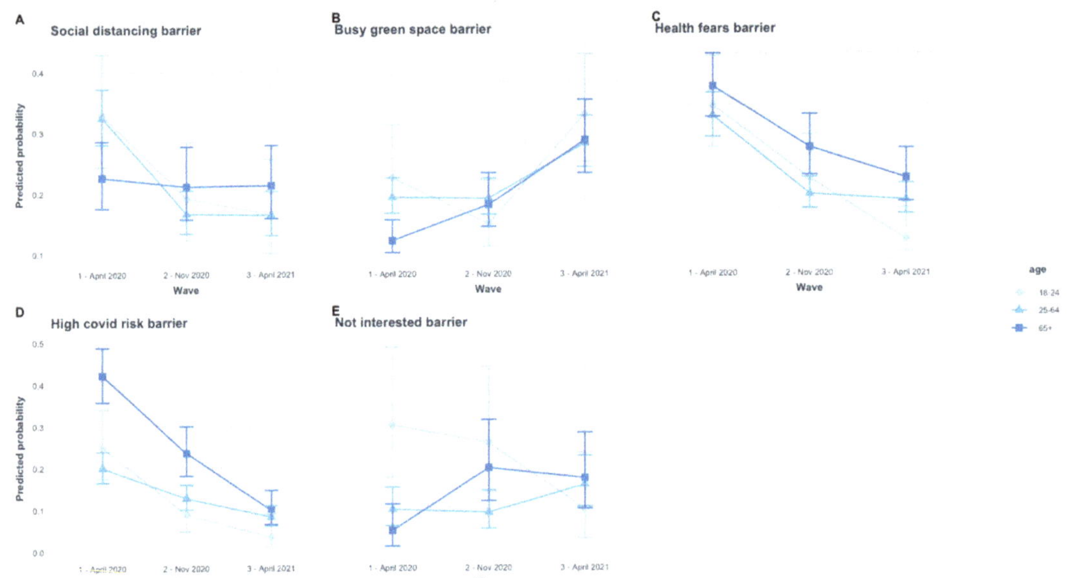

Figure 4. Predicted probabilities from logistic regression models with significant interactions between wave and age-group for: (**A**) the barrier 'I am worried I will not be able to socially distance', (**B**) the barrier 'green spaces are too busy', (**C**) the barrier 'I fear for my health when I go outdoors', (**D**) the barrier 'I/a member of my household is at higher risk of being severely affected by coronavirus', and (**E**) the barrier 'I am not interested in visiting green spaces'.

Reporting high risk of severe effects from COVID-19 as a barrier to the use of green spaces fell over time among all age groups, but a substantially higher starting level among those age 65+ and a relatively steeper decline produced a significant interaction (Figure 4—panel D). Finally, there were quite complex changes in the relationship between age groups and reporting a lack of interest in visiting green spaces (Figure 4—panel E). Perhaps the strongest signal from these was that the lack of interest fell sharply among the youngest age group, in contrast to rises among the older age groups.

4. Discussion

This study aims to describe the variation and change over time in green space visits and reported barriers to visiting during the COVID-19 pandemic in the UK. We drew on three waves of repeat cross-sectional data from April 2020, November 2020, and April 2021.

Visits to green spaces during the 4 weeks prior to the survey increased across the three waves—from 49% of respondents in April 2020 to 68% a year later. This increase was echoed in the regression results, confirming a significant association between survey wave and visiting. There were no other directly comparable surveys (in terms of timing) of green space use in the UK, but several other cohort studies and surveys did take place. They largely echo our findings, suggesting substantial increases in visits to natural environments well beyond what had been achieved in the years prior to the pandemic [29,41,42]. If we assume that contact with nature is generally good for health and wellbeing, this marks a positive impact of COVID-19. However, the extent of the increase in visits shown by our data is partly driven by the timing of the surveys. We were fortunate to conduct wave 1 during the strictest lockdown the UK experienced. Outdoor exercise was strongly limited during this time. The data we present here capture visits to green spaces, but measures of time spent there point to an inevitable reduction as a consequence of the restrictions. By waves 2 and 3, restrictions on outdoor exercise were reduced and this probably partly explains the sharp rise [37]. It will be important to continue monitoring visits to green spaces as the pandemic wanes to see if a permanent shift in behaviour has taken place.

Although the proportion of people visiting green spaces increased, socio-economic inequality in visits also rose and this raises concerns about environmental justice. The environment itself did not alter, but behavioural response to it did. Throughout the study period, higher social grade respondents were more likely to have visited green spaces than the lower social grade respondents. The gap between these groups in terms of the predicted probability of visiting rose by 33% over time. Several other studies have noted socio-economic inequalities in the use of green spaces both pre- and inter-pandemic [19,41–44]. One likely contributor is inequality in access—a classic issue of environmental injustice. For example, a study found that British people with an annual household income lower than £15,000 were less likely to live within a 5-min walk of green spaces, to live somewhere where the streets are green, and to report good walking routes where they live, compared to households with an annual income of £35,000 or £70,000 [41]. However, another contributor could be the risk of infection when using green spaces. A recent study found that the boroughs in London with a higher risk of infection shared common characteristics. These included the low accessibility of green spaces, high covid case concentration, and high vulnerability to virus transmission (calculated using the Indices of Multiple Deprivation) [45]. Lower social grade respondents may have felt at more risk of contracting coronavirus when visiting neighbourhood green spaces compared to higher social grade respondents.

It is difficult to directly compare countries' changes in green space use and barriers to using green spaces during the COVID-19 pandemic. This is primarily due to international differences in the nature and timing of mobility restrictions at different stages of the pandemic and also to the dates of survey data collection in relation to these. Despite limitations to direct comparison, the evidence does suggest that there were substantial before and after changes in green space use in many countries following COVID-19, but the nature of the changes differed. Decreases in use were described in Saudi Arabia [46], Italy, and Spain [47], for example, whilst increases were described in Belgium [48] and Norway [49]. In New York, equal numbers of respondents reported that they increased (15%) and decreased (14%) their green space visits during the pandemic, which was influenced by Covid-related barriers (N = 1145). Individuals with greater concerns about crowded green spaces and lack of social distancing visited green spaces less often during the pandemic compared to before, while those who considered green spaces to be more important for their health visited more frequently [28]. Our study may be unique in having assessed changes in green space use at three different time points through the pandemic and in finding that increases in visits were sustained but COVID-19 specific barriers were not important in later waves.

Our study focused largely on barriers to visiting green spaces that could plausibly be created and mitigated by the progress of the pandemic. Results suggested a shift in the population's perceptions of risk. Being worried about social distancing, fearing for health, and respondents perceiving a household member or themselves as at risk of being severely affected by COVID-19 were all less likely to be reported by respondents in waves 2 and 3 compared to wave 1. This echoes studies from other countries. Research undertaken in Canada in 2020, for example, reported that some participants noted that seeing more people outside coincided with the easing of restrictions. They also stated that people viewed outdoor activities as permissible because it was occurring outdoors instead of indoors, and there was less risk of spreading coronavirus [50]. The relationships between reported barriers also shed some light on these changes. For example, by wave 3, respondents were more likely to report green spaces being too busy as a barrier than they had been in wave 1. This corroborates our finding that visits to green spaces increased over the year; the spaces really were busier. Furthermore, by wave 3, respondents were less likely to report being worried about social distancing as a barrier to using green spaces than in wave 1. This suggests concerns over crowding may be more connected to accessing the green space and having an enjoyable experience than worries about getting too close to others outdoors. This finding links with several other studies from around the world [32,51,52]. Research undertaken in Palestine, for example, found that respondents were more likely to visit green spaces alone, but less likely to 'relax' or 'socialise' in green spaces after the pandemic occurred compared to before [51]. It is possible that encouraging more people into green spaces had the unintended consequence of putting off others. Since these are repeat cross-sectional rather than panel data, we were unable to assess whether/who stopped visiting as a result.

Overall, the reduction in reporting of barriers relating to health and contracting COVID-19 may be due to both increased knowledge/understanding of the risk of contracting COVID-19 outdoors and progress with vaccination from December 2020 in the UK [52–54]. In late 2020 and early 2021, for example, studies identified a very low likelihood of COVID-19 transmission outdoors and these were widely reported in the UK [54–56]. However, both this new knowledge and the beginnings of the UK's vaccination programme would have likely only impacted the barriers reported in wave 3 and these were similar to those reported in wave 2. Further evidence that the public understood COVID-19 risks quite well comes from between-group differences in reporting barriers. The oldest group (65+) were most likely to report fearing for their health when outdoors and avoiding green spaces because they or a member of their household were at high risk of being severely affected by COVID-19. Given that those aged 80 years or older were seventy times more likely to die following a positive COVID-19 test compared with those under 40 [57], it seems reasonable older respondents reported these barriers to a greater extent.

Although we focused on barriers plausibly related to COVID-19, our inclusion of 'lack of interest' as a barrier revealed important trends. Across the whole year of study, respondents aged 18–24 years old were more likely than other age groups to report a lack of interest as a barrier to visiting green spaces. This echoes other research, including studies in England before and during the pandemic, which found that younger age groups (16–34 and 16–24) were more likely to report a lack of interest than older age groups [19,58]. However, in our study, there was a sharp reduction during the year in this lack of interest among young people, and one somewhat in contrast to the older age groups. An apparent increase in interest among young adults could be connected to the relaxing of restrictions from the end of March 2021. The change in restrictions meant outdoor spaces provided the only location for socialising with non-household members [37]. Perhaps younger groups became oriented to green spaces as a place to meet and socialise [59].

Strengths and Limitations

The study had several strengths including a nationally representative sample of the UK population when weightings were applied. To the best of our knowledge, our study provided the only data covering the UK population's change in green space use and barriers over one year of the COVID-19 pandemic starting in the period of maximum restrictions. The breadth of data enabled us to explore both changes in visits over time and reasons for not visiting. The latter is particularly poorly covered in the wider literature.

However, the survey data were self-reported and therefore liable to the usual biases, including recall and confirmation bias. The study was also repeated cross-sectionally rather than in a panel. The changes we observed and explored were therefore at a population or group level, not within individuals. There were some small numbers when exploring subgroups who exhibited particular behaviours or attitudes. Data on the youngest group were particularly prone to this problem and it seems likely that bias will have been introduced. These problems highlight the need for further research into barriers to green space use for the wider population, as well as more specifically for young adults across the UK. The information on barriers was somewhat crude and we would encourage further research to investigate barriers to green space use in more detail, perhaps through qualitative methods or open-ended survey questions. In particular, the survey data on barriers analysed for this paper focused specifically on the influence of the COVID-19 pandemic on green space use. There may have been other barriers associated with the pandemic but not captured here, such as changes in behaviours imposed by home schooling and home working. These are also likely to have been unequal.

5. Conclusions

At a time when the COVID-19 pandemic is ongoing and restrictions are continuing to change, our study provides novel findings on how the UK population's visits to green spaces, and barriers to visiting, changed over the first year of the pandemic. Inequalities in the use of green spaces by social grade were sustained and widened over the year, emphasising that environmental injustices were exacerbated. There is a need for further investigation and action to ensure equal access to good quality green spaces for all demographic groups and communities. The reporting of Covid-related barriers to green space use, such as fear for their health and worrying about social distancing, fell over time, in parallel perhaps with greater knowledge of the virus, how it spreads, and perceptions of risk. Levels of disinterest in visiting green spaces also changed over the year, particularly for 18–24 years old; the pandemic seems to have made changes to their perceptions of green spaces. Our findings suggest that, despite the benefits to health and wellbeing that green spaces use can provide, there are still barriers in place that restrict some of the population from using them. These findings also indicate that the barriers in place are not felt equally across socio-demographic groups, particularly by age. The pandemic has affected green space use in both positive and negative ways; it will be important to determine if these are permanent changes. We encourage further research exploring young people's use and perceptions of green spaces, with a particular focus on interest in green spaces and how the patterns relating to a 'lack of interest' may change as all lockdown restrictions are removed. Our study has highlighted that inequalities in the use of green spaces remain and future research should be focused on reasons for a lack of green space use that is not related to COVID-19, which remains a key gap in the existing literature.

Supplementary Materials: The following are available online at https://www.mdpi.com/article/10.3390/land11040503/s1, Table S1: Weighted counts and percentages of demographic variables, Table S2: Significant interaction results by wave and socio-demographic variables; Predicted Probabilities (95% Confidence Intervals>), Table S3: Significant interaction results by wave and socio-demographic variables, each interaction model was adjusted for wave, sex, age, and social grade; Interaction Odds Ratios (95% Confidence Intervals).

Author Contributions: Conceptualization, H.B., J.R.O. and R.M.; methodology, H.B., J.R.O. and R.M.; software, H.B.; validation, H.B., J.R.O. and R.M.; formal analysis, H.B.; data curation, H.B.; writing—original draft preparation, H.B.; writing—review and editing, H.B., J.R.O. and R.M.; visualization, H.B. All authors have read and agreed to the published version of the manuscript.

Funding: The authors (H.B., J.R.O. and R.M.) are part of the Places and Health Programme at the MRC/CSO Social and Public Health Sciences Unit (SPHSU), University of Glasgow, supported by the Medical Research Council (MC_UU_00022/4) and the Chief Scientist Office (SPHSU19). H.B. is also funded by a Medical Research Council and University of Glasgow College of Medical, Veterinary and Life Sciences PhD studentship (MC_ST_U18004).

Institutional Review Board Statement: Ethical review and approval were waived for this study due to the data collection being conducted by YouGov, who use a double opt-in procedure across all Panels. All data were anonymized.

Informed Consent Statement: Informed consent was obtained by YouGov from all subjects involved in the Panel.

Data Availability Statement: We have made our research dataset publicly available (Datacite DOI: 10.5525/gla.researchdata.1038 and 10.5525/gla.researchdata.1091—embargoed until 31 October 2023).

Conflicts of Interest: The authors declare no conflict of interest.

References

1. World Health Organisation. Coronavirus Disease (COVID-19). World Health Organ. Available online: https://www.who.int/health-topics/coronavirus#tab=tab_1 (accessed on 17 January 2022).
2. Johnson, B.; Prime Minister's Statement on Coronavirus (COVID-19): 23 March 2020. GOV.UK. 2020. Available online: https://www.gov.uk/government/speeches/pm-address-to-the-nation-on-coronavirus-23-march-2020 (accessed on 7 July 2020).
3. Evandrou, M.; Falkingham, J.; Qin, M.; Vlachantoni, A. Changing living arrangements and stress during COVID-19 lockdown: Evidence from four birth cohorts in the UK. *SSM-Popul. Health* **2021**, *13*, 100761. [CrossRef]
4. Office for National Statistics. A "New Normal"? How People Spent Their Time after the March 2020 Coronavirus Lockdown. ONS. 2020. Available online: https://www.ons.gov.uk/peoplepopulationandcommunity/healthandsocialcare/conditionsanddiseases/articles/anewnormalhowpeoplespenttheirtimeafterthemarch2020coronaviruslockdown/2020-12-09 (accessed on 20 January 2022).
5. Office for Health Improvement & Disparities. COVID-19 Mental Health and Wellbeing Surveillance: Report, 2. Important Findings. UK Gov. 2021. Available online: https://www.gov.uk/government/publications/COVID-19-mental-health-and-wellbeing-surveillance-report/2-important-findings-so-far (accessed on 17 January 2022).
6. Kontis, V.; Bennett, J.E.; Rashid, T.; Parks, R.M.; Pearson-Stuttard, J.; Guillot, M.; Asaria, P.; Zhou, B.; Battaglini, M.; Corsetti, G.; et al. Magnitude, demographics and dynamics of the effect of the first wave of the COVID-19 pandemic on all-cause mortality in 21 industrialized countries. *Nat. Med.* **2020**, *26*, 1919–1928. [CrossRef] [PubMed]
7. Peters, A.; Rospleszcz, S.; Greiser, K.H.; Dallavalle, M.; Berger, K. The Impact of the COVID-19 Pandemic on Self-Reported Health. *Dtsch. Arztebl. Int.* **2020**, *117*, 861–867. [CrossRef]
8. Yao, Y.; Lu, Y.; Guan, Q.; Wang, R. Can parkland mitigate mental health burden imposed by the COVID-19? A national study in China. *Urban For. Urban Green.* **2022**, *67*, 127451. [CrossRef] [PubMed]
9. Giuntella, O.; Hyde, K.; Saccardo, S.; Sadoff, S. Lifestyle and mental health disruptions during COVID-19. *Proc. Natl. Acad. Sci. USA* **2021**, *118*, e2016632118. [CrossRef]
10. Sia, A.; Tan, P.Y.; Wong, J.C.M.; Araib, S.; Ang, W.F.; Er, K.B.H. The impact of gardening on mental resilience in times of stress: A case study during the COVID-19 pandemic in Singapore. *Urban For. Urban Green.* **2022**, *68*, 127448. [CrossRef] [PubMed]
11. Welsh, C.E.; Albani, V.; Matthews, F.E.; Bambra, C. The effects of the first national lockdown in England on geographical inequalities in the evolution of COVID-19 case rates: An ecological study. *medRxiv* **2021**. [CrossRef]
12. Bambra, C.; Lynch, J.; Smith, K.E. *The Unequal Pandemic: COVID-19 and Health Inequalities*; Policy Press: Bristol, UK, 2021.
13. McCarthy, M. Nature Got Us through Lockdown. Here's How It Can Get Us through the Next One. Guard. 2020. Available online: https://www.theguardian.com/books/2020/oct/03/nature-got-us-through-lockdown-heres-how-it-can-get-us-through-the-next-one (accessed on 18 February 2022).
14. Briggs, H. Nature 'More Important than ever during Lockdown'. *BBC News*. 2021. Available online: https://www.bbc.co.uk/news/science-environment-56889322 (accessed on 18 February 2022).
15. Twohig-Bennett, C.; Jones, A. The health benefits of the great outdoors: A systematic review and meta-analysis of greenspace exposure and health outcomes. *Environ. Res.* **2018**, *166*, 628–637. [CrossRef]
16. Geng, D.C.; Innes, J.; Wu, W.; Wang, G. Impacts of COVID-19 pandemic on urban park visitation: A global analysis. *J. For. Res.* **2021**, *32*, 553–567. [CrossRef]
17. Burnett, H.; Olsen, J.R.; Nicholls, N.; Mitchell, R. Change in time spent visiting and experiences of green space following restrictions on movement during the COVID-19 pandemic: A nationally representative cross-sectional study of UK adults. *BMJ Open* **2021**, *11*, e044067. [CrossRef] [PubMed]
18. Cheng, Y.; Zhang, J.; Wei, W.; Zhao, B. Effects of urban parks on residents' expressed happiness before and during the COVID-19 pandemic. *Landsc. Urban Plan.* **2021**, *212*, 104118. [CrossRef]
19. Boyd, F.; White, M.P.; Bell, S.L.; Burt, J. Who doesn't visit natural environments for recreation and why: A population representative analysis of spatial, individual and temporal factors among adults in England. *Landsc. Urban Plan.* **2018**, *175*, 102–113. [CrossRef]
20. Morris, J.; O'Brien, E.; Ambrose-Oji, B.; Lawrence, A.; Carter, C.; Peace, A. Access for all? Barriers to accessing woodlands and forests in Britain. *Local Environ.* **2011**, *16*, 375–396. [CrossRef]

21. Ode Sang, Å.; Knez, I.; Gunnarsson, B.; Hedblom, M. The effects of naturalness, gender, and age on how urban green space is perceived and used. *Urban For. Urban Green.* **2016**, *18*, 268–276. [CrossRef]

22. Verheij, J. *Urban Green Space as a Matter of Environmental Justice: The Case of Lisbon's Urban Greening Strategies*; KTH Royal Institute of Technology: Stockholm, Sweden, 2019. Available online: http://kth.diva-portal.org/smash/get/diva2:1342024/FULLTEXT01. pdf (accessed on 14 March 2022).

23. Wolch, J.R.; Byrne, J.; Newell, J.P. Urban green space, public health, and environmental justice: The challenge of making cities 'just green enough'. *Landsc. Urban Plan.* **2014**, *125*, 234–244. [CrossRef]

24. Pallathadka, A.; Pallathadka, L.; Rao, S.; Chang, H.; Van Dommelen, D. Using GIS-based spatial analysis to determine urban greenspace accessibility for different racial groups in the backdrop of COVID-19: A case study of four US cities. *GeoJournal* **2021**, 1–21. [CrossRef] [PubMed]

25. Mitchell, R. What Is Equigenesis and How Might It Help Narrow Health Inequalities? CRESH. 2013. Available online: https: //cresh.org.uk/2013/11/08/what-is-equigenesis-and-how-might-it-help-narrow-health-inequalities/ (accessed on 24 July 2019).

26. Mitchell, R.J.; Richardson, E.A.; Shortt, N.K.; Pearce, J.R. Neighborhood Environments and Socioeconomic Inequalities in Mental Well-Being. *Am. J. Prev. Med.* **2015**, *49*, 80–84. [CrossRef]

27. Moran, M.R.; Bilal, U.; Dronova, I.; Ju, Y.; Gouveia, N.; Caiaffa, W.T.; de Friche, A.A.L.; Moore, K.; Miranda, J.J.; Rodríguez, D.A. The equigenic effect of greenness on the association between education with life expectancy and mortality in 28 large Latin American cities. *Health Place* **2021**, *72*, 102703. [CrossRef]

28. Lopez, B.; Kennedy, C.; Field, C.; McPhearson, T. Who benefits from urban green spaces during times of crisis? Perception and use of urban green spaces in New York City during the COVID-19 pandemic. *Urban For. Urban Green.* **2021**, *65*, 127354. [CrossRef]

29. Natural England. The People and Nature Survey for England: Data and Publications from Adults Survey Year 1 (April 2020– March 2021) (Official Statistics) Main Findings. 2021. Available online: https://www.gov.uk/government/statistics/the-people- and-nature-survey-for-england-data-and-publications-from-adults-survey-year-1-april-2020-march-2021-official-statistics/ the-people-and-nature-survey-for-england-data-and-publications-from-adults-survey-year-1-april-2020-march-2021-official- statistics-main-finding (accessed on 17 January 2022).

30. Office for National Statistics. How Has Lockdown Changed Our Relationship with Nature? 2021. Available online: https://www. ons.gov.uk/economy/environmentalaccounts/articles/howhaslockdownchangedourrelationshipwithnature/2021-04-26 (accessed on 18 January 2022).

31. Heo, S.; Desai, M.U.; Lowe, S.R.; Bell, M.L. Impact of changed use of greenspace during COVID-19 pandemic on depression and anxiety. *Int. J. Environ. Res. Public Health* **2021**, *18*, 5842. [CrossRef] [PubMed]

32. Korpilo, S.; Kajosaari, A.; Rinne, T.; Hasanzadeh, K.; Raymond, C.M.; Kyttä, M. Coping with Crisis: Green Space Use in Helsinki Before and During the COVID-19 Pandemic. *Front. Sustain. Cities* **2021**, *3*, 99. [CrossRef]

33. YouGov. About YouGov Company. Available online: https://yougov.co.uk/about/ (accessed on 8 June 2020).

34. Prime Minister's Office. Prime Minister Announces New National Restrictions. GOV.UK. 2020. Available online: https://www.gov. uk/government/news/prime-minister-announces-new-national-restrictions (accessed on 19 January 2022).

35. Drakeford, M. Written Statement: New National Covid Measures for Wales (2 November 2020). Welsh Gov. 2020. Available online: https://gov.wales/written-statement-new-national-covid-measures-wales (accessed on 19 January 2022).

36. Sturgeon, N. Coronavirus (COVID-19) Update: First Minister's Statement Tuesday 17 November 2020. Scottish Gov. 2020. Available online: https://www.gov.scot/publications/coronavirus-COVID-19-update-first-ministers-statement-tuesday-17 -november-2020/ (accessed on 19 January 2022).

37. The Institute for Government. Timeline of UK Government Coronavirus Lockdowns and Restrictions. Inst. Gov. 2021. Available online: https://www.instituteforgovernment.org.uk/charts/uk-government-coronavirus-lockdowns (accessed on 19 January 2022).

38. The Market Research Society (MRS). Definitions Employed in Social Grading. MRS. Available online: https://www.mrs.org.uk/ pdf/Definitions%20used%20in%20Social%20Grading%20based%20on%20OG7.pdf (accessed on 2 July 2020).

39. R Core Team. R: A Language and Environment for Statistical Computing. 2018. Available online: https://www.r-project.org/ (accessed on 19 January 2022).

40. YouGov. Panel Methodology. YouGov. Available online: https://yougov.co.uk/about/panel-methodology/ (accessed on 19 January 2022).

41. Ramblers. *The Grass Isn't Greener for Everyone: Why Access to Green Space Matters*; Ramblers: London, UK, 2020.

42. Holland, F. *Out of Bounds Equity in Access to Urban Nature*; Groundwork: Birmingham, UK, 2021.

43. Hoffimann, E.; Barros, H.; Ribeiro, A.I. Socioeconomic Inequalities in Green Space Quality and Accessibility-Evidence from a Southern European City. *Int. J. Environ. Res. Public Health* **2017**, *14*, 916. [CrossRef] [PubMed]

44. Spotswood, E.N.; Benjamin, M.; Stoneburner, L.; Wheeler, M.M.; Beller, E.E.; Balk, D.; McPhearson, T.; Kuo, M.; McDonald, R.I. Nature inequity and higher COVID-19 case rates in less-green neighbourhoods in the United States. *Nat. Sustain.* **2021**, *4*, 1092–1098. [CrossRef]

45. Pan, J.; Bardhan, R.; Jin, Y. Spatial distributive effects of public green space and COVID-19 infection in London. *Urban For. Urban Green.* **2021**, *62*, 127182. [CrossRef]

46. Addas, A.; Maghrabi, A. How did the COVID-19 pandemic impact urban green spaces? A multi-scale assessment of Jeddah megacity (Saudi Arabia). *Urban For. Urban Green.* **2022**, *69*, 127493. [CrossRef] [PubMed]

47. Ugolini, F.; Massetti, L.; Calaza-Martínez, P.; Cariñanos, P.; Dobbs, C.; Ostoic, S.K.; Marin, A.M.; Pearlmutter, D.; Saaroni, H.; Šaulienė, I.; et al. Effects of the COVID-19 pandemic on the use and perceptions of urban green space: An international exploratory study. *Urban For. Urban Green.* **2020**, *56*, 126888. [CrossRef] [PubMed]

48. da Schio, N.; Phillips, A.; Fransen, K.; Wolff, M.; Haase, D.; Ostoić, S.K.; Živojinović, I.; Vuletić, D.; Derks, J.; Davies, C.; et al. The impact of the COVID-19 pandemic on the use of and attitudes towards urban forests and green spaces: Exploring the instigators of change in Belgium. *Urban For. Urban Green.* **2021**, *65*, 127305. [CrossRef]

49. Venter, Z.S.; Barton, D.N.; Gundersen, V.; Figari, H.; Nowell, M. Urban nature in a time of crisis: Recreational use of green space increases during the COVID-19 outbreak in Oslo, Norway. *Environ. Res. Lett.* **2020**, *15*, 104075. [CrossRef]

50. McCormack, G.R.; Petersen, J.; Naish, C.; Ghoneim, D.; Doyle-Baker, P.K. Neighbourhood environment facilitators and barriers to outdoor activity during the first wave of the COVID-19 pandemic in Canada: A qualitative study. *Cities Health* **2022**, 1–13. [CrossRef]

51. Dawwas, E.B.; Dyson, K. COVID-19 Changed Human-Nature Interactions across Green Space Types: Evidence of Change in Multiple Types of Activities from the West Bank, Palestine. *Sustainability* **2021**, *13*, 3831. [CrossRef]

52. NHS England. Landmark Moment as First NHS Patient Receives COVID-19 Vaccination. 2020. Available online: https://www.england.nhs.uk/2020/12/landmark-moment-as-first-nhs-patient-receives-COVID-19-vaccination/ (accessed on 18 January 2022).

53. The Institute for Government. Coronavirus Vaccine Rollout. 2021. Available online: https://www.instituteforgovernment.org.uk/explainers/coronavirus-vaccine-rollout (accessed on 18 January 2022).

54. Jones, N.R.; Qureshi, Z.U.; Temple, R.J.; Larwood, J.P.J.; Greenhalgh, T.; Bourouiba, L. Two metres or one: What is the evidence for physical distancing in COVID-19? *BMJ* **2020**, *370*, m3223. [CrossRef] [PubMed]

55. Public Health England (PHE) Transmission Group. *Factors Contributing to Risk of SARS-CoV-2 Transmission Associated with Various Settings*; Public Health England: London, UK, 2020.

56. Covid: Can You Catch the Virus Outside? *BBC News.* 2021. Available online: https://www.bbc.co.uk/news/explainers-55680305 (accessed on 25 January 2022).

57. Public Health England. *Disparities in the Risk and Outcomes of COVID-19*; Public Health England: London, UK, 2020. Available online: https://assets.publishing.service.gov.uk/government/uploads/system/uploads/attachment_data/file/908434/Disparities_in_the_risk_and_outcomes_of_COVID_August_2020_update.pdf (accessed on 18 January 2022).

58. Natural England and Kantar Public. *Impact of COVID-19 on Engagement with Green and Natural Spaces*; Natural England: York, UK, 2021.

59. Ramblers. Empowering the Next Generation to Explore the Great Outdoors. Ramblers. 2021. Available online: https://www.ramblers.org.uk/outthere (accessed on 19 January 2022).

Article

Barriers Affecting Women's Access to Urban Green Spaces during the COVID-19 Pandemic

Carolina Mayen Huerta * and Ariane Utomo

School of Geography, Earth & Atmospheric Sciences, University of Melbourne, Melbourne, VIC 3010, Australia; ariane.utomo@unimelb.edu.au
* Correspondence: cmayenhuerta@student.unimelb.edu.au

Abstract: During the COVID-19 pandemic, urban green spaces (UGS) have gained relevance as a resilience tool that can sustain or increase well-being and public health in cities. However, several cities in Latin America have seen a decrease in their UGS use rates during the health emergency, particularly among vulnerable groups such as women. Using Mexico City as a case study, this research examines the main barriers affecting women's access to UGS during the COVID-19 pandemic in Latin America. We applied a sequential mixed-methods approach in which the results of a survey distributed via social media in June 2020 to women aged 18 and older were used to develop semi-structured interviews with 12 women during October 2020. One year later, in November 2021, the continuity of the themes was evaluated through focus groups with the same group of women who participated in the interviews. Our results suggest that (1) prohibiting access to some UGS during the first months of the pandemic negatively impacted UGS access for women in marginalized neighborhoods; (2) for women, the concept of UGS quality and safety are intertwined, including the security level of the surrounding streets; and (3) women who live in socially cohesive neighborhoods indicated using UGS to a greater extent. Our findings highlight that while design interventions can affect women's willingness to use UGS by improving their perceived safety and comfort, they remain insufficient to fully achieve equity in access to UGS.

Keywords: geography of fear; gender; green spaces; violence; Latin America; quality; safety; mixed-methods; social equity; fear of crime

Citation: Mayen Huerta, C.; Utomo, A. Barriers Affecting Women's Access to Urban Green Spaces during the COVID-19 Pandemic. *Land* **2022**, *11*, 560. https://doi.org/10.3390/land11040560

Academic Editors: Francesca Ugolini and David Pearlmutter

Received: 11 March 2022
Accepted: 7 April 2022
Published: 10 April 2022

1. Introduction

A growing body of evidence suggests that exposure to urban green spaces (UGS) is critical to maintaining or increasing public health in cities, especially during emergencies such as the recent COVID-19 pandemic [1–7]. Access to UGS is associated with a multiplicity of benefits, including fostering social cohesion, improving air quality, enhancing the livability of neighborhoods, reducing overall stress and anxiety, and, more recently, lessening the feeling of isolation experienced during confinements [1,8,9]. Nevertheless, access to UGS is rarely neutral, predominantly in cities of the Global South [10–12]. Women, low-income individuals, people with disabilities, and ethnic minorities often encounter barriers, such as an absence of UGS in close proximity, fear of violence, poor quality or inadequate facilities, smaller available UGS, and high density, which during the pandemic was a critical hindrance to using these spaces [7,13–15]. The barriers mentioned above influence the possibility of visiting and enjoying UGS, which are essential to take full advantage of their restorative capabilities and improve people's health and overall well-being [3,16,17].

In Latin American cities, women are often underrepresented in public spaces, including green spaces [10,18]. Even so, research that focuses on the barriers preventing women from accessing UGS in the cities of this continent is scarce [19], particularly that which explores the restricted use of green spaces by women during the COVID-19 pandemic [7,11,15,20,21]. Unlike countries in the Global North such as the United States,

Norway, Italy, or England, where evidence on the use of UGS after the COVID-19 outbreak showed a significant increase for most segments of the population [22,23], the current data on women's use of UGS in countries such as Mexico, Peru, Brazil, and Argentina shows the opposite trend [20,21,24–27]. This is particularly concerning since access to UGS is of paramount importance for women, with far-reaching repercussions on women's mental health compared to men [17], as evidenced by Roe et al.'s [28] study on the implications of low exposure to UGS, which emphasizes the need to target this demographic. More broadly, beyond the context of the pandemic, cities aiming to become more inclusive should evaluate whether all population subgroups can access the existing green infrastructure, adapting the built environment to account for and promote the well-being of all residents [12,29–32].

Using Mexico City as a case study, this study aims to identify and understand the main barriers that have affected women's UGS access in Latin America and, therefore, UGS use during the COVID-19-induced crisis. Two central questions guide our paper: (1) What have been the main barriers to women's UGS access during the pandemic? (2) What factors could help increase women's use of UGS? These two questions are critical to understanding women's perspectives on UGS use and how to improve their experience while using these spaces. Furthermore, they have the potential to inform evidence-based policymaking designed to improve existing green areas and make access to these spaces more equitable in Latin American cities, which is of great relevance during a prolonged public health emergency such as the COVID-19 pandemic. To answer these two questions, we use a sequential mixed-methods approach, encompassing three stages in its data collection process and strategy of analysis, one quantitative and two qualitative [33,34]. Our research's qualitative components helped further explore the statistical results of the quantitative component, a technique widely used in exploratory studies [35,36]. First, we analyzed quantitative data from an online survey (*n* = 1245) launched in June 2020. Second, through the analysis of survey results, we developed a qualitative instrument, semi-structured in-depth interviews with twelve women in October 2020. Finally, through the interviews, we were able to identify critical themes that required further exploration, which led us to conduct two focus groups with the same participants in November 2021.

1.1. The COVID Pandemic and UGS

In March 2020, the novel coronavirus outbreak was categorized as a global pandemic, prompting governments worldwide to take drastic measures to contain the spread of the disease and subjecting almost half of the world's population to stay-at-home orders [2,6]. Several cities began confinement periods during which mobility and contact between people were restricted [37]. Preliminary evidence on the effects of lockdown measures on the population has demonstrated that containment policies, such as stay-at-home orders and the closure of non-essential businesses and schools, had adverse effects on people's well-being by introducing psychological stress, anxiety, feelings of loneliness, anger, frustration, and in some cases, depression, especially for women [38–41]. In particular, as numerous reports have demonstrated, once SAR-CoV-2 was categorized as a pandemic, women experienced rising rates of domestic violence, loss of jobs, deterioration in mental health, and increased anxiety about paying for health services [42,43], which negatively impacted their fear of contracting the disease compared to men [44,45].

In the vast majority of affected cities, urban residents identified UGS as one of the only means to avoid a sedentary lifestyle, cope with the stress of the situation, and improve their physical health while maintaining distance from others, as most sports and recreational facilities closed during the first months of the emergency [14,46–50]. For example, during a study conducted in Oslo in March 2020, Venter et al. [22] showed that the use of green spaces had increased, becoming a measure of resilience to the pandemic that helped sustain the well-being of users. Pouso et al. [2] conducted an online survey in nine countries during the early months of the pandemic and found that people who accessed UGS reported more positive emotions and used these spaces to cope with lockdown measures. Furthermore, research in Belgium by da Schio et al. [23] revealed that respondents also placed more importance on UGS after the pandemic began, showcasing the benefits perceived after their use.

The studies mentioned above support the claim that maintaining access to UGS and encouraging their use are essential measures for amplifying positive health behaviors in urban populations [51]. Even more so, considering that the growing body of evidence on the transmission of COVID-19 shows that outdoor spaces can be categorized as safe options for recreation and leisure [14,47,52]. However, despite the diversity of benefits UGS can provide, not all cities observed an increased UGS use during the health crisis [53]. For example, in Argentina, a survey administered in Buenos Aires ($n = 1740$) demonstrated that despite respondents associating UGS use with feelings of calmness and tranquility, they did not increase the frequency of use or time spent in UGS after the pandemic began [26]. In Mexico City, Google's COVID-19 Community Mobility Report revealed a decrease of nearly 34% in the use of parks during 2020 compared to 2019, with women significantly decreasing their use of UGS compared to men [24,25]. The latter is worrisome, as limited use of UGS during the COVID-19 pandemic has been associated with a deterioration in health, primarily mental health [54]. It is still unclear whether the fear of contracting or spreading COVID-19, the closure of UGS, restrictions in mobility, or additional factors—such as changes in the perception of UGS—could have potentially discouraged UGS visitation expressly for women.

1.2. Barriers Affecting Women's Access to UGS

The barriers that affect women's UGS access are varied and, in some cases, depend on the specific context of their location. For example, Manyani, Shackleton, and Cocks [18] have cited social norms in South Africa as obstacles that prevent women from using green spaces on ceremonial occasions. Another context-specific factor is extreme weather conditions, which can discourage the use of these spaces [55]. For instance, in Hyderabad, India, green spaces that lack adequate vegetation (i.e., tree cover) are not commonly used in the summer when the temperature exceeds 40 °C [56]. The racial or ethnic composition of neighborhoods can also affect the frequency of UGS use by minority women [57]. Comparable to these examples, there are several others in which barriers associated with specific social norms or geographies limit women's UGS access.

In perspective, aspects such as the aforementioned barriers are difficult to modify. Regardless, additional obstacles restrict women's access to UGS, such as distance, limited size, poor quality, perceived insecurity [13], and more recently, COVID-19 restrictions or fears [37]. Urban planners and policymakers' influence can modify these impediments to promote greater UGS use by women, thereby improving their health and well-being, particularly during a public health emergency [58–60].

1.2.1. Distance and Size of UGS

The consensus in the literature has suggested that the distance to UGS is the main factor that enables or prevents UGS access because use is bounded by geography [16,61,62]. Distance plays a significant role in women's access to UGS, as evidenced by a study in two Chilean cities, which indicates that women walk shorter distances to access green spaces [63]. In particular, given that pandemic restrictions caused the closure of UGS or restrained residents' mobility in several cities, this factor was reported to have significantly influenced UGS use during COVID-19-induced lockdowns by reducing or eradicating nearby options and eliminating the possibility of visiting distant green spaces [4,43,46]. As an example, the results of a survey in La Palma and Zaragoza, Spain, where strict home confinement measures were put in place to prevent the spread of COVID-19, showed that people who live closer to green areas reported higher UGS use compared with those who live far away [64]. Furthermore, the results of another study based on a survey conducted in six countries suggested that, during the COVID-19 confinement period, people tended to visit green spaces at closer distances [37]. Given the extensive evidence on distance and its relationship to UGS access, particularly women's UGS access, it is safe to imply that, in contexts with COVID-19 mobility restrictions, the lack of nearby UGS could be a critical barrier to women's UGS use [4].

In addition to distance, size is another factor behind the gendered patterns of UGS use. Studies carried out in densely populated cities, such as New York, revealed that users reported feeling calmer and accessing more often spaces that allowed social distancing, which is linked to the size of UGS [5]. Overall, larger spaces allowed residents to engage in activities while maintaining adequate distance from others, which was generally associated with a propensity to use them [20]. Nevertheless, pre-pandemic evidence shows that women tend to avoid larger spaces where they are not visible to others [19].

1.2.2. UGS Quality and Safety

For women, safety and quality—which are strongly correlated and often used to explain one another—are generally more significant barriers to UGS access than for men, resulting in higher UGS use for women who have access to UGS of higher perceived quality or who do not fear violence when using these spaces [12,65–69]. For instance, Ode Sang et al.'s [17] study in Gothenburg, Sweden, and a second study conducted in the 40 major cities in China by Carli [21] showed that gender has a strong effect on the perception and use of green spaces, with women placing more emphasis on UGS quality characteristics, such as cleanliness, maintenance, order, and beauty, than men. Similarly, survey results from the Spanish city of Carmona have indicated that women attribute a higher value to quality and security features in green spaces than men, leading to increased use by women in safer and higher quality UGS [70]. Furthermore, a study conducted in South Africa found that women felt more discouraged than men to use green spaces because of the lack of maintenance, the arrangement of vegetation, and safety concerns [18]. Knapp et al. [71], who explored the relationship between UGS use and quality in 31 green spaces within low-income Black neighborhoods in New Orleans, United States, found that signs of concerns, which are commonly associated with safety-related feelings, and attractiveness were significant predictors of use for female users, while these same variables were insignificant among men. These results are similar to those of a study conducted by Williams et al. [68], who noted that women of color are exceedingly affected by the perceived lack of quality in UGS; in this study, safety (measured by crime rates) was the sole indicator used to define UGS quality.

Valentine [72] has defined the "geography of fear" as how feelings of vulnerability affect women's choices, mobility, and use of public spaces. To that end, perceptions or feelings are more frequently used to measure UGS quality and safety than objective measures [73]. Specifically, preconceived notions of a neighborhood or the particular experiences of women in a space may have a more significant effect on UGS use than objective crime rates [74]. In Latin America, such geographies of fear result from structural inequalities [75], which are especially alarming at a time when collective anxiety has risen as a result of the COVID-19 pandemic, and when measures to counteract the negative impacts of the pandemic are desperately needed [9,32,43,76,77].

2. Materials and Methods

2.1. Phase 1: Quantitative Analysis

In June 2020, we conducted an online survey to assess people's use and perception of UGS during the start of the COVID-19 pandemic. The survey was launched through social media platforms, promoting it only to adults (18 years and older). The survey was completely anonymous and consisted of four sections containing multiple-choice and open-ended questions related to (1) sociodemographic characteristics, (2) UGS use and frequency of visits before and after the COVID-19 pandemic began, (3) rating of neighborhood and UGS characteristics' importance, and (4) health-related questions. For the current paper, we use a subset of the survey sample consisting of only respondents who identify as women. Of the four sections in the questionnaire, we used the information corresponding to the first three to carry out the subsequent analysis.

A total of 1914 women started the survey, and 1245 completed it. We received responses from all municipalities and income groups. All the questionnaires that exhibited 100 percent progress were considered, including those questionnaires in which the participants decided for some reason not to answer one question but continued to answer the rest. This decision was made following the reasoning behind Lopez, Kennedy, and McPhearson's [5] analysis of park use in NYC during the early stages of the COVID-19 pandemic. The authors decided to evaluate the responses of those who had completed most of the survey due to the added value of their answers for the study's findings. To achieve a higher level of participation, we avoided asking for specific information. For example, asking about the respondent's income was ruled out. Instead, we requested respondents to indicate their income brackets. This approach is standard for data collection over the internet, where people may be more reluctant to provide information [40].

First, descriptive statistics were used to obtain information on the sociodemographic characteristics of the women participating in the study's first phase. Second, given that the evidence indicates that the quality of UGS is a determining barrier to women's access to UGS [19], to understand whether the quality of UGS has influenced access during the pandemic, we ran three logistic regression models to explore the association between UGS access and UGS quality. UGS use was adopted as a proxy for UGS access based on the study by Van Herzele and Wiedemann [13] that identifies access as a precondition for the use of UGS. The first model included only the association between UGS use and respondents' perception of UGS quality in their neighborhood (good quality = 1, not good quality = 0). The second model accounted for potential confounders to calculate adjusted odds ratios, exploring if they were significantly associated with UGS use. Finally, the third model incorporated neighborhood characteristics, including the respondents' opinion of whether there are enough UGS in their neighborhood and whether their neighborhood is considered quiet.

Finally, we carried out a logit regression model to address the second research question and examine which factors are associated with UGS quality and could help increase women's use of UGS if present. The model assessed the effect of ten binary variables (cleanliness and maintenance, good lighting, walls, low noise, markets, toilets, ample size, events, police presence, and playgrounds) on the likelihood of respondents perceiving UGS in their neighborhood as them being of good quality or not. The significant variables were later explored in the qualitative components of the study. The predictor variables were tested a priori to validate that there was no violation of the assumption of linearity of the logit.

2.2. Phases 2 and 3: Qualitative Analysis

The findings of the first stage of the analysis were used to develop the second phase of the study, which consisted of semi-structured interviews that delved into women's barriers and enablers related to UGS access. The introduction of qualitative components makes it possible to recognize and articulate cultural contexts into policymaking, achieving higher levels of effectiveness in the analysis [78]. The rationale for including two sequential qualitative components in the study (QUAN → QUAL → QUAL) was to deepen the reasons behind the survey's key findings, enriching the robustness of the study's insights by ensuring that the follow-up qualitative data provided a better understanding of the survey results [79]. By integrating qualitative data, we were able to provide a depth and breadth that the quantitative approach lacked by itself [80].

We invited survey respondents to participate in in-depth interviews that would take place in October 2020. Twelve women (ages 20 to 59) were selected to participate in the interviews, which lasted between 45 and 80 min via telephone. Building on the work of Sargeant [81], who defines how to ensure the quality of participants in qualitative studies, we selected participants who could best inform our research questions and enhance our understanding of the barriers that affected women's access to UGS during the pandemic. Aligned with grounded theory, we used a maximum variation sampling strategy, selecting women from different social strata living in different city municipalities; some lived alone while others with their families, partners, or friends [82]. Additionally, we asked interested parties if they had young children or lived with older adults to assess whether attitudes changed depending on their care responsibilities. The selection of participants intended to provide multiple perspectives since the selected women came from different backgrounds and had diverse living conditions.

Comparable studies, such as Noël et al. [48] in Belgium and McCormack et al. [83] in Canada, have also used in-depth interviews to better understand the association between UGS use and other variables. The type of analysis used in this stage was thematic, designed to find, review, and name common themes regarding the factors that encourage or discourage women's UGS use. Thematic analyses have been identified as a useful tool to investigate the patterns and commonalities among group participants [83].

Finally, in November 2021, this same group of women was divided and invited to participate in two 90-min focus group sessions via Zoom. The objective of the third stage of the data collection and analysis was to investigate the continuity of the previously identified themes once the restrictions associated with COVID-19 eased. Additionally, we inquired about characteristics related to the quality of UGS in the survey to comprehend whether changes or enhancements in these factors are conducive to improving women's geographies of fear while using green spaces. All participants gave verbal and written consent to audio-record their interview and focus group session, with the guarantee that their identity would remain anonymous. The ethical standards of this study were approved by the University of Melbourne's Psychology Health and Applied Sciences Human Ethics Sub-Committee (2056618).

3. Results

Table 1 presents a summary of the characteristics of the survey respondents.

Table 1. Summary of the sociodemographic characteristics of online survey respondents, women living in Mexico City (n = 1245) June 2020.

Age Group	n	%
18–24	412	33.1%
25–29	154	12.4%
30–34	109	8.8%
35–39	88	7.1%
40–44	84	6.7%
45–49	76	6.1%
50–54	94	7.6%
55–59	86	6.9%
60–64	77	6.2%
65+	63	5.1%
# of other people living in the house		
Living alone	44	3.5%
Living with 1 person	242	19.4%
Living with 2 people	268	21.5%
Living with 3 people	297	23.9%
Living with 4 people	180	14.5%
Living with 5 or more	214	17.2%

Table 1. *Cont.*

Socio-economic status		
Low-income	507	40.7%
Middle-income	284	22.8%
High-income	442	35.5%
Education		
High school or less	201	16.1%
Technical degree	128	10.3%
Undergraduate	615	49.4%
Graduate school	298	23.9%

3.1. Characteristics of Survey Respondents and UGS Use (Access) during the Pandemic

Before Mexico City's government introduced restrictions to prevent the spread of COVID-19 on March 21, 2020, 1189 (95.5%) women indicated using UGS. This figure dropped to 700 (56.2%) once restrictions came into place, a decrease of almost 40%. Out of the 700 women who used UGS after restrictions were introduced, the majority (78.4%) chose to use the green space closest to home. Interestingly, 79% of women indicated not having enough UGS in their neighborhood. Of users, almost 10% reported staying at UGS for less than 15 min per visit, 34% between 16 to 30 min per visit, and most users, 56%, indicated using UGS for more than 30 mins per visit on average.

Notably, while only 39% of women living alone reported using UGS during the restrictions, this percentage increased to 57% for women living with someone else. It is also interesting that 62% of low-income women indicated using UGS after restrictions began, compared with 54% of middle-income women and 51% of high-income women. Regarding age, 53% of women aged 65 and older reported having stopped using UGS once the pandemic began, the largest decrease of any group. This result was expected given the vulnerability of this group to complications in the event of contracting COVID. Surprisingly, women aged 30 to 34 years were the second-largest group of lost UGS users, with 46% revealing that they stopped using UGS after restrictions started. In third place were women aged 60 to 64, at 44%. The rest of the age groups suffered a reduction of 36 to 42% in the number of users.

When we asked the 44% of women who specified not using UGS after the pandemic began about their motives, 88% of those women indicated that, due to the health emergency, they preferred to stay at home, either for fear of catching COVID, for fear of infecting someone at home, or, in some cases, because they were ill and needed to quarantine. Safety concerns occupied the second spot, with 26% of respondents who did not use UGS expressing fear of suffering some type of violence while using these spaces. Meanwhile, not having UGS close to home and them being of poor quality was the third (23%) and fourth (14%) most cited reasons. Proximity issues included women's responses that their closest UGS had closed due to pandemic regulations, leaving them with no other nearby options.

When we asked the women who said they had used UGS at some point (either before or after the COVID restrictions began) what activities they carried out within UGS, the activity reported as most common was walking (83%), followed by passive engagement or relaxing (66.2%), playing sports (20.8%), cycling (19.6%), socializing (14.4%), and buying or selling things in the local markets located inside the green spaces (7%). Notably, out of the 1033 women who indicated they liked to walk while visiting UGS, just over one quarter (262) mentioned that they walked with their pets.

3.2. Association between UGS Use (Access) and Perceived Quality of UGS in the Neighborhood

Next, we present the logistic regression results of Models 1, 2, and 3 (see Table 2), which show that the perceived quality of UGS in the neighborhood affected UGS use (access) after COVID restrictions were introduced. The positive association between perceived quality and access to UGS during mobility restriction is slightly reduced—but remains significant—when additional potential confounders were added to the model. In Model 2, after controlling for age group, living arrangement, and income group, the odds of using UGS after the restrictions are about 1.59 times higher among those who reported good quality UGS in their neighborhood relative to those who reported otherwise (95% CI [1.26–2.00]). Moreover, in Model 3, the positive association between perceived quality of UGS in the neighborhood and use of UGS persists after introducing neighborhood factors: the perceived quietness of the neighborhood and perceptions of adequate UGS in the area. The Hosmer–Lemeshow test for Model 2 yielded a χ^2 (8) of 6.09, and for Model 3, a χ^2 (8) of 10.01 and were both insignificant ($p > 0.05$), suggesting that the models fit the data well.

Table 2. Odds ratios and 95% confidence intervals for predictor variables associated with UGS use (access) after COVID-19 restrictions were introduced (21 March 2020), online survey of women living in Mexico City ($n = 1245$), June 2020.

	M1				M2				M3			
	OR	*p*		95% CI	OR	*p*		95% CI	OR	*p*		95% CI
Quality of UGS in the neighborhood												
Not good	1				1				1			
Good	1.62	0.00	***	(1.30–2.04)	1.59	0.00	***	(1.26–2.00)	1.44	0.00	***	(1.13–1.83)
Age group												
18–24					1				1			
25–34					0.97	0.87		(0.70–1.34)	1.00	1.00		(0.72–1.39)
35–44					1.31	0.16		(0.90–1.93)	1.30	0.19		(0.88–1.91)
45–54					1.23	0.30		(0.84–1.80)	1.24	0.28		(0.84–1.83)
55–64					1.12	0.59		(0.75–1.66)	1.07	0.75		(0.71–1.60)
65+					0.86	0.61		(0.49–1.53)	0.78	0.40		(0.43–1.40)
Living arrangement												
Living alone					1				1			
Living with 1 person					1.45	0.27		(0.75–2.82)	1.51	0.23		(0.77–2.95)
Living with more than 1 person					2.04	0.03	**	(1.08–3.86)	2.05	0.03	**	(1.08–3.89)
Income group												
Low-income					1				1			
Middle-income					0.70	0.02	**	(0.52–0.96)	0.67	0.01	***	(0.49–0.91)
High-income					0.66	0.01	***	(0.49–0.88)	0.61	0.00	***	(0.45–0.82)
Quiet neighborhood												
No									1			
Yes									1.42	0.04	**	(1.03–2.16)
Enough UGS in the neighborhood												
Strongly disagree									1			
Disagree									1.20	0.25		(0.88–1.65)
Neither agree nor disagree									1.22	0.24		(0.87–1.72)
Agree									1.48	0.06	*	(0.99–2.21)
Strongly agree									1.61	0.03	**	(1.04–2.49)

Notes: Significance levels *** $p < 0.01$, ** $p < 0.05$, * $p < 0.1$. Model 1: Unadjusted. Model 2: Model 1 adjusted for respondents' age group, living arrangement and income group. Model 3: Model 2 adjusted for neighborhood characteristics. Mc Fadden's R-squared, Model 1:0.010 ($p = 0.000$). Mc Fadden's R-squared, Model 2: 0.025 ($p = 0.000$). Mc Fadden's R-squared, Model 3: 0.035 ($p = 0.000$).

3.3. Features Associated with UGS Quality

What are particular features of UGS associated with women reporting having UGS of "good quality" in their neighborhood? Table 3 suggests that cleanliness and maintenance, good lighting, the presence of walls, and having playgrounds or sports facilities are significantly and positively associated with the likelihood of perceiving UGS as having good quality. Interestingly, the presence of police, a standard safety indicator, is not significantly associated with perceived good UGS quality. The Hosmer–Lemeshow test for Model 4 yielded a χ^2 (8) of 6.81 and was insignificant ($p > 0.05$), suggesting that the model fit the data well.

Table 3. Model 4: Odds ratios and 95% confidence intervals for predictor variables associated with perceived good quality of UGS in the neighborhood after COVID-19 restrictions were introduced, online survey of women living in Mexico City (n = 1245), June 2020.

Predictors	OR	p		95% CI	
Clean and well-maintained					
No	1				
Yes	3.79409	0.000	***	2.032134	7.083748
Good lighting					
No	1				
Yes	4.386839	0.000	***	2.672711	7.200311
Presence of walls					
No	1				
Yes	2.85302	0.000	***	2.137672	3.80775
Low noise					
No	1				
Yes	1.316099	0.225		0.844746	2.050459
Presence of markets					
No	1				
Yes	0.74423	0.106		0.520042	1.065064
Toilets					
No	1				
Yes	1.225118	0.173		0.91493	1.64047
Size (ample)					
No	1				
Yes	1.264952	0.269		0.833822	1.919001
Presence of events					
No	1				
Yes	0.989428	0.948		0.716886	1.365583
Presence of police					
No	1				
Yes	0.86368	0.310		0.65088	1.146053
Playgrounds or sports facilities					
No	1				
Yes	5.078593	0.000	***	3.824721	6.743527

McFadden's R^2: 0.043 (p = 0.000), McKelvey and Zavoina's R^2: 0.003 (p = 0.000). *** $p < 0.01$.

3.4. Qualitative Analysis Results

Through the in-depth interviews carried out in October 2020, we identified three central themes, which were further explored during the November 2021 focus groups to observe changes in participants' behavior. Below are the key reflections from the interviews and focus groups on each central theme and the changes in perceptions once restrictions were eased.

Theme 1. *The availability of UGS in the neighborhood was a critical hindrance to women's UGS access (use) due to COVID-19 concerns.*

One of the main barriers affecting women's use of UGS during the pandemic was the lack of available UGS in the neighborhood, either because those neighborhoods currently lack UGS or due to the restrictions associated with the COVID-19 pandemic, which included the closure of some public spaces. When reviewing our initial results, we noticed that UGS closures were associated with the size of spaces, with several survey respondents stating that small UGS were often closed. However, the interviews highlighted that residents of upper-middle or upper-income neighborhoods, where most green areas are located, were not restricted in their access to UGS. On the contrary, in neighborhoods with high UGS availability, women were more inclined to use these spaces because they remained open, were not saturated, and adherence to physical distancing protocols was easier.

> *"I often used the parks nearby, especially to walk with my dog and clear my head for a while. The parks near our building did not close. Sometimes, police officers make sure that there are not too many people and that those who come to the park wear face masks." (Participant 10, age 33)*

Conversely, in marginalized areas, which already have little UGS availability and where UGS are often small, restrictions on UGS access were introduced to prevent the spread of COVID-19 due to overcrowding concerns. Consequently, it is safe to assume that access was not strictly limited by UGS size but by neighborhoods' UGS availability.

> *"In my neighborhood, their [UGS] use is not allowed at the moment." (Participant 1, age 20)*

Up to October 2020, most interviewees who used UGS limited themselves to visiting the green space closest to their homes, avoiding spending much time outside. The lack of availability of a nearby green space did not appear to be an incentive to use distant spaces. Even when restrictions began to ease, some women decided to walk around the block instead of walking long distances to access UGS in other neighborhoods due to the fear of contagion. These women continued to engage in this practice even though they reported feeling happier when they saw greenery, indicating a greater urgency to increase UGS availability in underserved neighborhoods.

> *"From September [2020] onwards, I started going for a walk because I was desperate to avoid being at home all day, but not to the park because it is closed. I usually just walk around the block." (Participant 4, age 27)*

In November 2021, most participants communicated using public spaces more frequently than the year before, including UGS. Nevertheless, focus group discussions revealed that the use of these spaces was still not as frequent as before the pandemic began. In areas with low UGS availability, overcrowding and a lack of adherence to physical distancing profoundly inhibited the use of some UGS for fear of becoming infected or spreading the disease to other household members.

> *"I live with someone who is at higher risk of serious complications if they get sick with COVID-19, and I don't want to expose them." (Participant 6, age 23)*

> *"I still don't go out much. I go out much more than a few months ago, but I am still not entirely comfortable since some people don't wear masks. . . . The park in front [of the house] is sometimes crowded, and there are new variants." (Participant 2, age 57)*

Theme 2. *For women, the concept of UGS quality and safety, including street safety, are intertwined.*

Similar to the survey results, the findings from the interviews suggest a strong association between safety and perceived UGS quality. Interviewees indicated that they avoided those spaces in which they did not feel safe (those of low quality) or waited to have company to use them, with "company" referring to male friends or relatives. A year later, the focus groups showed that women's propensity to avoid unsafe or poor quality spaces continued once the restrictions decreased and UGS use had grown.

> *"I cannot say that a place is of good quality if it is not safe. What is more, I am not particularly eager to go if I consider it unsafe." (Participant 3, age 45)*

Interestingly, not only the quality and safety of UGS prevented women from using them, but also the perceived safety of the surrounding streets. In particular, the interviewees emphasized that walking on inadequate roads increased their feelings of vulnerability and discomfort, already exacerbated by the pandemic, which sometimes led them to avoid using UGS. Those participants who described living in low-income neighborhoods, where minimal infrastructure is common, conveyed more intense feelings of discomfort while walking around their respective neighborhoods.

> *"The road leading to the park is ugly; I do not feel safe walking there." (Participant 6, age 23)*

In turn, when focus group members discussed what factors could improve the generalized feeling of vulnerability, good lighting was accentuated as a central factor for encouraging UGS use, especially for women who used green areas at dawn or night. It is important to point out that, during the focus groups, a distinction was made that referring to good lighting as an enabler encompassed the green space and the adjacent streets. For example, there are instances where, despite having a green area of perceived good quality nearby, if the streets leading to it are poorly lit, the participant preferred not to use it due to safety concerns.

> *"The neighborhood park is not ugly, it's okay, but I hardly use it because I only have time in the mornings, and the street is very dark at that time." (Participant 7, age 28)*

Consistent with the survey results, the interviews showed that maintenance and cleanliness were fundamental to defining good UGS quality. Indeed, if the space was neglected or had a lot of garbage, it was considered a deterrent to its use, predominantly for those women who visited UGS with children. The focus groups stressed that with the removal of some restrictions, and as the spaces became more crowded, the perception of cleanliness and maintenance of various UGS decreased, negatively affecting their use and women's perception of safety.

> *"I feel that if the place looks neglected, half abandoned, it is a dangerous area where there may be gangs or drugs are sold since no one is going to monitor or clean the place." (Participant 12, age 59)*

Interviewees also indicated that their perception of UGS quality was associated with the activities that could be carried out in a given space. For example, UGS with running tracks, playgrounds, or basketball courts were generally perceived as high-quality spaces. Through the focus groups, the topic of exercise equipment also emerged. In specific green areas, the city government has installed fitness equipment. The participants mentioned being in favor of the initiative, as it allows them to participate in a wider variety of exercises without having to go to a gym, incentivizing them to be more active. This type of feature is especially appreciated in the case of small UGS that might not be large enough for people to run or engage in other kinds of sports that require larger spaces. The installation of exercise equipment is also linked to feeling safe since it provides a perception of care and maintenance to the area.

> *"In the greenway across the street, they put exercise machines, and I use them with my sister and nephews from time to time; they keep us active. It looks nice because they are well cared for and functional." (Participant 5, age 25)*

An issue that was not evaluated in the survey or interviews but that came up in the discussion groups is the presence of homeless people or street vendors in the parks or the streets neighboring UGS. Their appearance seems to be a common hindrance to UGS use, increasing the geographies of fear.

In contrast to findings from earlier studies, such as Navarrete-Hernandez, Vetro, and Concha [74]—where the presence of walls deters UGS use and is negatively related to safety, our survey indicated a positive relationship with quality, which was further explored in the qualitative analysis. The interviews and focus groups established that for women who visit medium or large UGS, such as Chapultepec Park, especially those accompanied by children or pets, walls provide the feeling of protection against possible road incidents. However, this observation differed for small parks or gardens, where easy access was preferred.

"I like that I can see my boys play … even if they run, they are in a closed space where they will not be able to run into the street." (Participant 9, age 35)

Theme 3. *Social cohesion increased women's propensity to use UGS in their neighborhoods despite heightened fears of insecurity experienced during the pandemic.*

In general, the interviews conveyed a heightened feeling of insecurity beyond UGS use. The closure of businesses, the low presence of people in the streets, and the uncertainty of the situation increased women's feelings of vulnerability when using any public space. In addition, seven participants expressed that these feelings did not improve from March 2020 to October 2020 due to a widespread perception that crime rates had increased because of the economic instability resulting from the pandemic.

"In these months, there have been more assaults. Nothing has happened to me yet, but I watch the news, and the situation looks bad." (Participant 11, age 33)

Remarkably, eleven out of the twelve women indicated concern about gender violence in the city, with three of them explicitly addressing the number of femicides per day, eleven, a number constantly repeated in the press. News about gender-based violence had increased women's collective anxiety about using public spaces even before the pandemic began, which was expressed in the focus groups. For instance, one of the interviewees mentioned participating in a march to protest gender-based violence, which was attended by thousands of women in March 2020, a few days before the COVID-19 restrictions came into force.

Both focus groups debated the perception of generalized insecurity in the city. In particular, it is important to mention that this perception only seems to affect women when they use public spaces alone—when accompanied by their partner, family, or friends, the anxiety or fear of being attacked decreases considerably. The latter is consistent with the survey results, which showed higher use of UGS among women who live with others. In this sense, participants agreed that public life is lived in a very different way for women, specifying that as a woman, one must live in a state of constant alertness, especially when going out at night, at dawn, or visiting spaces where there are few people. Geographies of fear seem to be a constant in Mexico City, although it is critical to emphasize that the pandemic has intensified those feelings.

"I do not go out to walk my dog at night unless my boyfriend is home, and we go together. It is a deal we have, and that way, I feel calmer and enjoy the walk." (Participant 8, age 38)

Finally, women living in communities with a sense of social cohesion also described using UGS in their neighborhoods more frequently. In this context, participants articulated social cohesion as having a relationship with their neighbors and a strong sense of solidarity between community members. This finding is especially noteworthy since distant spaces, where the participants do not know other users, do not stimulate feelings of trust and, therefore, use. Thus, a sense of community seems crucial to diminish women's geographies of fear. Social cohesion also affects how women perceive the police. On the one hand, women living in neighborhoods with police assigned to that area, who know police officers personally and have a cordial relationship with them, feel more comfortable and safer having officers around. For instance, women living in gated communities recognized police officers as part of their community.

"On the block, there are two policemen who make the rounds on their bicycles and also look after people in the park nearby. We know them well; I greet them whenever I am outside watering my plants." (Participant 10, age 33)

On the other hand, police presence increases stress levels for women who do not see officers as part of the community, as they associate police with corruption, crime, and mistreatment.

"It makes me uncomfortable to encounter cops anywhere. A patrol began to guard our neighborhood after COVID-19 to check that everything was fine. It makes me very nervous that cops are around here, especially because sometimes they go into the buildings to use the bathrooms, and I fear they might steal something." (Participant 9, age 35)

4. Discussion

In Latin American cities, the COVID-19 pandemic has exposed systemic inequalities and accentuated pre-existing urban challenges with significant gender dimensions [42,43]. Specifically, in Mexico City, the pandemic has widened gender disparities by increasing women's sense of vulnerability, affecting their opportunities, motivation, and access to UGS. This effect is evidenced by the results of our survey, which show a significant decrease in the use of UGS by women since the pandemic began. Our quantitative and qualitative analyses illustrate that anxiety about contracting or spreading COVID, UGS closures, and increased geographies of fear have been core barriers to women's access to UGS during this period, predominantly for those women living in underserved areas with low availability of UGS and poor street infrastructure. Markedly, the closure of UGS in marginalized neighborhoods during the first phase of the pandemic negatively impacted the mental health of women in these areas by further restricting their exercise and relaxation options, amplifying health disparities [49,84].

Similar to Dunckel Graglia's [75] empirical evidence on the use of public spaces such as the metro in Mexico City, our findings reveal that the fear of violence hinders women's belief in their right to the city and, therefore, their access to UGS. To this end, research on gender inequalities in Latin America suggested that the growing number of femicides in the region, associated with a fear of violence, could be undermining women's desire to leave the house by influencing their perception of safety in public spaces [10,85]. The latter coincides with observations from our interviews and focus groups, where participants expressed a heightened sense of vulnerability linked to a perception of increased criminality. Although the fear of violence when using public spaces is a barrier experienced by women worldwide, gender-based violence is more prevalent in the cities of the Global South, which translates into lower participation rates among women in the urban public sphere [86,87]. For instance, while a recent study of access to public parks and gardens in urban areas of England and Wales revealed that women preferred to visit UGS that were less crowded due to COVID-19 contagion concerns [88], our study uncovered that women in Mexico City, conversely, tend to avoid parks with few people due to the fear of crime. Thus, to increase women's access to UGS both on a day-to-day basis and during a public health emergency, it is necessary to consider expanding not only equity in UGS availability but also equity in social access [69], or what Mazumder [12] has defined as experiential equity.

To achieve experiential equity, it is essential to understand why access to infrastructure is restricted beyond the distribution of public resources [12]. In societies with structural inequities, such as Mexico, the fear of objectification, harassment, assault, and rape restrict women's power to participate in society by excluding them from public spaces or limiting their ability to benefit from urban infrastructure, including environmental amenities [21,73,74]. Having no access to open green spaces to exercise or decompress during the pandemic confinement has been a negative repercussion of gendered geographies in Latin America [75], where women cannot use and enjoy UGS due to safety concerns. Accordingly, exploring pathways to ameliorate the perception of geographies of fear can be a tool for differentiating redistribution in cities and reducing health disparities during catastrophic events [16,76]. If green spaces are not well lit, do not have adequate conditions such as clear pathways, have enclosed areas with blind spots, or seem unwelcoming, they can be perceived as hazardous rather than places that stimulate a healthy urban environment [3,19,89].

In this sense, our results suggest that since a better sense of belonging to one's neighborhood can boost women's participation in the public sphere [90,91], it is imperative to promote programs aimed at improving social cohesion to increase women's confidence in using public spaces and improve their trust in institutions such as the police [19,92,93]. Access to social goods in communities with high levels of social fragmentation tends to greatly disadvantage marginalized groups, such as women [94]. Consequently, strengthening neighborhood belonging and identity, and creating networks that enrich social capital are necessary to guarantee women's right to the city [95].

Our findings also indicate that to expand women's UGS access in Mexico City and, thereby, improve their well-being and enhance their feeling of comfort, it is critical to incorporate quality characteristics to existing UGS, such as exercise equipment or playgrounds [17]. Even more so, to reduce women's feeling of vulnerability, UGS and the surrounding streets must have similar quality elements, such as good lighting, cleanliness, and maintenance. This finding is consistent with the growing body of behavioral research showing that routes to access public spaces are especially important for incorporating women into the public sphere and increasing their opportunities to access health and recreation spaces [69,96]. In the case of cleanliness, maintenance, and good lighting, evidence from other countries, such as Australia and the U.K., highlights these two characteristics as essential to intensify feelings of comfort and safety among women; otherwise, spaces evoke fear of crime [19,62].

It is essential to highlight that the suggestions presented in this study are specific to Mexico City. Although the city shares similar characteristics to other Latin American cities [19], and it is reasonable to argue that some observations might be comparable to women's experiences elsewhere, the results cannot be extrapolated to other settings without additional research. For example, given that the literature on geographies of fear indicates that women in other Latin American cities share a fear of violence when using public spaces [97], policies aimed at improving public lighting might be expected to increase the use of UGS by women in other Latin American cities. However, more evidence is needed to test these hypotheses and assess the extent to which the recommendations of this study are relevant to cities in other Latin American countries. This limitation is related to the study's design, which in its second phase, uses a select number of women to deepen the understanding of women's lack of access to UGS instead of seeking representativeness [98].

Finally, although women's access to UGS during a health crisis can be improved by modifying the existing infrastructure to accommodate their needs [99], access to UGS will not be entirely equitable unless structural inequalities that reinforce gender discrimination are eliminated [72]. Accordingly, studies like this that examine potential design interventions that can affect women's willingness to use UGS in specific contexts, ameliorating their perceived safety and comfort, are necessary to develop evidence-based policies that will increase cities' inclusiveness and environmental justice. However, these efforts alone are not sufficient to eliminate the problem of the geographies of fear [100].

5. Conclusions

The present paper investigated women's barriers to UGS access during the COVID-19 pandemic in Latin America, using a mixed-methods approach that included a survey, in-depth interviews, and focus groups. The use of various analysis methods made it possible to incorporate sociocultural perceptions into the data analyses, enriching the robustness of the study findings. Using Mexico City as a case study, our results suggest that, in Latin America, women stopped using UGS or decreased their frequency of use since the COVID restrictions took effect. The three main barriers affecting women's UGS access during the pandemic have been the limited availability of green spaces in some neighborhoods, the poor quality of UGS and surrounding streets, and the lack of social cohesion, which has deeply hurt women's willingness to use UGS during the health emergency.

Notably, the interviews and focus groups conveyed that the quality of UGS and their safety are intrinsically connected to women. In turn, the safety of the route to access UGS is of the utmost importance to encourage their use, with elements such as good lighting, cleanliness, and maintenance promoting a lesser sense of vulnerability. Moreover, the study results point to a need for better guidelines that enforce physical distance during a disease outbreak or pandemic rather than prohibiting access to UGS altogether, particularly in marginalized areas, as evidence shows that restricting access in underserved neighborhoods only contributes to increasing health disparities.

Studies like this one, which capture the perceptions and motivations of women, are instruments of citizen participation necessary to improve cities' infrastructure and environmental justice. However, the equity of access to UGS cannot be entirely achieved until the structural inequalities that lead to gender violence are eliminated. Ensuring women's social, economic, and political inclusion remains a persistent challenge in cities of Latin America.

Author Contributions: Conceptualization, C.M.H. and A.U.; methodology, C.M.H. and A.U.; software, C.M.H.; validation, C.M.H. and A.U.; formal analysis, C.M.H.; investigation, C.M.H.; data curation, C.M.H.; writing—original draft preparation, C.M.H.; writing—review and editing, C.M.H. and A.U.; supervision, A.U.; project administration, C.M.H.; funding acquisition, C.M.H. All authors have read and agreed to the published version of the manuscript.

Funding: This research was supported by the School of Geography, Earth, & Atmospheric Sciences at the University of Melbourne.

Institutional Review Board Statement: The study was conducted according to the guidelines of the Declaration of Helsinki and approved by the Psychology Health and Applied Sciences Human Ethics Sub-Committee of the University of Melbourne (Ethics ID 2056618, approved 8 May 2020).

Informed Consent Statement: Informed consent was obtained from all subjects involved in the study.

Acknowledgments: Special thanks to the women who contributed to this study.

Conflicts of Interest: The authors declare no conflict of interest.

References

1. Vidal, D.G.; Teixeira, C.P.; Dias, R.C.; Fernandes, C.O.; Filho, W.L.; Barros, N.; Maia, R.L. Stay close to urban green spaces: Current evidence on cultural ecosystem services provision. *Eur. J. Public Health* **2021**, *31* (Suppl. S3), 31. [CrossRef]
2. Pouso, S.; Borja, A.; Fleming, L.E.; Gomez-Baggethun, E.; White, M.P.; Uyarra, M.C. Contact with blue-green spaces during the COVID-19 pandemic lockdown beneficial for mental health. *Sci. Total Environ.* **2021**, *756*, 143984. [CrossRef] [PubMed]
3. Wang, R.; Feng, Z.; Pearce, J.; Liu, Y.; Dong, G. Are greenspace quantity and quality associated with mental health through different mechanisms in Guangzhou, China: A comparison study using street view data. *Environ. Pollut.* **2021**, *290*, 117976. [CrossRef] [PubMed]
4. Burnett, H.; Olsen, J.R.; Nicholls, N.; Mitchell, R. Change in time spent visiting and experiences of green space following restrictions on movement during the COVID-19 pandemic: A nationally representative cross-sectional study of UK adults. *BMJ Open* **2021**, *11*, e044067. [CrossRef]
5. Lopez, B.; Kennedy, C.; McPhearson, T. Parks are Critical Urban Infrastructure: Perception and Use of Urban Green Spaces in NYC during COVID-19. *Preprints* **2020**, 2029918. [CrossRef]
6. Grima, N.; Corcoran, W.; Hill-James, C.; Langton, B.; Sommer, H.; Fisher, B. The importance of urban natural areas and urban ecosystem services during the COVID-19 pandemic. *PLoS ONE* **2020**, *15*, e0243344. [CrossRef]
7. Borkenhagen, D.; Grant, E.; Mazumder, R.; Negami, H.R.; Srikantharajah, J.; Ellard, C. The effect of COVID-19 on parks and greenspace use during the first three months of the pandemic—A survey study. *Cities Health* **2021**, *10*, 1–10. [CrossRef]
8. Wang, C.; Li, C.; Wang, M.; Yang, S.; Wang, L. Environmental justice and park accessibility in urban China: Evidence from Shanghai. *Asia Pac. Viewp.* **2021**. [CrossRef]
9. Yessoufou, K.; Sithole, M.; Elansary, H.O. Effects of urban green spaces on human perceived health improvements: Provision of green spaces is not enough but how people use them matters. *PLoS ONE* **2020**, *15*, e0239314. [CrossRef]
10. Wright Wendel, H.E.; Zarger, R.K.; Mihelcic, J.R. Accessibility and usability: Green space preferences, perceptions, and barriers in a rapidly urbanizing city in Latin America. *Landsc. Urban Plan.* **2012**, *107*, 272–282. [CrossRef]
11. Rigolon, A.; Browning, M.; Lee, K.; Shin, S. Access to Urban Green Space in Cities of the Global South: A Systematic Literature Review. *Urban Sci.* **2018**, *2*, 67. [CrossRef]
12. Mazumder, R. Experiential equity: An environmental neuroscientific lens for disparities in urban stress. In *Global Reflections on COVID-19 and Urban Inequalities*; Bristol University Press: Bristol, UK, 2021; Volume 1, pp. 187–194.

13. Van Herzele, A.; Wiedemann, T. A monitoring tool for the provision of accessible and attractive urban green spaces. *Landsc. Urban Plan.* **2003**, *63*, 109–126. [CrossRef]
14. Pipitone, J.M.; Jović, S. Urban green equity and COVID-19: Effects on park use and sense of belonging in New York City. *Urban For. Urban Green.* **2021**, *65*, 127338. [CrossRef]
15. Bolte, G.; Nanninga, S.; Dandolo, L. Sex/Gender Differences in the Association between Residential Green Space and Self-Rated Health-A Sex/Gender-Focused Systematic Review. *Int. J. Environ. Res. Public Health* **2019**, *16*, 4818. [CrossRef]
16. Curtis, D.S.; Rigolon, A.; Schmalz, D.L.; Brown, B.B. Policy and Environmental Predictors of Park Visits During the First Months of the COVID-19 Pandemic: Getting out While Staying in. *Environ. Behav.* **2021**, *54*, 487–515. [CrossRef]
17. Ode Sang, Å.; Knez, I.; Gunnarsson, B.; Hedblom, M. The effects of naturalness, gender, and age on how urban green space is perceived and used. *Urban For. Urban Green.* **2016**, *18*, 268–276. [CrossRef]
18. Manyani, A.; Shackleton, C.M.; Cocks, M.L. Attitudes and preferences towards elements of formal and informal public green spaces in two South African towns. *Landsc. Urban Plan.* **2021**, *214*, 104147. [CrossRef]
19. Sreetheran, M.; van den Bosch, C.C.K. A socio-ecological exploration of fear of crime in urban green spaces—A systematic review. *Urban For. Urban Green.* **2014**, *13*, 1–18. [CrossRef]
20. Mayen Huerta, C.; Cafagna, G. Snapshot of the Use of Urban Green Spaces in Mexico City during the COVID-19 Pandemic: A Qualitative Study. *Int. J. Environ. Res. Public Health* **2021**, *18*, 4304. [CrossRef]
21. Carli, L.L. Women, Gender equality and COVID-19. *Gend. Manag. Int. J.* **2020**, *35*, 647–655. [CrossRef]
22. Venter, Z.S.; Barton, D.N.; Gundersen, V.; Figari, H.; Nowell, M. Urban nature in a time of crisis: Recreational use of green space increases during the COVID-19 outbreak in Oslo, Norway. *Environ. Res. Lett.* **2020**, *15*, 104075. [CrossRef]
23. da Schio, N.; Phillips, A.; Fransen, K.; Wolff, M.; Haase, D.; Ostoić, S.K.; Živojinović, I.; Vuletić, D.; Derks, J.; Davies, C.; et al. The impact of the COVID-19 pandemic on the use of and attitudes towards urban forests and green spaces: Exploring the instigators of change in Belgium. *Urban For. Urban Green.* **2021**, *65*, 127305. [CrossRef]
24. Google. Google COVID-19 Community Mobility Reports. 2020. Available online: https://www.google.com/covid19/mobility/ (accessed on 6 April 2022).
25. Mayen Huerta, C.; Utomo, A. Evaluating the association between urban green spaces and subjective well-being in Mexico city during the COVID-19 pandemic. *Health Place* **2021**, *70*, 102606. [CrossRef] [PubMed]
26. Marconi, P.L.; Perelman, P.E.; Salgado, V.G. Green in times of COVID-19: Urban green space relevance during the COVID-19 pandemic in Buenos Aires City. *Urban Ecosyst.* **2022**, 1–13. [CrossRef] [PubMed]
27. Viezzer, J.; Biondi, D. The influence of urban, socio-economic, and eco-environmental aspects on COVID-19 cases, deaths and mortality: A multi-city case in the Atlantic Forest, Brazil. *Sustain. Cities Soc.* **2021**, *69*, 102859. [CrossRef] [PubMed]
28. Roe, J.J.; Thompson, C.W.; Aspinall, P.A.; Brewer, M.J.; Duff, E.I.; Miller, D.; Mitchell, R.; Clow, A. Green space and stress: Evidence from cortisol measures in deprived urban communities. *Int. J. Environ. Res. Public Health* **2013**, *10*, 4086–4103. [CrossRef]
29. Pereira, R.H.M.; Schwanen, T.; Banister, D. Distributive justice and equity in transportation. *Transp. Rev.* **2016**, *37*, 170–191. [CrossRef]
30. Lennon, M. Green space and the compact city: Planning issues for a 'new normal'. *Cities Health* **2020**, 1–4. [CrossRef]
31. Bell, S.; Montarzino, A.; Travlou, P. Mapping research priorities for green and public urban space in the UK. *Urban For. Urban Green.* **2007**, *6*, 103–115. [CrossRef]
32. Richardson, E.A.; Mitchell, R. Gender differences in relationships between urban green space and health in the United Kingdom. *Soc. Sci. Med.* **2010**, *71*, 568–575. [CrossRef]
33. Creswell, J.W.; Fetters, M.D.; Ivankova, N.V. Designing A Mixed Methods Study in Primary Care. *Ann. Fam. Med.* **2004**, *2*, 7–12. [CrossRef] [PubMed]
34. Ivankova, N.V. Implementing Quality Criteria in Designing and Conducting a Sequential QUAN → QUAL Mixed Methods Study of Student Engagement With Learning Applied Research Methods Online. *J. Mix. Methods Res.* **2013**, *8*, 25–51. [CrossRef]
35. O'Cathain, A.; Murphy, E.; Nicholl, J. Three techniques for integrating data in mixed methods studies. *BMJ* **2010**, *341*, c4587. [CrossRef] [PubMed]
36. Creswell, J.W.; Tashakkori, A. Developing publishable mixed methods manuscripts. *J. Mix. Methods Res.* **2007**, *1*, 107–111. [CrossRef]
37. Ugolini, F.; Massetti, L.; Calaza-Martinez, P.; Carinanos, P.; Dobbs, C.; Ostoic, S.K.; Marin, A.M.; Pearlmutter, D.; Saaroni, H.; Sauliene, I.; et al. Effects of the COVID-19 pandemic on the use and perceptions of urban green space: An international exploratory study. *Urban For. Urban Green.* **2020**, *56*, 126888. [CrossRef]
38. Ettman, C.K.; Abdalla, S.M.; Cohen, G.H.; Sampson, L.; Vivier, P.M.; Galea, S. Prevalence of Depression Symptoms in US Adults Before and During the COVID-19 Pandemic. *JAMA Netw. Open* **2020**, *3*, e2019686. [CrossRef]
39. Dozois, D.J.A. Anxiety and depression in Canada during the COVID-19 pandemic: A national survey. *Can. Psychol. Psychol. Can.* **2021**, *62*, 136–142. [CrossRef]
40. Heo, S.; Desai, M.U.; Lowe, S.R.; Bell, M.L. Impact of Changed Use of Greenspace during COVID-19 Pandemic on Depression and Anxiety. *Int. J. Environ. Res. Public Health* **2021**, *18*, 5842. [CrossRef]
41. Cindrich, S.L.; Lansing, J.E.; Brower, C.S.; McDowell, C.P.; Herring, M.P.; Meyer, J.D. Associations Between Change in Outside Time Pre- and Post-COVID-19 Public Health Restrictions and Mental Health: Brief Research Report. *Front. Public Health* **2021**, *9*, 619129. [CrossRef]

42. Burki, T. The indirect impact of COVID-19 on women. *Lancet Infect. Dis.* **2020**, *20*, 904–905. [CrossRef]
43. Triguero-Mas, M.; Anguelovski, I.; Cole, H.V.S. Healthy cities after COVID-19 pandemic: The just ecofeminist healthy cities approach. *J. Epidemiol. Community Health* **2021**, *76*, 354–359. [CrossRef] [PubMed]
44. Giordani, R.C.F.; Zanoni da Silva, M.; Muhl, C.; Giolo, S.R. Fear of COVID-19 scale: Assessing fear of the coronavirus pandemic in Brazil. *J. Health Psychol.* **2022**, *27*, 901–912. [CrossRef] [PubMed]
45. Broche-Perez, Y.; Fernandez-Fleites, Z.; Jimenez-Puig, E.; Fernandez-Castillo, E.; Rodriguez-Martin, B.C. Gender and Fear of COVID-19 in a Cuban Population Sample. *Int. J. Ment. Health Addict.* **2020**, *20*, 83–91. [CrossRef] [PubMed]
46. Marino, R.; Vargas, E.; Flores, M. Impacts of COVID-19 lockdown restrictions on housing and public space use and adaptation: Urban proximity, public health, and vulnerability in three Latin American cities. *Taylor Francis* **2021**, *22*, 363.
47. Hanzl, M. Urban forms and green infrastructure—The implications for public health during the COVID-19 pandemic. *Cities Health* **2020**, *10*, 1–5. [CrossRef]
48. Noël, C.; Rodriguez-Loureiro, L.; Vanroelen, C.; Gadeyne, S. Perceived Health Impact and Usage of Public Green Spaces in Brussels' Metropolitan Area During the COVID-19 Epidemic. *Front. Sustain. Cities* **2021**, *3*, 3. [CrossRef]
49. Slater, S.J.; Christiana, R.W.; Gustat, J. Recommendations for Keeping Parks and Green Space Accessible for Mental and Physical Health During COVID-19 and Other Pandemics. *Prev. Chronic Dis.* **2020**, *17*, E59. [CrossRef]
50. Larson, L.R.; Mullenbach, L.E.; Browning, M.; Rigolon, A.; Thomsen, J.; Metcalf, E.C.; Reigner, N.P.; Sharaievska, I.; McAnirlin, O.; D'Antonio, A.; et al. Greenspace and park use associated with less emotional distress among college students in the United States during the COVID-19 pandemic. *Environ. Res.* **2022**, *204*, 112367. [CrossRef]
51. Douglas, M.; Katikireddi, S.V.; Taulbut, M.; McKee, M.; McCartney, G. Mitigating the wider health effects of covid-19 pandemic response. *BMJ* **2020**, *369*, m1557. [CrossRef]
52. Chu, D.K.; Akl, E.A.; Duda, S.; Solo, K.; Yaacoub, S.; Schunemann, H.J.; on behalf of the COVID-19 Systematic Urgent Review Group Effort (SURGE) study authors. Physical distancing, face masks, and eye protection to prevent person-to-person transmission of SARS-CoV-2 and COVID-19: A systematic review and meta-analysis. *Lancet* **2020**, *395*, 1973–1987. [CrossRef]
53. Freeman, S.; Eykelbosh, A. COVID-19 and outdoor safety: Considerations for use of outdoor recreational spaces. *Natl. Collab. Cent. Environ. Health* **2020**, *829*, 1–15.
54. Reyes-Riveros, R.; Altamirano, A.; De La Barrera, F.; Rozas-Vásquez, D.; Vieli, L.; Meli, P. Linking public urban green spaces and human well-being: A systematic review. *Urban For. Urban Green.* **2021**, *61*, 127105. [CrossRef]
55. Schipperijn, J.; Ekholm, O.; Stigsdotter, U.K.; Toftager, M.; Bentsen, P.; Kamper-Jørgensen, F.; Randrup, T.B. Factors influencing the use of green space: Results from a Danish national representative survey. *Landsc. Urban Plan.* **2010**, *95*, 130–137. [CrossRef]
56. Basu, S.; Nagendra, H. Perceptions of park visitors on access to urban parks and benefits of green spaces. *Urban For. Urban Green.* **2021**, *57*, 126959. [CrossRef]
57. Derose, K.P.; Han, B.; Williamson, S.; Cohen, D.A.; Corporation, R. Racial-Ethnic Variation in Park Use and Physical Activity in the City of Los Angeles. *J. Urban Health* **2015**, *92*, 1011–1023. [CrossRef]
58. van den Berg, M.; Wendel-Vos, W.; van Poppel, M.; Kemper, H.; van Mechelen, W.; Maas, J. Health benefits of green spaces in the living environment: A systematic review of epidemiological studies. *Urban For. Urban Green.* **2015**, *14*, 806–816. [CrossRef]
59. Hunter, R.F.; Christian, H.; Veitch, J.; Astell-Burt, T.; Hipp, J.A.; Schipperijn, J. The impact of interventions to promote physical activity in urban green space: A systematic review and recommendations for future research. *Soc. Sci. Med.* **2015**, *124*, 246–256. [CrossRef]
60. Vos, S.; Bijnens, E.M.; Renaers, E.; Croons, H.; Van Der Stukken, C.; Martens, D.S.; Plusquin, M.; Nawrot, T.S. Residential green space is associated with a buffering effect on stress responses during the COVID-19 pandemic in mothers of young children, a prospective study. *Environ. Res* **2022**, *208*, 112603. [CrossRef]
61. Wood, L.; Hooper, P.; Foster, S.; Bull, F. Public green spaces and positive mental health—Investigating the relationship between access, quantity and types of parks and mental wellbeing. *Health Place* **2017**, *48*, 63–71. [CrossRef]
62. Sugiyama, T.; Gunn, L.D.; Christian, H.; Francis, J.; Foster, S.; Hooper, P.; Owen, N.; Giles-Corti, B. Quality of Public Open Spaces and Recreational Walking. *Am. J. Public Health* **2015**, *105*, 2490–2495. [CrossRef]
63. Rojas, C.; Páez, A.; Barbosa, O.; Carrasco, J. Accessibility to urban green spaces in Chilean cities using adaptive thresholds. *J. Transp. Geogr.* **2016**, *57*, 227–240. [CrossRef]
64. Jato-Espino, D.; Moscardo, V.; Vallina Rodriguez, A.; Lazaro, E. Spatial statistical analysis of the relationship between self-reported mental health during the COVID-19 lockdown and closeness to green infrastructure. *Urban For. Urban Green.* **2022**, *68*, 127457. [CrossRef] [PubMed]
65. Adinolfi, C.; Suárez-Cáceres, G.P.; Cariñanos, P. Relation between visitors' behaviour and characteristics of green spaces in the city of Granada, south-eastern Spain. *Urban For. Urban Green.* **2014**, *13*, 534–542. [CrossRef]
66. Bell, S.L.; Phoenix, C.; Lovell, R.; Wheeler, B.W. Green space, health and wellbeing: Making space for individual agency. *Health Place* **2014**, *30*, 287–292. [CrossRef] [PubMed]
67. Lee, A.C.; Jordan, H.C.; Horsley, J. Value of urban green spaces in promoting healthy living and wellbeing: Prospects for planning. *Risk Manag. Healthc. Policy* **2015**, *8*, 131–137. [CrossRef]
68. Williams, T.G.; Logan, T.M.; Zuo, C.T.; Liberman, K.D.; Guikema, S.D. Parks and safety: A comparative study of green space access and inequity in five US cities. *Landsc. Urban Plan.* **2020**, *201*, 103841. [CrossRef]

69. Talal, M.L.; Santelmann, M.V. Visitor access, use, and desired improvements in urban parks. *Urban For. Urban Green.* **2021**, *63*, 127216. [CrossRef]
70. Braçe, O.; Garrido-Cumbrera, M.; Correa-Fernández, J. Gender differences in the perceptions of green spaces characteristics. *Soc. Sci. Q.* **2021**, *102*, 2640–2648. [CrossRef]
71. Knapp, M.; Gustat, J.; Darensbourg, R.; Myers, L.; Johnson, C. The Relationships between Park Quality, Park Usage, and Levels of Physical Activity in Low-Income, African American Neighborhoods. *Int. J. Environ. Res. Public Health* **2018**, *16*, 85. [CrossRef]
72. Valentine, G. The Geography of Women's Fear. *Area* **1989**, *21*, 385–390.
73. Wesely, J.K.; Gaarder, E. The Gendered "Nature" of the Urban Outdoors. *Gend. Soc.* **2016**, *18*, 645–663. [CrossRef]
74. Navarrete-Hernandez, P.; Vetro, A.; Concha, P. Building safer public spaces: Exploring gender difference in the perception of safety in public space through urban design interventions. *Landsc. Urban Plan.* **2021**, *214*, 104180. [CrossRef]
75. Dunckel Graglia, A. Finding mobility: Women negotiating fear and violence in Mexico City's public transit system. *Gend. Place Cult.* **2015**, *23*, 624–640. [CrossRef]
76. Kumar, A.; Nayar, K.R. COVID 19 and its mental health consequences. *J. Ment. Health* **2021**, *30*, 1–2. [CrossRef]
77. Spotswood, E.N.; Benjamin, M.; Stoneburner, L.; Wheeler, M.M.; Beller, E.E.; Balk, D.; McPhearson, T.; Kuo, M.; McDonald, R.I. Nature inequity and higher COVID-19 case rates in less-green neighbourhoods in the United States. *Nat. Sustain.* **2021**, *4*, 1092–1098. [CrossRef]
78. Justice, J. The bureaucratic context of international health: A social scientist's view. *Soc. Sci. Med.* **1987**, *25*, 1301–1306. [CrossRef]
79. Mao, J. Social media for learning: A mixed methods study on high school students' technology affordances and perspectives. *Comput. Hum. Behav.* **2014**, *33*, 213–223. [CrossRef]
80. Burns, A. Mixed Methods. In *Qualitative Research in Applied Linguistics: A Practical Introduction*; Heigham, J., Croker, R.A., Eds.; Palgrave Macmillan UK: London, UK, 2009; pp. 135–161.
81. Sargeant, J. Qualitative research part II: Participants, analysis, and quality assurance. *J. Grad. Med. Educ.* **2012**, *4*, 1–3. [CrossRef]
82. Palinkas, L.A.; Horwitz, S.M.; Green, C.A.; Wisdom, J.P.; Duan, N.; Hoagwood, K. Purposeful Sampling for Qualitative Data Collection and Analysis in Mixed Method Implementation Research. *Adm. Policy Ment. Health* **2015**, *42*, 533–544. [CrossRef]
83. McCormack, G.R.; Petersen, J.; Naish, C.; Ghoneim, D.; Doyle-Baker, P.K. Neighbourhood environment facilitators and barriers to outdoor activity during the first wave of the COVID-19 pandemic in Canada: A qualitative study. *Cities Health* **2022**, *10*, 1–13. [CrossRef]
84. Volenec, Z.M.; Abraham, J.O.; Becker, A.D.; Dobson, A.P. Public parks and the pandemic: How park usage has been affected by COVID-19 policies. *PLoS ONE* **2021**, *16*, e0251799. [CrossRef] [PubMed]
85. Wilson, T.D. Violence against Women in Latin America. *Lat. Am. Perspect.* **2013**, *41*, 3–18. [CrossRef]
86. Salahub, J.E.; Gottsbacher, M.; De Boer, J. *Social Theories of Urban Violence in the Global South: Towards safe and Inclusive Cities*; Routledge: London, UK, 2018.
87. Sanchez, O.R.; Vale, D.B.; Rodrigues, L.; Surita, F.G. Violence against women during the COVID-19 pandemic: An integrative review. *Int. J. Gynaecol. Obs.* **2020**, *151*, 180–187. [CrossRef] [PubMed]
88. Shoari, N.; Ezzati, M.; Baumgartner, J.; Malacarne, D.; Fecht, D. Accessibility and allocation of public parks and gardens in England and Wales: A COVID-19 social distancing perspective. *PLoS ONE* **2020**, *15*, e0241102. [CrossRef] [PubMed]
89. Hamstead, Z.A.; Fisher, D.; Ilieva, R.T.; Wood, S.A.; McPhearson, T.; Kremer, P. Geolocated social media as a rapid indicator of park visitation and equitable park access. *Comput. Environ. Urban Syst.* **2018**, *72*, 38–50. [CrossRef]
90. Young, A.F.; Russell, A.; Powers, J.R. The sense of belonging to a neighbourhood: Can it be measured and is it related to health and well being in older women? *Soc. Sci. Med.* **2004**, *59*, 2627–2637. [CrossRef]
91. Jennings, V.; Bamkole, O. The Relationship between Social Cohesion and Urban Green Space: An Avenue for Health Promotion. *Int. J. Environ. Res. Public Health* **2019**, *16*, 452. [CrossRef]
92. Han, Z.; Sun, I.Y.; Hu, R. Social trust, neighborhood cohesion, and public trust in the police in China. *J. Polic. Int. J. Police Strateg. Manag.* **2017**, *40*, 380–394. [CrossRef]
93. Martínez-Ferrer, B.; Vera, J.A.; Musitu, G.; Montero-Montero, D. Trust in Police and Fear of Crime Among Young People from a Gender Perspective: The Case of Mexico. *Violence Gend.* **2018**, *5*, 226–232. [CrossRef]
94. Chakravarty, S.; Fonseca, M.A. The effect of social fragmentation on public good provision: An experimental study. *J. Behav. Exp. Econ.* **2014**, *53*, 1–9. [CrossRef]
95. Beauvais, C.; Jenson, J. *Social Cohesion: Updating the State of the Research*; CPRN: Ottawa, ON, Canada, 2002; Volume 62.
96. Viswanath, K.; Mehrotra, S.T. 'Shall we go out?' Women's safety in public spaces in delhi. *Econ. Polit. Wkly.* **2007**, *42*, 1542–1548.
97. Dammert, L. *Fear and Crime in Latin America*; Routledge: London, UK, 2012.
98. Queirós, A.; Faria, D.; Almeida, F. Strengths and limitations of qualitative and quantitative research methods. *Eur. J. Educ. Stud.* **2017**, *3*, 93131. [CrossRef]
99. Jorgensen, A.; Hitchmough, J.; Calvert, T. Woodland spaces and edges: Their impact on perception of safety and preference. *Landsc. Urban Plan.* **2002**, *60*, 135–150. [CrossRef]
100. Young, I.M. CHAPTER 8. City Life and Difference. In *Justice and the Politics of Difference*; Princeton University Press: Princeton, NJ, USA, 2012; pp. 226–256.

 *land*

Article

Urgent Biophilia: Green Space Visits in Wellington, New Zealand, during the COVID-19 Lockdowns

Maggie MacKinnon [1], Rebecca MacKinnon [2], Maibritt Pedersen Zari [3,*], Kain Glensor [4] and Tim Park [5]

[1] Wellington School of Architecture, Faculty of Architecture and Design Innovation, Te Herenga Waka Victoria University of Wellington, Wellington 6011, New Zealand; maggie.mackinnon@vuw.ac.nz
[2] Graduate School of Life Sciences, Faculty of Science, Utrecht University, Padualaan 8, 3584 CH Utrecht, The Netherlands; rebecca.mackinnon1@gmail.com
[3] School of Future Environments, Faculty of Design and Creative Technologies, Te Wānanga Aronui o Tāmaki Makau Rau Auckland University of Technology, Auckland 1010, New Zealand
[4] Analytics and Modelling, Ministry of Transport, Wellington 6011, New Zealand; k.glensor@transport.govt.nz
[5] Parks, Sport and Recreation, Wellington City Council, Wellington 6011, New Zealand; tim.park@wcc.govt.nz
* Correspondence: maibritt.pedersen.zari@aut.ac.nz

Abstract: Urgent biophilia describes the conscious desire of humans to seek interactions with nature during periods of stress. This study examines the changes in frequency and reason for visiting urban green spaces by residents of Wellington, New Zealand, to determine whether resident behavior during a stressful period exemplifies the principles of urgent biophilia. The COVID-19 pandemic and resulting lockdowns were used as the study period due to the significant physical and mental health stressors they triggered. Pedestrian and cyclist counters located in key urban green spaces in Wellington were used to collect data on visits pre- and post-pandemic. Two surveys were used to assess residents' reasons for visiting urban green spaces during lockdowns. Increased green space visits were seen during the strictest lockdowns, though there was some variation in visits depending on the location of the green space. The most frequently reported reason for visiting green spaces during lockdown was mental wellbeing, followed by recreation. These results suggest that Wellington residents used urban green spaces as a coping mechanism during stressful lockdown periods for wellbeing benefits, exemplifying the principles of urgent biophilia. Urban planners and policymakers must consider and implement urban green infrastructure as a public health resource.

Keywords: urgent biophilia; urban green space; nature-based coping; human wellbeing; pandemic; COVID-19; urban green infrastructure

Citation: MacKinnon, M.; MacKinnon, R.; Pedersen Zari, M.; Glensor, K.; Park, T. Urgent Biophilia: Green Space Visits in Wellington, New Zealand, during the COVID-19 Lockdowns. *Land* **2022**, *11*, 793. https://doi.org/10.3390/land11060793

Academic Editors: Francesca Ugolini and David Pearlmutter

Received: 1 May 2022
Accepted: 24 May 2022
Published: 27 May 2022

Publisher's Note: MDPI stays neutral with regard to jurisdictional claims in published maps and institutional affiliations.

1. Introduction

1.1. Urgent Biophilia and Human Wellbeing

The concept of biophilia, introduced by E. Fromm in 1964 and popularised by E.O. Wilson in 1984, describes the "innate human tendency to focus on and affiliate with life forms and life-like processes" [1]. This affiliation and desire to connect with nature, often termed the biophilia hypothesis, is said to be encoded in human genetics as a result of our evolutionary and historical dependence on other species and biological systems for survival and reproduction [2,3]. A growing body of quantitative and qualitative research provides evidence for the mental and physical advantages associated with biophilia and contact with nature, as well as the adverse effects of a lack of contact with nature [4,5].

A framework for incorporating biophilia into the built environment at the architectural scale, termed biophilic design, was introduced by S. Kellert, a colleague of E.O. Wilson, in 2008 [6]. Biophilic design frameworks related to urban scales have also been devised [7,8], and an international Biophilic Cities Network exists to facilitate a global network of partner cities "working to pursue a natureful city within their unique and diverse environments and cultures" [9].

Biophilic design utilises natural morphologies, materials, and spatial patterns and arrangements to provide more opportunities for humans to connect with nature, either directly or indirectly, in buildings and cities, thereby improving human wellbeing [6]. Russell et al. [10] categorised the contributions of non-material experiences of nature to the many facets of human wellbeing, including certainty and control, inspiration and fulfilment, sense of place and identity, and connectedness and belonging.

In 2012, K. Tidball [11] proposed the concept of urgent biophilia to describe nature's role in human resilience. In contrast to the biophilia hypothesis, which suggests that our innate affinity to nature is mostly subconscious, urgent biophilia suggests that humans consciously seek out contact with nature to strengthen their resilience during a crisis or disaster. Tidball's 2012 paper [11] reviews the therapeutic benefits of contact with nature and suggests that within the context of a disaster or crisis, individuals or communities may consciously seek out nature to reap those benefits and aid in their recovery. This hypersensitised manifestation of the human affinity for nature functions as a self-administered or doctor-prescribed [12] nature-based therapy that can improve our capacity to withstand and adapt to hardship [13].

1.2. Green Space and Human Wellbeing

Ecosystems and contact with nature contribute to human wellbeing through physical, psychological, philosophical, social, cultural, and spiritual pathways [14]. The term "wellbeing" goes beyond the meeting of basic needs and includes elements such as a positive physical and mental state, social cohesion and participation in society, and a sense of purpose and achievement [5]. These more intangible benefits of contact with nature are central to human values and preferences, such as cultural diversity and identity, cultural landscapes and heritage, inspiration, recreation, and tourism [10,14].

An extensive body of literature documents the mental and physical health benefits related to nature-based therapies and living in close proximity to nature [4,5]. Nature-based therapies for mental wellbeing include practices such as forest bathing, horticulture, and community gardening [15]. There is evidence for the positive impacts of nature on stress reduction [16], social cohesion [17], and improved mood [18]. Significant associations have been found between the proximity and accessibility of urban green spaces and positive physical and mental health outcomes [19]. However, McDonald et al. [20] found that only 13% of urban residents live near enough forest cover to confer significant wellbeing benefits. Due to age or financial constraints, residents with limited mobility are particularly impacted by a lack of access to good quality urban green space [21]. Urgent biophilia suggests that access to the physical and mental wellbeing benefits of urban green spaces is especially important during times of crisis. Therefore, this study examines the COVID-19 lockdowns and their impacts on green space visits by urban residents.

1.3. COVID-19 Pandemic and Human Wellbeing

The global public health crisis presented by the COVID-19 pandemic sent unprecedented regulations throughout countries worldwide. When the COVID-19 virus arrived in New Zealand in February 2020, the government implemented rapid, strict lockdowns to stop its spread. While this resulted in New Zealand having fewer COVID-19 cases and deaths, there were other impacts on wellbeing due to the economic, social, and health consequences of the border closure and lockdowns [22]. Such large and sudden disruptions to everyday life negatively impact the wellbeing of populations, particularly those in urban environments with limited access to green spaces [23]. Whether through direct contact with the COVID-19 virus or the indirect impacts of the local and global restrictions put into place to reduce its spread, the COVID-19 pandemic has had serious impacts on human wellbeing [24]. The COVID-19 period has been associated with significantly higher levels of depression and anxiety [25], and lockdown severity significantly impacted mental health [26]. The COVID-19 pandemic is a crisis scenario that contains no physical destruction or disaster but remains a threat to public health and social cohesion. The self-isolation and stay-at-home orders severely limited travel globally and locally and reduced the opportunities for coping mechanisms, such as social activities and time spent with loved ones. Without these conventional coping strategies, how did individuals cope with the stress and uncertainty brought on by the pandemic? Based on the concepts and evidence in the biophilia, human wellbeing, and urban green space literature, this study investigates if and why residents of Wellington, New Zealand, used their local green spaces for nature-based coping mechanisms during the COVID-19 lockdowns, as the theory of urgent biophilia might suggest.

2. Materials and Methods

2.1. COVID-19 Lockdowns in Wellington, New Zealand

New Zealand is a temperate island nation in the South Pacific with a population of approximately 5 million [27]. The first COVID-19 case in New Zealand was reported on 28 February 2020. The New Zealand border was closed to all but citizens and permanent residents on 19 March 2020. The New Zealand Government implemented a National Action Plan to manage the spread of COVID-19 and outlined four levels of restrictions [28]. The Alert Level system was introduced and put in place on 21 March 2020, and New Zealand was placed in Alert Level 2. A series of Alert Level changes and lockdowns followed over the next five months, with the most stringent isolation requirements (Alert Level 4) lasting four consecutive weeks. During Alert Level 4, people were to remain at home except to visit essential services. Outside recreation (e.g., walking, jogging) was allowed with social distancing but was limited to residents' local neighbourhoods. The COVID-19 Stringency Index rated New Zealand's Alert Level 4 protocol the strictest in the world (96.30 out of 100), followed by Italy (93.52), France (87.96), and the UK (79.63) [29]. During Alert Level 3, people were required to work and study from home, if possible. There was limited domestic travel, and public venues were closed, other than most urban green spaces. During Alert Level 2, alternative ways of working were encouraged to limit the virus spread. There were reduced capacities in public venues, other than urban green spaces, and restrictions on mass gatherings. During Alert Level 1, most restrictions were removed, except for physical distancing, contact tracing, and restrictions on mass gatherings. Self-isolation and testing were required for anyone symptomatic [28]. Alert levels varied in different parts of New Zealand, particularly in 2021.

Wellington, a coastal city on the North Island, is the capital of New Zealand and has a population of approximately 216,505 [30]. The city's steep hilly topography constrains development and affords residents a high proportion of visible green and blue space [31]. As a result of the preservation of a series of interconnected urban green spaces (known as the Town Belt) and other key native habitat reserves [32], central Wellington has an average of 20 m^2 of urban green space per person, but this varies considerably between different suburbs [33]. This is more than twice the WHO recommendation of at least 9 m^2 of urban green space per person [34]. Because most residents have access to nearby green space, Wellington is a good case study for examining green space usage by urban residents, particularly during the COVID-19 period (Figure 1) when travelling to larger natural areas on the urban periphery was restricted. Wellington is one of the cities in the International Biophilic Cities Movement [35].

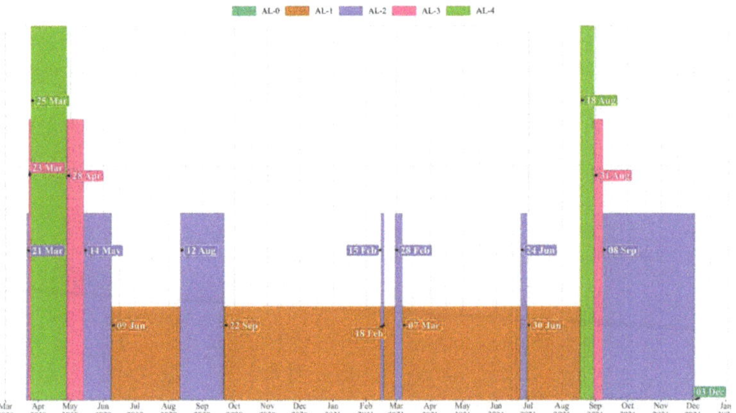

Figure 1. Wellington, New Zealand, COVID-19 Alert Level restrictions and dates from 2019 to 2021. Alert Level (AL) restrictions increase in strictness, with AL-4 being the strictest (everything except essential services closed). The start dates of the Alert Level periods are shown in the labels.

2.2. Wellington Pedestrian and Cyclist Counter Data

Wellington City Council has a large network of pedestrian and cyclist counters throughout its green spaces to track trail visits. For this analysis, data from counters in six urban green space locations in the Wellington region were used: Gilberd Bush, Hataitai to City Walkway, Southern Walkway (and "Super D" mountain bike trail), Berhampore Golf Course, Mount Kaukau, and Waimapihi (Polhill) (Figure 2). These counters produced the most reliable and consistent data over the study period of January 2019 to March 2022. The counters used by Wellington City Council are from Eco-Counter [36]. The PYRO Evo counter is a passive infrared sensor sitting inside a wooden post beside the trail that counts pedestrians only. The MULTI Nature counter uses a passive infrared sensor and an inductive loop sensor buried under the trail to distinguish between pedestrians and cyclists. Data were collected via cellular transmission using the Eco-Visio data analysis platform from Eco-Counter.

Figure 2. Pedestrian and cyclist counter locations in Wellington green spaces. Berhampore Golf Course (GC) and Mount Kaukau had pedestrian-only counters. The other four locations had MULTI counters. The base satellite image is from Landcare Research [37].

The data from these counters were processed using the statistical analysis package "R". Trail visits in the various Alert Level periods in Wellington throughout 2020, 2021, and 2022 were compared to the (pre-COVID) trail visits in 2019. The process for each counter is as follows:

1. The daily aggregate of counts are calculated.
2. The daily mean count on weekends and weekdays is calculated for each month.
3. A correction factor for each month's weekends/weekdays is calculated by comparing the monthly means with a reference month (March 2019). The month is arbitrarily chosen.
4. Each day's aggregate count is adjusted by the correction factor.

This process corrects for differences in green space visits in Wellington between weekends and weekdays and across seasons. Extreme values (those far more or less than expected) were investigated to identify faulty counters and were not included in the analysis.

2.3. Tanera Park Survey

Tanera Park in Aro Valley is one of the green spaces that make up the Wellington Town Belt. During the first national lockdown, labelled glass jars were tied to a fence at one of the park's viewpoints, and small stones were piled near them to create an interactive survey installation (Figure 3). The survey was carefully designed so that it would not increase virus transmission potential. Posters next to the jars asked people to place a stone in the jar that best described the reason for their visit to the park on a given day (Figure 4). The stones were removed from the jars and counted at the end of each day. The installation was active from 19 April to 29 May 2020, during which Wellington transitioned from Alert Level 4 to Alert Level 2. During the first phase of the survey, there were five jars, each corresponding to a known green space benefit: mental wellbeing, education, beauty, inspiration, and recreation. After an initial assessment of the data collected, phase 2 of the survey expanded the categories related to mental wellbeing and recreation, linking them to attributes of biophilic design. Seven jars were installed in groups related to mental wellbeing, recreation, and aesthetics. There were three jars in the mental wellbeing category (upliftment, stress relief, and hope), two jars in the recreation category (sport and leisure), and two jars in the aesthetics category (beauty and inspiration).

Figure 3. The location of the jar survey in Tanera Park in Aro Valley, Wellington. The base satellite image is from Landcare Research [37].

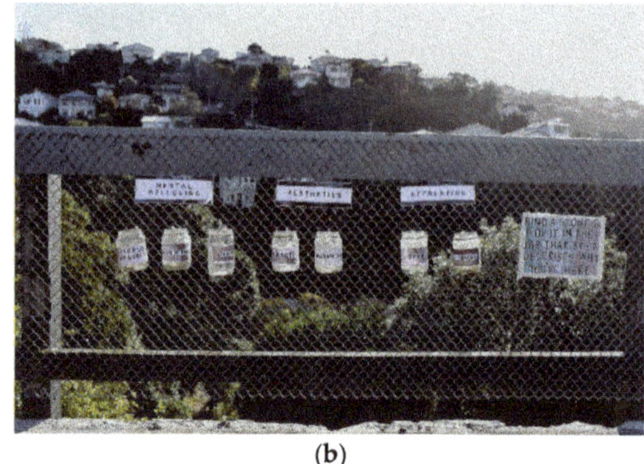

(a)   (b)

Figure 4. Images of the Tanera Park jar survey: (**a**) jar labels for phase 1 of the survey; (**b**) jar labels for phase 2 of the survey.

In order to better understand if people's behavior was exemplifying the principles of urgent biophilia, an online survey was created using Qualtrics and was active from 27 May to 30 June 2020. The online survey was distributed via a QR code installed next to the jars, and through popular Facebook groups, community noticeboards, and flyers. However, it was limited to people who had visited and/or interacted with the jar survey in Tanera Park. The survey consisted of 27 questions that expanded upon the jar survey in the park and further investigated people's frequency and reasons for visiting the park during COVID-19 lockdowns. The survey also allowed open text answers for the reason for visiting the park to capture responses that were not covered by the jar categories, and it also provided a list of options for participants to describe their emotional state.

3. Results

3.1. Wellington Pedestrian and Cyclist Counter Analysis

In total, the six Wellington green space locations registered 538,000 visits in 2019, compared to 564,000 (+4.8%) in 2020 and 554,000 (+3.0%) in 2021; however, some individual stations registered larger changes, such as Gilberd Bush in 2020 with a 67.3% increase or the Hataitai to City Walkway with a decrease of 22.1% in 2020 (Table 1).

Table 1. Summary of green space visits in 2019 (pre-COVID), 2020, and 2021.

Counter	2019 Visits	2020 Visits (Change from 2019)	2021 Visits (Change from 2019)
Berhampore GC	18,172	21,484 (+18.2%)	25,059 (+37.9%)
Gilberd Bush	12,438	20,803 (+67.3%)	16,849 (+35.5%)
Hataitai to City Walkway	134,103	104,467 (−22.1%)	111,419 (−16.9%)
Mount Kaukau	156,067	160,990 (+3.2%)	145,395 (−6.8%)
Southern Walkway/Super D	161,711	182,479 (+12.8%)	192,556 (+19.1%)
Waimapihi (Polhill)	55,739	74,091 (+32.9%)	63,129 (+13.3%)
Total	538,230	564,314 (+4.8%)	554,407 (+3.0%)

Figures 5 and 6 are frequency distribution graphs of daily visits at the six green space sites during high Alert Levels (3–4) and low Alert Levels (0–2) for each post-COVID-19 year, each compared to the pre-COVID-19 (2019) period. These graphs show the proportion of days (y-axis) against the number of visits per day (x-axis). A narrow peak on these graphs results from the number of visits per day being relatively consistent. In contrast, low, flat curves result from the number of visits being highly variable.

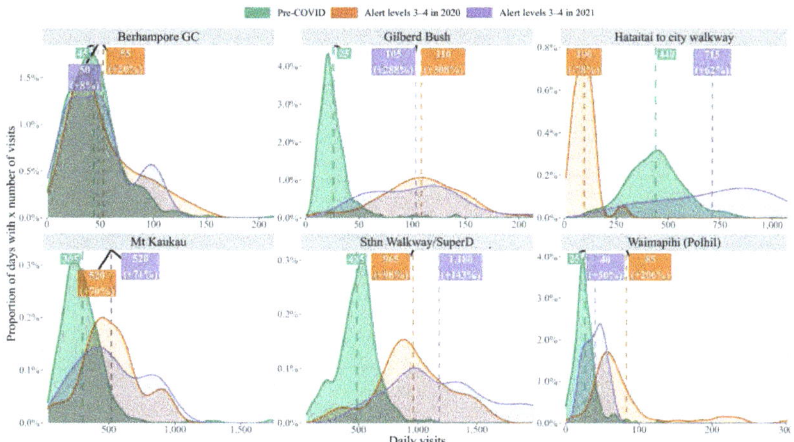

Figure 5. The distribution of daily green space visits during high Alert Levels (3–4) in 2020 (orange) and 2021 (purple) compared to the pre-COVID-19 (2019) periods. The graphs show the proportion of days (y-axis) with a number of daily visitors (x-axis), and thus show how consistent the number of visits is. The mean daily visitors are also marked with a dashed line and label with this compared to 2019. All values have been corrected for mean monthly and weekend/weekday differences.

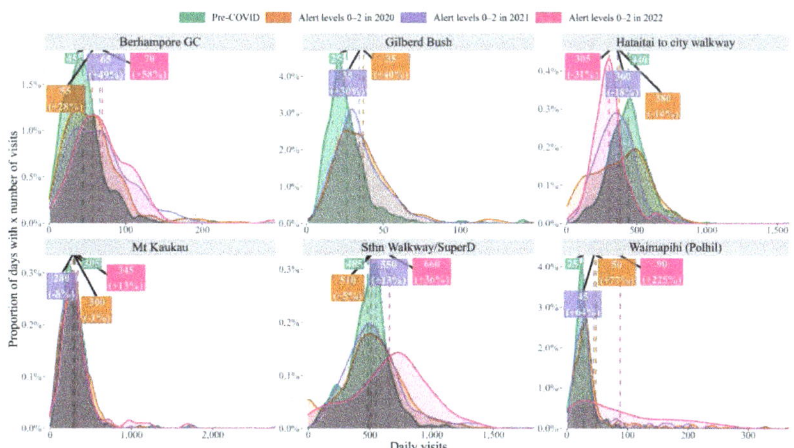

Figure 6. The distribution of daily visits during low Alert Levels (0–2) in 2020 (orange), 2021 (purple), and 2022 (pink) compared to the pre-COVID-19 (2019) periods. The graphs show the proportion of days (y-axis) with a number of daily visitors (x-axis), and thus show how consistent the number of visits is. The mean daily visitors are also marked with a dashed line and label with this compared to 2019. All values have been corrected for mean monthly and weekend/weekday differences.

At high Alert Levels, there were also variations in the visit patterns of different green space trails. For example, the Hataitai to City Walkway is often used for people's commute to work; however, with business and school closures and people working and studying from home, there was a decrease in the number of people on that trail, −78% in 2020, compared to an increase of 62% in 2021. Recreational bush trails, namely Waimapihi (Polhill) (+206% in 2020), Gilberd Bush (+308% in 2020 and +288% in 2021), and the Southern Walkway/Super D (+143% in 2021), saw dramatic increases in visits during the lockdowns. Several of the trails saw large increases in the day-to-day variance in the number of visits, especially Gilberd Bush and the Haitaitai to City Walkway.

At low Alert Levels (Figure 6), the differences from pre-COVID-19 visits were, as expected, more modest compared to high alert levels. Especially Mount Kaukau had little variation (−8% to +13%). The visit profile for both the Southern Walkway/Super D and Waimapihi (Polhill) became notably more variable compared to the other years for each. The Hataitai to City Walkway has seen decreased use in all years, suggesting the increase in use in 2021 high Alert Levels (Figure 5) may have been the result of increased leisure use of this trail while commuting use may have decreased overall.

3.2. Tanera Park Survey Analysis

On average, 60 stones were collected every day from the jars. During the first phase of the jar survey, which coincided with Alert Level 4, the jars that received the most stones were recreation and mental wellbeing, each accounting for a third of the total stones counted (Figure 7). During the second phase of the jar survey, which expanded the jar options and grouped them into three categories, mental wellbeing and recreation remained the main reasons for visiting the park. Stress relief and leisure were the jars that received the most stones during Alert Level 3, while the beauty and leisure jars became the most popular during Alert Level 2.

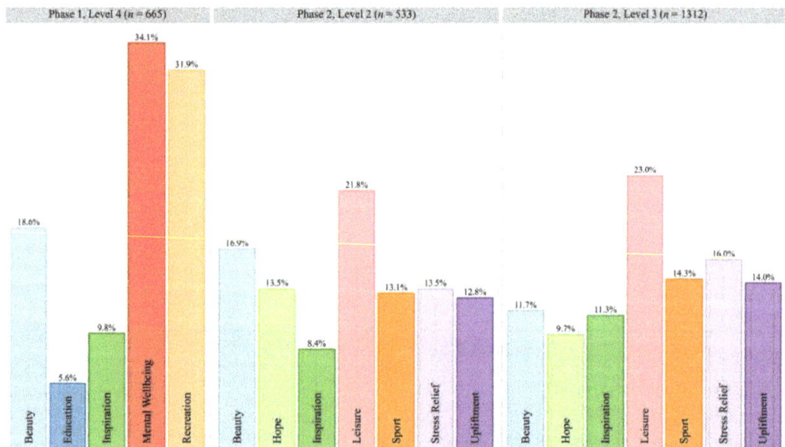

Figure 7. Reasons for visiting Tanera Park during the COVID-19 lockdowns in Wellington. Phase 1 of the jar survey included five options. These options were expanded in phase 2 of the survey and grouped into the categories of mental wellbeing (upliftment, stress relief, and hope), recreation (sport and leisure), and aesthetics (beauty and inspiration).

There were 134 responses to the online survey, of which 58 met the selection criteria of having visited and interacted with the jar installation in Tanera Park. The majority of the respondents lived within a 10 min walk from the green space. Over half (58%) of the respondents said they were visiting the park more frequently during the lockdowns than they did prior to them, with 19% stating they visited the same amount and 23% stating they visited less (Figure 8). The reasons for this could be grouped into four main themes: routine change, more time, lack of other options, and feeling confined indoors or cabin fever. With the removal of daily commutes, errands, and social activities, some respondents stated that visiting the park provided an opportunity to break up the monotony of their lockdown experience. With fewer commitments and more flexible working or schooling from home options, some respondents stated that they had more time in their day to walk and spend time in the park. The lack of other safe options for exercise and outdoor gatherings that abided by the social distancing guidelines was another reason some respondents visited the park. The proximity to their home and the spaciousness of the paths made Tanera Park more attractive than some of the neighbouring green spaces that had narrower paths. The most common theme in visitor responses was that of cabin fever. Respondents expressed a need for fresh air or to escape from the indoor confinement and isolation imposed by the lockdowns.

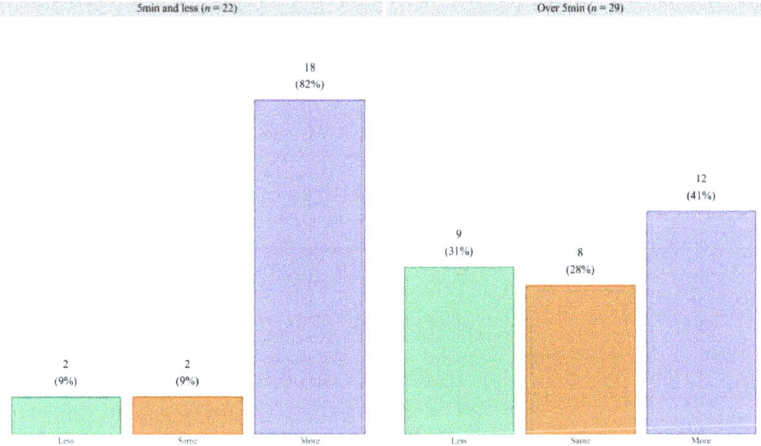

Figure 8. Respondents' respective walking distance from Tanera Park and their change in frequency of park visits during the COVID-19 lockdowns in Wellington.

In response to the questions relating to their emotional states during the lockdowns, many respondents selected uncertainty and gratitude, followed closely by relaxation, stress, and anxiety. The varied and sometimes contradictory emotions selected reflect the complexity of the impacts of the lockdowns on mental wellbeing. However, the majority (77%) of the respondents answered that visiting the park helped alleviate some of their negative emotions. When asked how they thought the park helped them cope with negative emotions, some respondents answered that it helped them overcome uncertainty by providing a space for them to gain perspective and reflect. Others stated it helped them alleviate stress by providing an opportunity to escape from media and clear their mind. As one of the few options for social distancing, some respondents said the park helped combat loneliness by allowing safe social interactions. Some respondents said that the park offered a change of scenery and activity to alleviate cabin fever. Though many respondents recognised their need for the wellbeing benefits of nature, when asked to elaborate whether this need was more during the pandemic than it was before, the majority responded that their need for it remained the same (Figure 9). This indicates that many respondents recognise and rely on nature in their daily lives and not only in times of crisis and that other factors, such as more time and fewer other options, could have been the driver behind their increase in visits to the park.

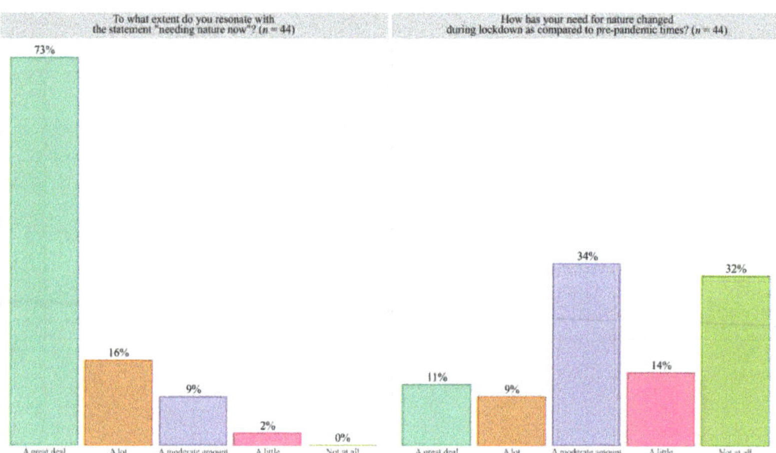

Figure 9. The responses to two questions of the online survey. The questions were: to what extent do you resonate with the statement "needing nature now", and how has your need for nature changed during the lockdown compared to the pre-pandemic period?

4. Discussion

The Wellington City Council pedestrian and cyclist counter data showed that there was an increase in urban green space visits during the COVID-19 lockdowns, with visit increases being the highest during the strictest Alert Level (3 and 4) periods. Biophilia is generally described as a subconscious affinity [1]; however, the Tanera Park surveys showed that the majority of visitors were conscious of and made use of nature-based coping mechanisms to alleviate negative emotions and lifestyle changes during the COVID-19 lockdowns, exemplifying the principles of urgent biophilia [11]. However, some important limitations to the data should be noted. The counters used recorded only the numbers of pedestrians or cyclists passing and did not track the patterns of individual users. Further research using individual tracker data could help isolate the changes in the total number of park visitors and individuals' frequency of visits to the park during the COVID-19 lockdowns. Individual tracker data could also supplement the surveys conducted in this research, which were limited by small sample sizes and self-selected participants. No visitor sociodemographic information was collected by the counters or surveys. Future research using individual tracker data could help capture differences in green space usage or engagement with the surveys between subgroups of Wellington's population.

Despite some data limitations, the results of this research support other surveys of New Zealand residents and their interactions with nature during the COVID-19 pandemic. The New Zealand Department of Conservation conducted qualitative interviews from November 2020 to January 2021 to understand how visitors' participation and perspective towards the outdoors had changed as a result of the pandemic [38]. They found similar drivers, such as coping with anxiety, uncertainty, and change, behind people's participation in the outdoors. An online panel survey of three standardised wellbeing measures found that New Zealanders experienced higher levels of severe psychological distress and anxiety during the COVID-19 lockdowns, particularly in young adults who reported the highest levels of suicidality [39]. However, compared to a cross-sectional study conducted in the UK, Gasteiger et al. [40] found that levels of anxiety and stress were significantly lower in New Zealand overall during the pandemic. A similar result was found in a study comparing mental health and wellbeing in the UK, Ireland, New Zealand, and Australia, with New Zealanders reporting the highest levels of mental health and wellbeing [41]. Another study by Sibley et al. [21] found that there was a small increase in psychological distress in New Zealanders as a result of the COVID-19 lockdowns; however, they also found that participants felt a greater sense of community. With limited indoor options for social gathering, this may suggest that the safe social gathering opportunities provided by urban green spaces contributed to wellbeing during stressful lockdown periods in addition to the restorative benefits of nature itself. Though perhaps skewed as a result of the travel restrictions during lockdowns, the majority of respondents lived within a 10 min walk from the park, demonstrating the importance of local green space accessibility for nature-based coping mechanisms. A study of New Zealanders' physical activity during the March to May 2020 lockdown found that only half of the respondents were able to maintain their usual level of physical activity as a result of gyms closing [42]. A survey by Faulkner et al. [41] found that those who reported doing less exercise as a result of the COVID-19 lockdowns had reduced mental health and wellbeing. Because green spaces can be important locations for recreation and outdoor physical exercise, having accessible, evenly distributed green spaces in cities could increase opportunities for physical activity during closures of public and private exercise facilities. This is an additional reason why urban green space contributes to mental health and wellbeing and why it must be strategically incorporated across urban landscapes in relation to the spatial distribution of human population densities.

The increase in green space visits by urban residents for wellbeing benefits during the COVID-19 lockdowns found in this study also corroborates international research findings on green space usage during the pandemic. Using mobile tracking data, Venter et al. [43] found that recreational green space usage increased by 291% in Oslo, Norway, in comparison to a 3-year average for the same days after adjusting for other factors, such as weather. As with the results of this study, these findings demonstrate the importance of urban green space as a critical piece of resilience infrastructure during crises. Marconi et al. [44] found that the meaning of urban green space to survey respondents in Buenos Aires, Argentina, evolved from just a place to experience nature to an important piece of infrastructure in the city during the COVID-19 restrictions. Spatial analyses combined with online questionnaires were used by Robinson et al. [45] to find that respondents in England, UK, spent more time in nature during the pandemic, most stating it was for health and wellbeing benefits. They also found that higher amounts of green land cover within a 250 m radius from respondents' homes were an important predictor of increased mental wellbeing [45]. A survey conducted by Berdejo-Espinola et al. [46] in Brisbane, Australia, found that nearly 80% of respondents stated their reason for visiting green spaces during the lockdowns was for physical and mental wellbeing. In their international survey, Ugolini et al. [47] found that pre-pandemic, the most popular reasons for visiting urban green space were relaxing and physical exercise. When the lockdowns limited access to urban green space, many respondents stated that it was the outdoor environment and social activity elements of green space that they missed the most. Pouso et al. [26] found that having contact with nature especially helped those under strict lockdown. Their international European survey found that individuals with accessible outdoor space and visible blue and green space had more positive emotions. Reid et al. [25] found significant reductions in depression and anxiety were associated with spending time in green space as well as residents' perceived abundance of green space, after adjusting for sociodemographic and pandemic-related stressors.

5. Conclusions

This study combined quantitative data from pedestrian and cyclist counters with qualitative data from surveys to determine if and why the COVID-19 lockdowns changed Wellington residents' green space visits. Our findings show that urban residents did actively seek out green space for restorative mental wellbeing benefits during the uncertainty and stress of the COVID-19 lockdowns, exemplifying the principles of urgent biophilia. In order to enhance the resilience of urban residents in the face of global crises, it is crucial to provide accessible green spaces and green infrastructure in cities to provide opportunities for residents to engage in nature-based coping mechanisms and thus positively affect mental health and wellbeing and resilience. Though urban green space accessibility discourse often focuses on mobility-restricted populations, the COVID-19 pandemic showed that accessibility is important for everyone during periods of crisis that restrict movements (lockdowns, natural disasters, etc.). Urban planners and policymakers must consider urban green infrastructure and spaces as a public health necessity and ensure a strategic distribution of green spaces across urban environments that correlates to where people live. Increasing the opportunities for individuals to engage in nature-based coping mechanisms during periods of normalcy and in times of crisis is essential for the current and future wellbeing of urban populations.

Author Contributions: Conceptualization, M.P.Z., R.M. and T.P.; methodology, R.M., K.G. and T.P.; writing—original draft preparation, M.M. and R.M.; writing—review and editing, M.M., M.P.Z., R.M. and K.G.; visualization, K.G. All authors have read and agreed to the published version of the manuscript.

Funding: This research received no external funding.

Institutional Review Board Statement: The study was conducted in accordance with the Declaration of Helsinki, and was approved by the Ethics Committee of Te Herenga Waka Victoria University of Wellington (application 28552, approved 26/05/20).

Informed Consent Statement: Informed consent was obtained from all subjects involved in the study, as per the approved 28552 Ethics application.

Data Availability Statement: Data supporting reported results can be found by contacting the authors.

Acknowledgments: We acknowledge the support of our various intuitions that enabled us to undertake this collaborative work. We also acknowledge the support of Utrecht University supervisors who contributed to interpreting Tanera Park results.

Conflicts of Interest: The authors declare no conflict of interest.

References

1. Wilson, E.O.; Kellert, S.R. *The Biophilia Hypothesis*; Island Press: Washington, DC, USA, 1993.
2. Browning, W.; Ryan, C.; Clancy, J. *14 Patterns of Biophilic Design*; Terrapin Bright Green LLC: New York, NY, USA, 2014.
3. Joye, Y.; de Block, A. Nature and I are two: A critical examination of the biophilia hypothesis. *Env. Values* **2011**, *20*, 189–215. [CrossRef]
4. Roberts, A.B.L.; Kerr, G.; Lambert, S.; McWilliam, W.; Moore, K.; Quinn, J.; Simmons, D.; Thrush, S.; Townsend, M.; Blaschke, P.; et al. *The Nature of Wellbeing: How Nature's Ecosystem Services Contribute to the Wellbeing of New Zealand and New Zealanders*; Department of Conservation: Wellington, New Zealand, 2015.
5. Summers, J.K.; Smith, L.M.; Case, J.L.; Linthurst, R.A. A Review of the Elements of Human Well-Being with an Emphasis on the Contribution of Ecosystem Services. *Ambio* **2012**, *41*, 327–340. [CrossRef]
6. Kellert, S.R.; Heerwagen, J.; Mador, M. *Biophilic Design: The Theory, Science and Practice of Bringing Buildings to Life*; John Wiley & Sons: Hoboken, NJ, USA, 2008; ISBN 9780470163344.
7. Haase, D. Reflections on urban landscapes, ecosystems services and nature-based solutions in cities. *Plan. Theory Pract.* **2016**, *17*, 295–300. [CrossRef]
8. Zari, M.P. Understanding and designing nature experiences in cities: A framework for biophilic urbanism. *Cities Health* **2019**, 1–12. [CrossRef]
9. Biophilic Cities. Our Vision: Connecting Cities and Nature. 2022. Available online: https://www.biophiliccities.org/our-vision (accessed on 1 May 2022).
10. Russell, R.; Guerry, A.D.; Balvanera, P.; Gould, R.K.; Basurto, X.; Chan, K.M.A.; Klain, S.; Levine, J.; Tam, J. Humans and Nature: How Knowing and Experiencing Nature Affect Well-Being. *Annu. Rev. Environ. Resour.* **2013**, *38*, 473–502. [CrossRef]
11. Tidball, K.G. Urgent Biophilia: Human-Nature Interactions and Biological Attractions in Disaster Resilience. *Ecol. Soc.* **2012**, *17*, 113–131. [CrossRef]
12. Kondo, M.C.; Oyekanmi, K.O.; Gibson, A.; South, E.C.; Bocarro, J.; Hipp, J.A. Nature Prescriptions for Health: A Review of Evidence and Research Opportunities. *Int. J. Environ. Res. Public Health* **2020**, *17*, 4213. [CrossRef]
13. Tidball, K.G.; Krasny, M.E. *Greening in the Red Zone: Disaster, Resilience and Community Greening*, 1st ed.; Springer: Amsterdam, The Netherlands, 2014; ISBN 90-481-9947-6.
14. Millennium Ecosystem Assessment. *Ecosystems and Human Well-Being: Synthesis*; Island Press: Washington, DC, USA, 2005.
15. Wendelboe-Nelson, C.; Kelly, S.; Kennedy, M.; Cherrie, J.W. A Scoping Review Mapping Research on Green Space and As-sociated Mental Health Benefits. *Int. J. Environ. Res. Public Health* **2019**, *16*, 2081. [CrossRef]

16. Kilpatrick, A.M.; Salkeld, D.J.; Titcomb, G.; Hahn, M.B. Conservation of biodiversity as a strategy for improving human health and well-being. *Philos. Trans. R. Soc. B Biol. Sci.* **2017**, *372*, 20160131. [CrossRef]

17. de Vries, S.; van Dillen, S.; Groenewegen, P.; Spreeuwenberg, P. Streetscape greenery and health: Stress, social cohesion and physical activity as mediators. *Soc. Sci. Med.* **2013**, *94*, 26–33. [CrossRef]

18. van den Berg, M.; Wendel-Vos, W.; van Poppel, M.; Kemper, H.; van Mechelen, W.; Maas, J. Health benefits of green spaces in the living environment: A systematic review of epidemiological studies. *Urban Urban Green.* **2015**, *14*, 806–816. [CrossRef]

19. Nutsford, D.; Pearson, A.; Kingham, S. An ecological study investigating the association between access to urban green space and mental health. *Public Health* **2013**, *127*, 1005–1011. [CrossRef]

20. McDonald, R.I.; Beatley, T.; Elmqvist, T. The green soul of the concrete jungle: The urban century, the urban psychological penalty, and the role of nature. *Sustain. Earth* **2018**, *1*, 3. [CrossRef]

21. Dennis, M.; Cook, P.A.; James, P.; Wheater, C.P.; Lindley, S.J. Relationships between health outcomes in older populations and urban green infrastructure size, quality and proximity. *BMC Public Health* **2020**, *20*, 626. [CrossRef]

22. Officer, T.N.; Imlach, F.; McKinlay, E.; Kennedy, J.; Pledger, M.; Russell, L.; Churchward, M.; Cumming, J.; McBride-Henry, K. COVID-19 Pandemic Lockdown and Wellbeing: Experiences from Aotearoa New Zealand in 2020. *Int. J. Environ. Res. Public Health* **2022**, *19*, 2269. [CrossRef]

23. Samuelsson, K.; Barthel, S.; Colding, J.; Macassa, G.; Giusti, M. Urban nature as a source of resilience during social distancing amidst the coronavirus pandemic. *OSFPreprints* **2020**. [CrossRef]

24. Sibley, C.G.; Greaves, L.M.; Satherley, N.; Wilson, M.S.; Overall, N.C.; Lee, C.H.J.; Milojev, P.; Bulbulia, J.; Osborne, D.; Milfont, T.L.; et al. Effects of the COVID-19 Pandemic and Nationwide Lockdown on Trust, Attitudes Toward Government, and Well-Being. *Am. Psychol.* **2020**, *75*, 618–630. [CrossRef]

25. Reid, C.E.; Rieves, E.S.; Carlson, K. Perceptions of green space usage, abundance, and quality of green space were associated with better mental health during the COVID-19 pandemic among residents of Denver. *PLoS ONE* **2022**, *17*, e0263779. [CrossRef]

26. Pouso, S.; Borja, Á.; Fleming, L.E.; Gómez-Baggethun, E.; White, M.P.; Uyarra, M.C. Contact with blue-green spaces during the COVID-19 pandemic lockdown beneficial for mental health. *Sci. Total Environ.* **2021**, *756*, 143984. [CrossRef]

27. Stats NZ. Estimated Population of NZ. Available online: https://www.stats.govt.nz/indicators/population-of-nz (accessed on 4 April 2022).

28. National Crisis Management Centre. National Action Plan. 2020. Available online: https://covid19.govt.nz/assets/resources/legislation-and-key-documents/COVID-19-national-action-plan-2-issued-1-April.pdf (accessed on 29 April 2022).

29. Global Data Lab. Our World in Data: COVID-19 Stringency Index. Available online: https://ourworldindata.org/explorers/coronavirus-data-explorer?uniformYAxis=0&hideControls=true&Interval=7-day+rolling+average&Relative+to+Population=true&Color+by+test+positivity=false&country=USA~{}ITA~{}CAN~{}DEU~{}GBR~{}FRA&Metric=Stringency+index (accessed on 18 April 2022).

30. id (Informed Decisions) for Wellington City Council. Population and Household Forecasts 2013 to 2043. Available online: https://forecast.idnz.co.nz/wellington/home (accessed on 23 July 2021).

31. Rastandeh, A.; Pedersen Zari, M.; Brown, D.K.; Vale, R. Analysis of landform and land cover: Potentials for urban biodiversity conservation against rising temperatures. *Urban Policy Res.* **2019**, *37*, 338–349. [CrossRef]

32. Rastandeh, A.; Brown, D.K.; Pedersen Zari, M. Site selection of urban wildlife sanctuaries for safeguarding indigenous biodiversity against increased predator pressures. *Urban Urban Green* **2018**, *32*, 21–31. [CrossRef]

33. Blaschke, P.; Chapman, R.; Gyde, E.; Howden-Chapman, P.; Ombler, J.; Pedersen Zari, M.; Perry, M.; Randal, E. *Green Space in Wellington's Central City: Current Provision, and Design for Future Wellbeing*; New Zealand Centre for Sustainable Cities: Wellington, New Zealand, 2019.

34. World Health Organization (WHO). *Urban Green Spaces and Health*; WHO Regional Office for Europe: Copenhagen, Denmark, 2016.

35. Beatley, T. *Handbook of Biophilic City Planning and Design*; Island Press: Washington, DC, USA, 2016; pp. 75–84.

36. Eco-Counter. Products. Available online: https://www.eco-counter.com/products/ (accessed on 15 April 2022).

37. Manaaki Whenua—Landcare Research. Land Cover Database Version 5.0, Mainland New Zealand. Available online: https://lris.scinfo.org.nz/layer/104400-lcdb-v50-land-cover-database-version-50-mainland-new-zealand/ (accessed on 30 March 2022.).

38. Department of Conservation. Resilience through Nature: Qualitative Research Providing Insights into New Zealanders' Outdoors Participation during Lockdown Levels, September 2021. Available online: https://www.doc.govt.nz/globalassets/documents/about-doc/role/visitor-research/resilience-through-nature.pdf (accessed on 20 April 2022).

39. Every-Palmer, S.; Jenkins, M.; Gendall, P.; Hoek, J.; Beaglehole, B.; Bell, C.; Williman, J.; Rapsey, C.; Stanley, J. Psychological distress, anxiety, family violence, suicidality, and wellbeing in New Zealand during the COVID-19 lockdown: A cross-sectional study. *PLoS ONE* **2020**, *15*, e0241658. [CrossRef]

40. Gasteiger, N.; Vedhara, K.; Massey, A.; Jia, R.; Ayling, K.; Chalder, T.; Coupland, C.; Broadbent, E. Depression, anxiety and stress during the COVID-19 pandemic: Results from a New Zealand cohort study on mental well-being. *BMJ Open* **2021**, *11*, e045325. [CrossRef]

41. Faulkner, J.; O'Brien, W.J.; McGrane, B.; Wadsworth, D.; Batten, J.; Askew, C.D.; Badenhorst, C.; Byrd, E.; Coulter, M.; Draper, N.; et al. Physical activity, mental health and well-being of adults during initial COVID-19 containment strategies: A multi-country cross-sectional analysis. *J. Sci. Med. Sport* **2021**, *24*, 320–326. [CrossRef]

42. Meiring, R.M.; Gusso, S.; McCullough, E.; Bradnam, L. The Effect of the COVID-19 Pandemic Movement Restrictions on Self-Reported Physical Activity and Health in New Zealand: A Cross-Sectional Survey. *Int. J. Environ. Res. Public Health* **2021**, *18*, 1719. [CrossRef]
43. Venter, Z.S.; Barton, D.N.; Martinez-Izquierdo, L.; Langemeyer, J.; Baró, F.; McPhearson, T. Interactive spatial planning of urban green infrastructure—Retrofitting green roofs where ecosystem services are most needed in Oslo. *Ecosyst. Serv.* **2021**, *50*, 101314. [CrossRef]
44. Marconi, P.L.; Perelman, P.E.; Salgado, V.G. Green in times of COVID-19: Urban green space relevance during the COVID-19 pandemic in Buenos Aires City. *Urban Ecosyst.* **2022**, *25*, 941–953. [CrossRef]
45. Robinson, J.M.; Brindley, P.; Cameron, R.; MacCarthy, D.; Jorgensen, A. Nature's Role in Supporting Health during the COVID-19 Pandemic: A Geospatial and Socioecological Study. *Int. J. Environ. Res. Public Health* **2021**, *18*, 2227. [CrossRef]
46. Berdejo-Espinola, V.; Suárez-Castro, A.F.; Amano, T.; Fielding, K.S.; Oh, R.R.Y.; Fuller, R.A. Urban green space use during a time of stress: A case study during the COVID-19 pandemic in Brisbane, Australia. *People Nat.* **2021**, *3*, 597–609. [CrossRef]
47. Ugolini, F.; Massetti, L.; Calaza-Martínez, P.; Cariñanos, P.; Dobbs, C.; Ostoić, S.K.; Marin, A.M.; Pearlmutter, D.; Saaroni, H.; Šaulienė, I.; et al. Effects of the COVID-19 pandemic on the use and perceptions of urban green space: An international exploratory study. *Urban For. Urban Green.* **2020**, *56*, 126888. [CrossRef]

Article

Human-Nature Interactions during and after the COVID-19 Pandemic in Moscow, Russia: Exploring the Role of Contact with Nature and Main Lessons from the City Responses

Diana Dushkova [1,2,*], Maria Ignatieva [3], Anastasia Konstantinova [2], Viacheslav Vasenev [4], Elvira Dovletyarova [2] and Yury Dvornikov [2]

[1] Department of Urban and Environmental Sociology, Helmholtz-Centre for Environmental Research (UFZ), Permoser Str. 18, 04318 Leipzig, Germany

[2] Agrarian and Technological Institute, Peoples Friendship University of Russia (RUDN University), Miklukho-Maklaya Str. 8/2, 117198 Moscow, Russia; konstantinova-av@rudn.ru (A.K.); dovletyarova-ea@rudn.ru (E.D.); dvornikov_yua@pfur.ru (Y.D.)

[3] School of Design, The University of Western Australia (UWA), 35 Stirling Highway, Perth 6009, Australia; maria.ignatieva@uwa.edu.au

[4] Soil Geography and Landscape Group, Wageningen University, 6708 PB Wageningen, The Netherlands; slava.vasenev@wur.nl

* Correspondence: diana.dushkova@ufz.de

Citation: Dushkova, D.; Ignatieva, M.; Konstantinova, A.; Vasenev, V.; Dovletyarova, E.; Dvornikov, Y. Human-Nature Interactions during and after the COVID-19 Pandemic in Moscow, Russia: Exploring the Role of Contact with Nature and Main Lessons from the City Responses. *Land* **2022**, *11*, 822. https://doi.org/10.3390/land11060822

Academic Editors: Francesca Ugolini and David Pearlmutter

Received: 30 April 2022
Accepted: 28 May 2022
Published: 31 May 2022

Publisher's Note: MDPI stays neutral with regard to jurisdictional claims in published maps and institutional affiliations.

Abstract: Urban green spaces (UGS) as essential elements of the urban environment provide multiple ecosystem services including benefits for physical and mental health. Impacts of the COVID-19 pandemic and related restrictions have influenced human relationships with nature. Based on empirical research, this article explores the pathways and implications of human-nature interactions during and after COVID-19 and how human health and well-being could be supported by contact with nature. The article discusses the reasons that attract people to visit UGS (value of UGS, their perceptions, ways of contact with urban nature, etc.). It also analyses the effects of social isolation on the usage and perception of UGS during and after the COVID-19 pandemic. The research revealed current needs for UGS and their role in adaptation of urban development and greening strategy. For this purpose, an online questionnaire survey among residents of Moscow was conducted in April–July of 2020 when restrictive measures were imposed in the city in response to the COVID-19 pandemic. Additionally, non-participatory observations and photo documentation were used to supplement the data on UGS visitation and use. The GIS mapping method was applied to analyze the UGS provision (availability and accessibility of UGS). Moreover, expert interviews were conducted aiming to explore the implications of the COVID-19 pandemic on the urban fabric and life of the citizens. The aim was to reveal the main tendencies that can be used in the adaptation of urban development plans, especially regarding UGS and human-nature interactions. The results show that citizens (both survey respondents and experts) highly value urban nature as a tool for coping with COVID-19 challenges. They underlined a need for accessible UGS, most notably for breathing fresh air, reducing stress, relaxing, and observing and enjoying nature. The survey also revealed the particular health effects resulting from the reduction of UGS visitations due to COVID-19 restrictions. Several changes in human-nature interactions were also observed: many respondents especially missed spending time outdoors and meeting other people. That highlights the fact that while UGS normally provides places for social integration and socializing, during the COVID-19 isolation UGS were especially valued in regard to physical health and well-being (self-recovery). Both respondents and experts expressed their opinions regarding the future development of UGS network and how the UGS's structure and design should be adapted to the current challenges. The claimed interests/preferences included the need for providing all residents equal access to UGS in a time of pandemics and post pandemics. A set of limitations and directions for future research of UGS was suggested.

Keywords: green infrastructure; urban green space; human-nature relations; COVID-19 pandemic; green recovery in a (post) COVID world

---

1. Introduction

Urban green spaces (UGS) such as parks, gardens, forests, etc. are a vital element of the urban environment that provides healthy, sustainable living conditions and enhances the quality of life [1–3]. Defined as all urban land covered by vegetation of any kind [4], UGS include a diversity of ecosystems and their ecological processes, which support urban blue and green infrastructure and provide multiple ecosystem services (ES). Many recent studies confirmed that contact with nature in vegetated and water-rich urban areas improves people's physical and mental health by reducing stress in everyday life [3,5,6]. It also stimulates physical activity [7,8], enhances human wellbeing and mental state, and improves the quality of life [9–15]. Contact with nature facilitates social cohesion and inclusion [3,16–18], contributes to a sense of place, and shapes regional identity [19–21]. UGS increase resilience to climate change and environmental shocks by moderating the urban heat island effect, noise, chemical pollution, and excessive storm water run-off. UGS promote biodiversity conservation and sustainable lifestyles [1,22–25].

Accelerated urbanization and current societal challenges, especially during the COVID-19 pandemic and associated social isolation, increased stress and spatial recalibration of everyday life. It is also demonstrated that urban nature can play an essential role for the resilience of urban society by providing significant positive impacts on the physical and mental health of individuals and communities [2,26,27]. The extraordinary circumstances surrounding the worldwide coronavirus (COVID-19) pandemic, when billions of people over the globe were locked down with limited access to UGS, highlighted the importance of contact with urban nature for human well-being and shaping human-nature relationships and the recovery in the post-pandemic world [28–32]. In order to address the declaration of WHO measures to slow or prevent the spread of COVID-19, many cities worldwide had to implement various restrictions that affect individual mobility and public life, including limited access to services and facilities outside their neighborhoods, whilst lowering the intensity of their usual physical activity. In several cities, it also resulted in the closure of some public parks and other open spaces to help encourage physical distancing [29,33,34]. Thus, UGS within local neighborhoods has become more important than ever not only in hosting people's outdoor activities but also in allowing secure socializing [28,33,35]. In this regard, the quality of UGS and their availability, green planning and landscape design characteristics should receive special attention.

This study aims to explore the pathways and implications of human-nature interactions during and after the COVID-19 pandemic and how health problems and pandemic challenges can be mitigated through contact with nature (therapeutic value of nature). In particular, the study addresses the following research questions: (a) how COVID-19 impacted human life and personal well-being and what activities were undertaken by citizens to cope with COVID-related challenges; (b) what are the reasons that normally attract visitors to UGS (e.g., claimed as perceived personal benefits from UGS, value of UGS/role of nature, ways of contact with urban nature, etc.); (c) how COVID-19-related restrictions changed the human-nature interactions (visitation, usage, and availability of UGS); (d) what are the current preferences and needs for UGS and what are COVID-related adaptations of UGS to sustainable planning and management strategy. Following previous studies on effects of the COVID-19 pandemic on the use and perceptions of urban green space [2,26–28,31,33–35], UGS accessibility and availability [27,29,30,32], value of UGS from the planners' perspective [28,34], we conisder a wide diversity of human–nature interactions and how COVID-19 can affect their dynamics. Moreover, we aim to explore what novel actions and adaptation strategies can be established that can have positive outcomes for both humans and nature. There is an increasing number of studies on COVID-19 and UGS use published in 2020–2021; however, they are still dominated by European Union countries, followed by the USA. Insights from other countries, including Russia, remain underrepresented. In this sense, Moscow presents an interesting case, which can provide a different perspective. Moreover, this exploratory study adds to the literature by implementing an integrated approach of combining a social survey (as a pilot study) with expert interviews, GIS mapping and non-participatory observations providing the insights from UGS perception and use in a megacity demonstrating the impact of COVID-19 and how it is addressed in the city greening strategy. In particular, an online questionnaire survey among residents of Moscow has been conducted in May–July of 2020 when restrictive measures were imposed in response to the COVID-19 pandemic. In addition, non-participatory observations and photo-documentation were used to supplement the data on park visitation and the use of UGS, as well as GIS mapping to analyze the availability and accessibility of UGS. Other methods were semi-structured interviews with experts in January–March 2022 aiming to reveal the main tendencies in the adaptation of urban development plans to planning, design, and use of UGS during the pandemic.

Our prime hypothesis is that regular contact with nature and a variety of human-nature interactions is vital in ensuring resilience and sustainability under the (post)-COVID conditions and in times of continuous decline of nature in cities. The paper has the following structure: we begin by reflecting on the results of the questionnaire survey and non-participatory observation with photo-documentation regarding the human-nature interaction, e.g., the subjective values assigned to UGS and their importance for mental and physical well-being, especially during and after the COVID-19 pandemic. Then, we discuss the issues of UGS access and preferences, followed by suggestions for designing UGS. Finally, we compare the survey results with the UGS provisions and benefits maps based on the spatial proxies (e.g., availability of UGS, the total area of parks/maintained UGS, nearest distance to parks, averaged land surface temperature, number of tree/bird species, and urban heat stress index). We also elaborate on the role of UGS in adaptation to the new circumstances based on the analysis of expert interviews. This analysis also revealed the main tendencies in the adaptation of urban development plans regarding planning, design, and use of UGS. We also discussed some limitations and directions for future research, which are vital for creating resilient and livable cities promoting healthy lifestyles and habitats after the pandemic.

2. Materials and Methods

The study involved a three-step process (Figure 1) starting with the pre-investigation phase (preliminary research) aiming to identify the key issues related to research on human-nature interactions and identifying research gaps. That is followed by suggestions on how they can be covered by the presented research (identifying research questions and methodology) using a literature review. The next step refers to data collection and analysis. For this purpose, different methodologies were used (questionnaire survey, non-participatory observation, GIS mapping, and interviews with experts), and each of them refers to the particular research question (see main outcomes at Figure 1). The final step deals with data interpretation and comparison with previous studies as well reflecting on limitation and identifying directions for future research.

	RESEARCH STEP / METHODOLOGY	DESCRIPTION / ACTIVITIES	MAIN OUTCOMES
Preinvestigation	Defining research design / literature review	- Literature review in SCOPUS, ISI Web of Sciences, Google Scholar - Revealing the methods used in the related research and their main outcomes. Identifying research gap	Key issues related to research on human-nature interactions and therapeutic value of nature (also during pandemic) are identified
Preinvestigation	Premises for conducting research / literature review	- Defining the preliminary list of research gaps and suggestions how they can be covered by this research - Developing research framework/design and its main steps	Research goal and questions are formulated Research framework is elaborated
Data collection and analysis	Questionnaire survey	- Exploring: a) impact of COVID-19 on human life and well-being and activities which helped to cope with them; b) perceived personal benefits from UGS, value of UGS; c) ways of human-nature interactions and their changes due to COVID-19; d) needs for UGS.	Analysis of the respondents' answers on personal responses to COVID-19, activities to cope with them, UGS role/value, use, needs
Data collection and analysis	Non-participatory observations, photo-doc	- Applying the observational techniques to explore the use of UGS by people to supplement the questionnaire survey data	A set of photographs and recorded activities in UGS
Data collection and analysis	GIS mapping	- Geo-coding: extracting spatial data for each geo-coded point (e.g. availability of UGS, its area, heat stress, sealing, etc.)	Maps showing location of survey respondents, UGS availability, heat stress
Data collection and analysis	Inetrviews with experts	- As follow-up of questionnaire, semi-structured interviews with experts allowing to reveal the main tendencies in the adaptation o urban development plans regarding design and use of UGS	Linkages between claimed citizens' needs and plans for UGS adaptation and design
Finalizing	Discussion	- Comparing findings with the existing concepts and results - Reflection on the research hypothesis	Evaluated results and their relation to the reviewed literature and research questions
Finalizing	Conclusion	- Developing future directions for research	Contribution to the knowledge base

Figure 1. Research framework.

2.1. Study Area

The investigation was conducted in Moscow, Russia's capital city (Figure 2), which covers an area of 2500 square kilometers and is inhabited by over 12.5 million citizens (including the New Moscow district) (2018 census). It is the biggest Russian city and center of political, economic, and cultural life. The city is situated on the banks of the Moskva River. The climate is humid continental with long, cold winters usually lasting from mid-November to the end of March, and warm summers. The cold climatic conditions (snow is present for about five months a year, from mid-October to the beginning of April) and the recent change in Moscow's regional climate due to global warming (extreme heat is more frequent in the city and has reached 37.8–38.2 °C) act as limiting factors for outdoor recreation. Nevertheless, the city is characterized by a high percentage of green spaces (55% of the city territory). Green infrastructure framework dated back to 1935 when the initial City Master Plan was implemented. This Master Plan defined the protection and management of a Green Belt around the city, the establishment of seven major green zones stretching from the outskirts to the center, and a connecting system of boulevards and parks. During the Soviet period (1917–1991), UGS and their network were an important part of spatial planning. At the beginning of the post-Soviet time (1990's), many UGS were neglected, which is linked to the economic and political changes and instability. The administrative reform of 2012 resulted in a doubling of the urban area. Due to city limits, expansion arable lands occupy now the second place by size (12.43%). When comparing "Old" Moscow with the "New" Moscow, the total share of UGS is almost three times higher in New Moscow (75%) than in the "Old" part (28%). Moscow is an exceptional case of a European city where both inherited features of central planning from the Soviet time and modern methods of greening were integrated.

With 14 square meters of green spaces per capita, Moscow could be considered a city with a well-developed green infrastructure. The city includes the following types of UGS: (1) remnants of native vegetation (recreational forest and forest parks are the biggest UGS type making up 33% of the total area), (2) historical gardens and parks as the monuments of landscape architecture, (3) public district and local community parks and green spaces within local neighborhoods (residential block of houses), (4) green plazas and boulevards in neighborhoods, (5) alley and street green in all parts of the city, (6) rain gardens, swales, and other bioretention facilities to manage stormwater and rainwater run-off by planting native flora into depressions or channels; (7) natural and designed bogs in the nature parks and the residential districts, (8) rivers and canals as public waterways for recreational purposes, (9) ponds and creeks for recreational purposes, (10) specific blue spaces—artificially created water reservoirs that are used for recreational purposes. A great number of public parks and residential green spaces was established in the Soviet era. Moscow has an extensive forest-park zone for short-term recreational use based on native forests but containing planned and designed elements. Usually, natural forests in Moscow are part of specially designed nature-protected areas. The dominant protected areas in Moscow are natural and historical parks (10 parks in total) [33].

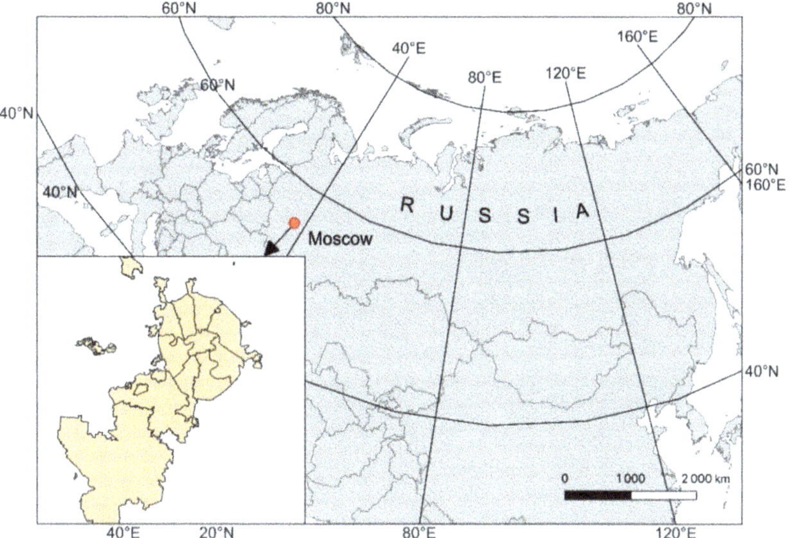

Figure 2. Study area.

2.2. Social Survey: Data Collection and Processing

Data on Moscow residents' perceptions of the personal impact of COVID-19-related restrictions and values of UGS was received via an online survey. An online survey tool Survio (survio.com (accessed on 20 May 2020)) was used to provide a fixed item, anonymous questionnaire. The questionnaire was composed of 25 questions written in the respective local language (Russian).

The survey focused on the use of UGS and their value augmented during and after the first wave of the COVID-19 in Moscow. The questionnaire was active from May to July 2020.

The general components of the questionnaire were composed according to the following logic. The first step was related to the area of residence (to confirm the study area—Moscow city) of the respondent and the duration of lockdown (number of days that the respondent spent at home). The second part of the questionnaire rated the subjective values assigned to UGS using multiple-choice questions with options that were derived from the scholarly literature [36,37]. The importance of UGS for mental and physical well-being was assessed using six-point Likert scale questions (from very unimportant to very important and additional options for answering "I don't know"). The third component included the subjective assessment of perceived impacts on daily routines and personal wellbeing of COVID-19 related restrictions such as lockdown, social distancing, working from home, and combining full-time remote (home) working and home schooling of children. In the fourth step, self-reported impacts of COVID-19 restrictions on access to UGS were examined. Three questions asked respondents to indicate from a list of options how frequently they accessed UGS before, during, and after COVID-19 restrictions respectively. The fifth step was to find out the abundance and quality of UGS in the respondents' local neighborhoods. A filter question with three options (agree, disagree, and don't know) was used to assess the sufficiency of UGS. Then, subsequent question requested the respondent's preferred additional UGS types by selecting options from a list (including "nothing" and "other") [38,39]. The final part consisted of demographic questions such as gender, age, level of education, current occupation, resident postcode, dwelling type, and the number of people living in the household. The survey template can be found in Questionnaire S1 of Supplementary Material.

To recruit participants, a targeted advertisement was created on the Vkontakte social network (vk.com (accessed on 20 May 2020)). Vkontakte is one of the most popular and frequently used social networks in Russia. According to the data presented by the network in 2020, 78% of Russian residents visit VKontakte at least once a month and 50% of residents visit the social network at least once a day (https://vk.com/press/q2-2020-results (accessed on 18 May 2022)). An important criterion is the representation of different groups of users by age categories that is more differentiated than on Instagram or Facebook. The target audience was limited only by the factor of residence in the Moscow agglomeration. This was important to ensure the participation in the survey of people with different socio-demographic characteristics, views, and life experiences. In total, 119,385 people have seen the advertisement, 988 of them visited the entry page of the survey, and 280 of them completed the survey. After reviewing the results, 59 respondents were excluded as they were not from Moscow. The final sample included 216 respondents.

Analysis of the data was conducted using the SPSS software (IBM SPSS v24). This analysis included descriptive statistics, χ^2 test and nonparametric tests (such as Mann–Whitney U and Kruskal–Wallis tests) for ordinal data.

2.3. Non-Participatory Observations and Photo-Documentation

In order to supplement the data on visitation and use of UGS, non-participatory observations and photo-documentation were conducted following the approach of Clark et al. [40] providing observational techniques for exploring the use of UGS by different social and activity groups. Non-participatory observation methods are widely used by experts to capture and assess human/local visitors' behavior without taking an active part in the situation under scrutiny [21,40–42]. During the observations, different types of UGS (public parks, gardens, urban forests, and small UGS such as playgrounds, alleys, boulevards, sport facilities, and front- and backyards of multi-story housing complexes) were visited randomly at different times of the day (between 09:00 and 19:00) in working days and weekends in two weeks in July 2019 and July 2020 (during COVID-19 lockdowns most of them were closed) to survey the types of visitors' activities that took place in these UGS. In total, five big urban parks and two forest-parks (Zaryadye, Gorky Park, Neskuchny Sad, Sokolniki, Bitzevsky Park, Timiryasevsky park, Kolomenskoe Park, Park VDNKh of All-Russian Exhibition Center) as well as selected alleys, boulevards, front- and backyards, sport facilities, and playgrounds of several housing complexes from different districts of Moscow were visited randomly.

The observers selected an observation point with a clear view of the entire place (where possible) in each UGS or selected several points in the big parts according to the functional zones of the parks and recorded all activities taking place in the spaces during observation hours using special fieldwork protocols.

2.4. GIS Mapping

Since the Survio service does not allow geotagging, we geo-located the respondents according to their postal codes. Moreover, some respondents have indicated their full home address within the questionnaire. We assigned the geospatial coordinates of the house in case the address has been provided. Otherwise, we assigned the coordinates of the postal office with the mentioned postal code to each questionnaire. The geo-coding was performed using python 9.3. The number of respondents that indicated their postal code/address was 147 out of 216. For each geo-coded point, we have extracted spatial data representing the environmental conditions/quality of the surroundings. These spatial data were averaged within the regular hexagonal grid (size ~ 48 ha) to better characterize the surrounding area. These spatial data include: availability of UGS (green stands, lawns, shrublands) per capita (m^2 pers^{-1}), the total area of parks/maintained UGS, nearest distance to parks, averaged land surface temperature, urban heat stress index, the proportion of the sealed area.

Spatial distribution of UGS/sealed surface was retrieved from satellite data. Land and cover classes were identified based on the mosaic of cloud-free Sentinel-2A MSI (level 2A), and optical satellite images were taken on 6 June 2019, available at ESA Copernicus Scientific Data Hub. The image pre-processing included resampling to 10-m spatial resolution of 10 bands with initial 10 and 20 m spatial resolution and mosaicking. The stepwise sub-pixel and per-pixel classification was performed to classify the surface into the following land cover classes: water, sealed areas, bare soil, lawns/grasslands, trees, and shrubs based on spectral signatures of these classes. The pre-processing and classification were performed within the Google Earth Engine cloud computing platform [43]. The population data were parsed from the open data portal of the Department of Housing and Utilities. The vector park contours were downloaded from the open data portal of the Moscow government (www.data.mos.ru (accessed on 15 April 2022)). The land surface temperature was calculated based on the mosaic of Landsat-8 (thermal infrared sensor, TIRS) satellite data (taken 6 July 2021). The spatial analysis was performed in the python 3.9 environment. Urban heat stress index was based on the physical equivalent temperatures [44,45] derived from the air temperature simulation by the COSMO-CLM model. The model was calibrated for the Moscow city environment [46,47] and projected heat stress for July 2010 (one of the hottest months in history of meteorological monitoring in Moscow) with 500 m spatial resolution [48].

2.5. Expert Interviews with Different Stakeholders

In addition to the questionnaire survey, which reflects the perspectives and opinions of citizens in general, the semi-structured interviews with the experts were conducted to allow a deeper exploration of specific aspects derived from the survey. In particular, the interviews aimed to reveal what are the main implications for urban planning and development resulting from the COVID-19. The interviews were also intended to explore specific urban planning strategies to address the current needs of citizens and issues of urban sustainability, resilience, and human-environmental interactions adapted to the new reality (post-COVID time). In the search for experts, the main selection criterion was a thorough knowledge of urban development, planning and environmental policy as well as practical implementation of greening projects in present and in the recent past. Using the snowball technique, 13 expert interviews were conducted (both face-to-face and online) with different stakeholders of Moscow in January–March 2022. From the experts who participated in the study, three experts were from public authority and decision making, four from academia and research institution, three from landscape design and architecture, and three from NGO and citizen groups. The interviews with experts ranged from 19 to 56 min, were recorded using a digital device and further transcribed; the process of manual inductive coding was applied. The experts according to Meuser and Nagel [49] were persons possessing institutionalized authority and knowledge with the potential of conditioning the actions of others in a meaningful way. Therefore, expert interviews facilitate gaining insights and context knowledge central to the research questions that cannot be deduced from other methodical approaches. A group of stakeholders (experts) was presented by representatives from public authorities and decision-makers (city administration, city planners), universities and research institutions (scientists), practitioners (architects and landscape architects), and civil society groups/organizations (representatives from NGOs, urban communities, volunteers).

In our interviews, we asked the experts to provide their perspectives on the following questions:

(a) what are the main lessons from the city's responses to COVID-19 regarding the development of the city and the change in the lifestyle of citizens?
(b) how COVID-19 pandemic influences the design and development of urban spaces in Moscow in general and regarding UGS and human-nature interactions?

The obtained data were analyzed to reveal the main tendencies in the adaptation of urban development plans regarding the use of UGS and to formulate the proposals for UGS management strategies.

3. Results

3.1. Impacts of COVID-19 on the Population (Citizens) Based on a Questionnaire Survey and Non-Participatory Observations

3.1.1. Respondents' Characteristics

The demographic characteristics of the survey respondents are presented in Table 1. The sample was dominated by women (73.0% vs. 27.0%). The age range distribution was characterized by two dominant categories 21–30 years old and 31–40 years old. The majority of respondents (80%) had a university diploma or higher obtained level of education. More than 40% were full-time employed. This was the dominant category along with studying at university (19.7%) and part-time employed (13.8%). More frequently, Moscow dwellers lived in a flat, apartment, or townhouse (81.4%). The prevalent number of families had no children and lived in the respondent household—63.9%.

Table 1. Demographic characteristic of respondents (*n* = 216 Moscow citizen).

Characteristics	Share of Respondents
Gender	27%—male 73%—female
Age	10.6%—less than 20 years old 31.9%—21–30 years old 26.4%—31–40 years old 12.0%—41–50 years old 10.2%—51–60 years old 8.8%—more than 60 years old
Highest obtained level of education	9.3%—School diploma 6.5%—Secondary special education 67.6%—University graduation (BSc., MSc., diploma) 13.4%—Post-university graduation (PhD., Dr., or similar) 3.2%—Other
Type of housing	13.9%—House 81.4%—Flat/Apartment 4.7%—Other
Current employment status	40.5%—Full time employed 13.8%—Part time employed 5.9%—Casual employment 4.5%—Stay at home (home duties) 3.7%—On parental leave 4.8%—Unemployed 2.2%—Temporarily laid off due to COVID-19 19.7%—Student 4.8%—Retired

Table 1. *Cont.*

Characteristics	Share of Respondents
Living/family circumstances	63.9%—Family without children 15.5%—Family with 0–6-year-old children 12.9%—Family with 7–12-year-old children 7.7%—Family with 13–17-year-old children

3.1.2. Impact of COVID-19 on Moscow Residents

The first wave of COVID-19 related restrictions in Moscow started on 26 March 2020. There were some measures in the first month of the pandemic where playgrounds, restaurants and cafes, and sports infrastructure were closed, universities were switched to remote mode, and social distancing and self-isolation for aged people were recommended. Self-isolation mode was prolonged until June 2020. However, in April and May, the measures in Moscow became stricter, including not only the closure of 55 public parks and banning UGS visits but a full lockdown for 1.5 months (Figure 3). The self-isolation regime (home quarantine) in Moscow lasted from 30 March to 9 June 2020. There were few exceptions to the restrictions that allowed going outside, for example, a necessity to be at the workplace, shopping at the nearest food store or pharmacy, walking pets within a distance of 100 m from the place of residence, taking out the garbage, seeking emergency medical help, or a direct threat to life and health. From 12 May 2020, wearing protective masks and gloves was mandatory in public transport and other public places. The majority of respondents answered that COVID-19 restriction measures such as lockdowns and working from home have not influenced their wellbeing: lockdowns (43.1%), working from home (44.4%) (Figure 4a). A slightly negative impact was identified in the case of social distancing (37.5%). At the same time, the impact of combining full-time remote (home) working and home schooling of children was statistically significant for families with 0–6-year-old children ($\chi^2 = 18.795$, $p = 0.002$) and with 7–12-year-old children ($\chi^2 = 28.282$, $p = 0.000$), while for families with 13–17-year-old children it was not ($\chi^2 = 1.969$, $p = 0.853$). For families with 0–6-year-old children, the impact of combining full-time remote (home) working and home schooling of children has two dominant categories: slightly negative (39.4%) and neutral (39.4%) (Figure 4b). For families with 7–12-year-old children, the most frequent answers were the same, but about a quarter of respondents (24.1%) defined the impact of COVID-19 restrictions as very negative.

Figure 3. *Cont.*

Figure 3. Closure of many parks and public spaces in Moscow (first line) and neighbourhood UGS (playgrounds, second line) due to COVID-19 restrictions, March–May 2020 (signboards indicating prohibiting visits to parks and playgrounds): Park Zaryadye (**top left**), Park VDNKh (**top right**), playgrounds in the multi-story housing complexes in the southern district of Moscow (**bottom images**). Photos: E. Kosenkov.

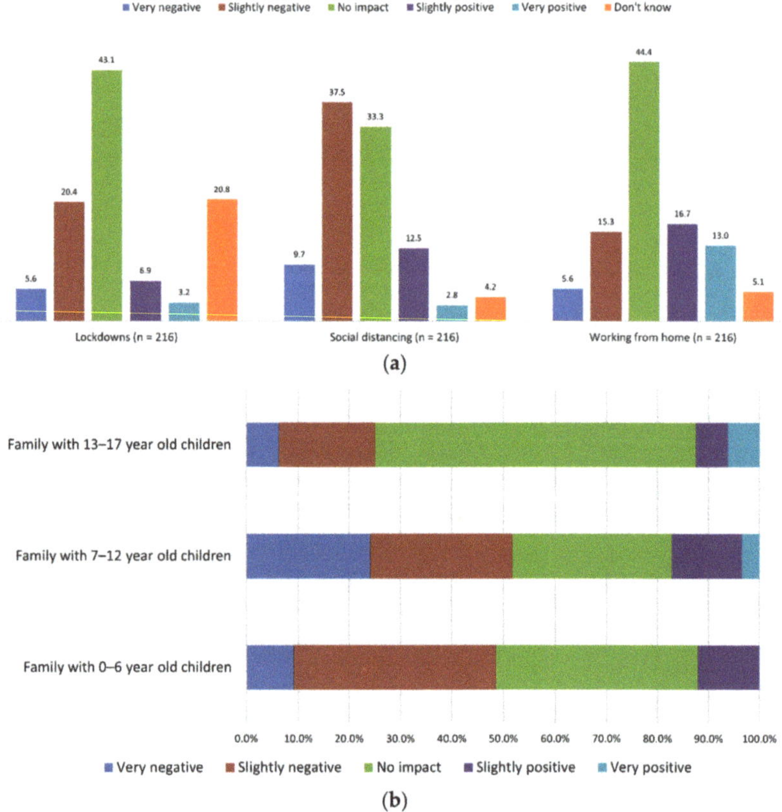

Figure 4. Impact on respondent's personal wellbeing of COVID-19 restrictions: (**a**) Lockdowns (*n* = 216), Social distancing (*n* = 216), and working from home (*n* = 216) and (**b**) combining full-time remote (home) working with home schooling of children (*n* = 171).

The situation with the impact of COVID-19 restrictions on daily routines was different. In the case of lockdowns, it was assessed as more negative, 42.6% of respondents called the impact slightly negative (Figure 5a). Social distancing on the contrary was identified as not so negative for daily life: neutral for 37.5% of respondents and slightly negative for 30.1%. The impact of working from home was in total more positive for 38.0% of respondents (slightly positive for 21.8% and very positive for 16.2%) or neutral for 34.7% of respondents. For families with 0–6-year-old children ($\chi^2 = 22.735$, $p = 0.000$) and with 7–12-year-old children ($\chi^2 = 22.653$, $p = 0.000$) the impact of combining full-time remote (home) working and home schooling (standing) of children was a real challenge. The majority of the respondents (50.0% and 36.7% respectively) assessed it as slightly negative and 20% of respondents with 7–12-year-old children have chosen the "very negative" category (Figure 5b). At the same time, for more than half (64.7%) of families with 13–17-year-old children it was not statistically significant ($\chi^2 = 4.411$, $p = 0.492$), and they have not seen any changes.

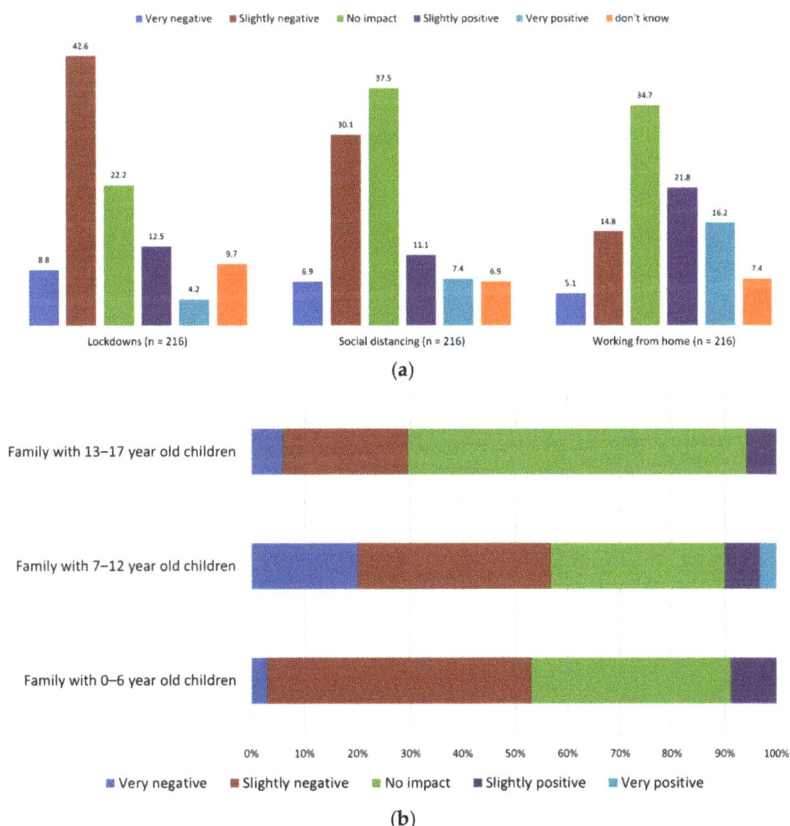

Figure 5. Impact on respondent's daily routings of COVID-19 restrictions: (**a**) Lockdowns ($n = 216$), Social distancing ($n = 216$), and working from home ($n = 216$) and (**b**) combining full-time remote (home) working with home schooling of children ($n = 171$).

There was not a significant difference in activities that helped to cope with COVID restrictions. They were depending on the respondents' age: I take a walk (if it is allowed) ($\chi^2 = 4.569$, $p = 0.471$); I do some outdoor physical/mental exercises ($\chi^2 = 3.406$, $p = 0.638$); I do some physical/mental exercises at home ($\chi^2 = 1.375$, $p = 0.927$); I am connected with the world via Internet ($\chi^2 = 3.554$, $p = 0.615$); I have developed/undertaken activity which helps me to reduce stress (χ^2 = 4.794, p = 0.442); Other (χ^2 = 2.530, p = 0.772). For two age categories (31–40 years old and 41–50 years old) the most frequently selected activity to cope with COVID-19 restrictions was taking a walk (33.0% and 29.3% respectively) (Table 2). For other age groups including respondents younger than 30 and older than 50, the dominant activity was doing indoor physical or mental exercises.

Table 2. Activities to cope with COVID restrictions according to the age of respondents (n = 216).

Activities	<20 Years Old	21–30 Years Old	31–40 Years Old	41–50 Years Old	21–60 Years Old	>60 Years Old
I take a walk (if it is allowed)	28.2%	28.7%	33.0%	29.3%	28.9%	24.1%
I do some outdoor physical/mental exercises	7.7%	7.0%	5.7%	7.3%	2.6%	0.0%
I do some physical/mental exercises at home	30.8%	33.9%	26.4%	29.3%	31.6%	37.9%
I am connected with the world via Internet	17.9%	13.0%	18.9%	14.6%	23.7%	20.7%
I 've developed/undertaken activity which help me to reduce stress	10.3%	9.6%	11.3%	9.8%	5.3%	3.4%
Other	5.1%	7.8%	4.7%	9.8%	7.9%	13.8%

3.1.3. Importance of Contact with Nature for Physical and Mental Well-Being

All respondents highly valued UGS for physical and mental well-being. The importance of contact with nature was not dependent on age group. Mental health and well-being (χ^2 = 25.311, p = 0.445) noted at least by 70% of the respondents (Figure 6a) and physical health and well-being (χ^2 = 17.439, p = 0.865) at least by 56% of the respondents (Figure 6b). At the same time, there is a tendency: the older the respondents, the more positive answers (related to the importance of nature) they choose. Participants from age groups of 51–60 and 60+ assessed the role of UGS exclusively positively, especially for mental health and well-being (73.7% of 60 + years old and 81.8% of age group 51–60 years old). Among the younger groups of respondents, the variability of the chosen answers increases. However, the total absence of impact ("Not at all important" category) is noted only by 1.4–4.3% of respondents younger than 40, and an insignificant impact ("Slightly important" category) by 3.5–13.0% of respondents of the same age.

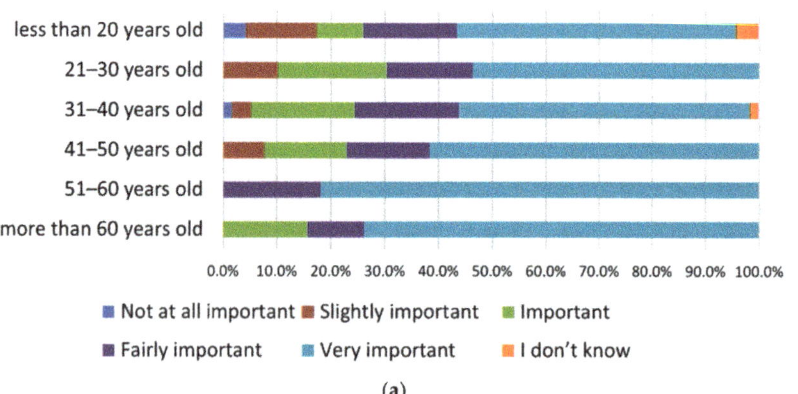

(**a**)

Figure 6. *Cont.*

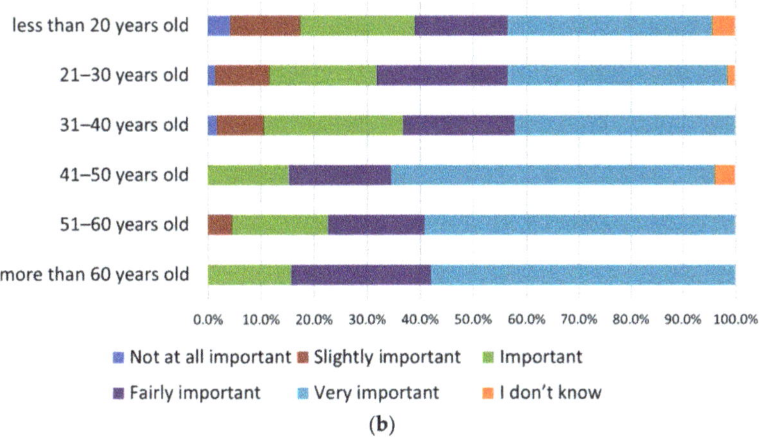

Figure 6. Importance of UGS for (**a**) mental health and well-being (n = 216) and (**b**) physical health and well-being (n = 216) according to subjective views of different age groups.

An analysis of responses on the value of nature showed that, depending on age, the results associated with social activity. "Meet other people" (friends, relatives, or simply to feel social inclusion) is statistically significant (χ^2 = 23.505, p = 0.000); "Have fun" (χ^2 = 19.861, p = 0.001). Additionally, the variable "Spend time with family and visiting playgrounds" is statistically significant and it is not only dependent on an age group (χ^2 = 12.904, p = 0.024), but also family status—having children (U = 3315, p = 0.000). Among perceived personal benefits of UGS use (Figures 7–10), the majority of respondents (12.3–16.8%) in all age groups named "breathe fresh air" (Table 3). The most frequently selected benefits are also "mental health benefits (to reduce stress etc.)" (12.2–12.8%), "enjoy scenic beauty" (11.3–13.6%), and "a place to relax and unwind" (11.1–14.4%). These categories did not depend on the age group of the respondents and rather expressed the unity of views of Moscow residents on UGS and their benefits for people.

Table 3. Perceived personal benefits of UGS by Moscow residents of different ages (n = 216).

Values to Be in Nature/ Urban Green Space	<20 Years Old	21–30 Years Old	31–40 Years Old	41–50 Years Old	21–60 Years Old	>60 Years Old
A place for physical exercise (jogging, running, etc.)	8.2%	8.4%	9.6%	7.2%	10.4%	11.2%
Breath fresh air	12.3%	14.9%	15.0%	16.8%	16.5%	16.3%
Sun bathing	4.1%	4.6%	5.4%	3.2%	6.1%	6.1%
Mental health benefits (to reduce stress etc.)	12.3%	12.2%	12.6%	12.8%	12.2%	12.2%
A place to relax	11.6%	11.9%	11.1%	14.4%	12.2%	14.3%
A safe place to be	4.8%	4.8%	4.2%	4.0%	5.2%	6.1%
Spend time with family and visiting playgrounds	2.1%	2.0%	5.7%	7.2%	3.5%	6.1%
Meet other people (friends, relatives, or simply to feel social inclusion)	9.6%	4.8%	3.3%	1.6%	4.3%	2.0%
Escape from the urban environment	8.9%	9.1%	8.7%	8.8%	7.8%	4.1%
Connect with nature	6.8%	9.6%	9.9%	8.8%	9.6%	8.2%
Have fun	6.8%	4.6%	2.4%	1.6%	0.9%	1.0%
Enjoy scenic beauty	12.3%	13.2%	12.3%	13.6%	11.3%	12.2%

Figure 7. UGS use for breathing fresh air, relaxing, and escaping from the crowded city (Photos: D. Dushkova, A. Konstantinova).

Figure 8. UGS for spending time with family and children (Photos: D. Dushkova).

Figure 9. UGS as a space for physical exercises and outdoor sport activities (Photos: D. Dushkova).

Figure 10. UGS for socializing (meeting other people) (Photos: D. Dushkova).

3.1.4. Changes in UGS Visitation

The change in UGS visitation before and during the pandemic was especially evident due to the strong lockdown and restriction of park visits and the use of playgrounds and sport facilities (Figures 11 and 12). Almost 30% of respondents have never visited UGS during the lockdown and 12.5% visited less than once per month (Figure 13). At the same time, some Moscow residents retained the frequency of visits despite the limitations; for example, 23.1% visited UGS a few times per week. After the pandemic, an increase in the number of everyday visits and a few times per week visits (almost 7%) was identified. It was higher than in the pre-pandemic period. Such visits to UGS respect hygienic norms and rules for social distance (Figure 14).

Figure 11. Closure of urban parks due to COVID-19 restrictions.

Figure 12. COVID-19 restrictions for use of playgrounds and sport facilities.

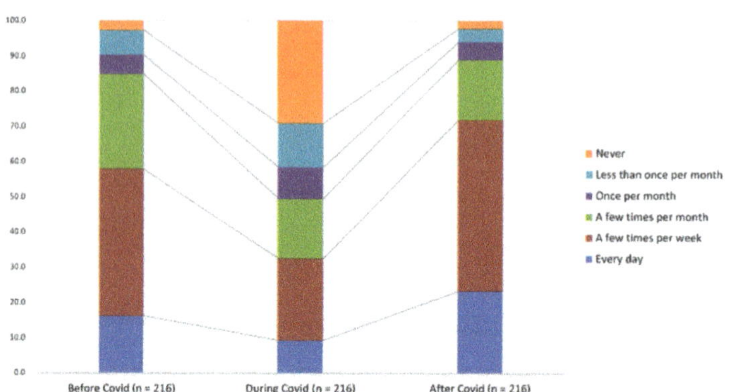

Figure 13. Frequency of UGS visits before, during, and after the COVID-19 restrictions (*n* = 216).

Figure 14. Social distancing (signboards indicating respect for social distancing of no less than 1.5 m) in Zaryadye park (**first line**) and Kolomenskoe Park of Moscow (**second line**). Photos: D. Dushkova, A.Konstantinova.

3.1.5. UGS Demand, Quality, and Preferences

In Moscow, the demand for UGS can be rated as rather high. More than 65% of respondents agreed that the city needs more UGS (Figure 15). At the same time, 47.2% assessed the quality of existing UGS as inadequate, and 15.3% could not assess UGS at all.

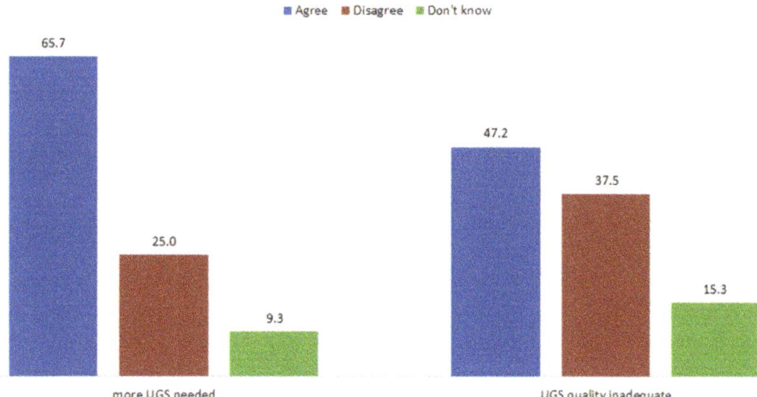

Figure 15. Demand for UGS (n = 216) and its quality (n = 216) according to Moscow residents' opinion.

Regarding the current needs for additional types of UGS, statistical significance was noted for children's playgrounds. The data is depending on the presence of children in the family (U = 4261, p = 0.000) and especially in families with 0–6-year-old children (χ^2 = 16.538, p = 0.000). For families with 7–12-year-old children, pocket parks and small gardens/squares for silent recreation are more desirable types of UGS (χ^2 = 4.633, p = 0.031). Otherwise, the dependencies are not determined. Most respondents made their choice in favor of greening in their neighborhood or district (Table 4). The most frequently selected types of UGS were "Green front- and backyards" (10.8–14.8% depending on the family type), "More green spaces near your house (trees, shrubs, etc.)" (one of the most popular answers for families with 0–6-year-old children–12.2%), "More pocket parks and small gardens/squares for silent recreation" (11.7–14.8%). At the same time, more street greening was indicated as the most desired measure (11.5–16.7%). Another frequently selected measure by respondents was to transform vacant lots and brownfields near their houses into more designed and managed green spaces.

Table 4. Additional required types of UGS mentioned by respondents (n = 216).

Types of Green Spaces	Family with 0–6 Year Old Children	Family with 7–12 Year Old Children	Family with 13–17 Year Old Children	Family without Children
Green front- and backyards	10.8%	12.2%	14.8%	12.2%
More green spaces near my house (trees, shrubs, etc.)	12.2%	7.8%	7.4%	11.3%
More playgrounds for children	7.9%	4.3%	3.7%	2.4%
More green space for active recreation (also with installation for physical exercises)	3.6%	7.0%	7.4%	6.6%
More public parks	10.8%	8.7%	7.4%	11.3%
More street greening	11.5%	13.0%	16.7%	13.3%
More community gardens	9.4%	9.6%	11.1%	11.1%
More school and kindergarten green spaces	8.6%	9.6%	5.6%	7.7%
More pocket parks and small gardens/squares for silent recreation	12.2%	14.8%	13.0%	11.7%
To transform vacant lots and brownfields which we have near our house	11.5%	11.3%	11.1%	9.7%
Other	1.4%	1.7%	1.9%	2.7%

3.2. UGS Provision (Availability and Assessibility of UGS) in Moscow Based on GIS Mapping

UGS cover more than half of the Moscow city area, whereas 38% of the city is sealed. The percentage of green spaces is comparable to such cities as Rostov-on-Don [50], Milan [51], and Madrid [52], but it is less than in Berlin [53]. The major part of the sealed areas is concentrated in the central and eastern parts of the city, which were historically occupied by industrial areas, whereas the north-west and south parts are sealed by less than 25% and considered more environmentally friendly and comfortable for life. Respondents live in areas that are 9.7–72.5% sealed (median $\pm$ sd = 41.6 $\pm$ 14.1%) (Figure 16). The level of soil sealing did not have a particular effect on the presence ($n = 16$) or absence ($n = 131$) of outside sport activity during lockdown (paired t-test, p-value = 0.614). Respondents that indicated the outside sport activity in their everyday life live closer to parks (181 m on average compared to those who do not—288 m, $p = 0.05$). Similarly, the permeable surfaces in the neighbourhoods have not affected answering positively ($n = 54$) or negatively ($n = 71$) concerning the demand for new UGS during the lockdown (paired t-test, p-value = 0.21). Respondents that indicated the need for additional UGS live closer to parks (190 m, $n = 36$) than those who answered negatively (278 m, $n = 100$), $p < 0.05$. Herewith, the overall UGS area in the surroundings did not affect answers of respondents. The nearest distance to parks had a minor influence on how often respondents visited these parks (Figure 17). Almost 90% of the respondents had a green space of at least 2 ha within 8 min walking distance. This fact indicates the high accessibility of green spaces in Moscow.

The availability of UGS in the central part of the city is lower compared to the suburbs and as a rule, it is below 50 m^2 per capita, which is considered in a health threshold/guideline for Moscow [54,55]. The highest availability of UGS was reported for New Moscow, south-western and north-eastern suburbs, where the close proximity to urban forests (e.g., Elk Island or Bitsevsky Forest) resulted in values of availability of UGS above 500 m^2 per capita. The availability of UGS (m^2 per capita) as well as its different components (trees/shrublands/lawns) did not influence how respondents indicate their potential needs for additional UGS ($p > 0.05$).

The locations of all respondents were within the highly urbanized area. Therefore, there was a little difference between extracted heat stress index and satellite-based land surface temperatures. There were no statistically significant effects of these hydrometeorological parameters on the willingness to go outside during the lockdown or the indicated need for additional UGS.

Although the availability and accessibility of UGS are widely accepted as indicators of the quality of life in cities. The composition and patchiness of UGS in Moscow are also highly important. For example, green spaces in the central area were dominated by small (less than 10 ha) parks with the vegetation composed mainly of lawns and shrubs with less than 30% covered by trees. Such green spaces are less beneficial for climate regulation compared to the larger patches dominated by tree stands [46–48]. It is clearly visible from the heat stress map, where the highest temperatures were shown for the city's centre, whereas the areas close to urban forests did not experience heat stress.

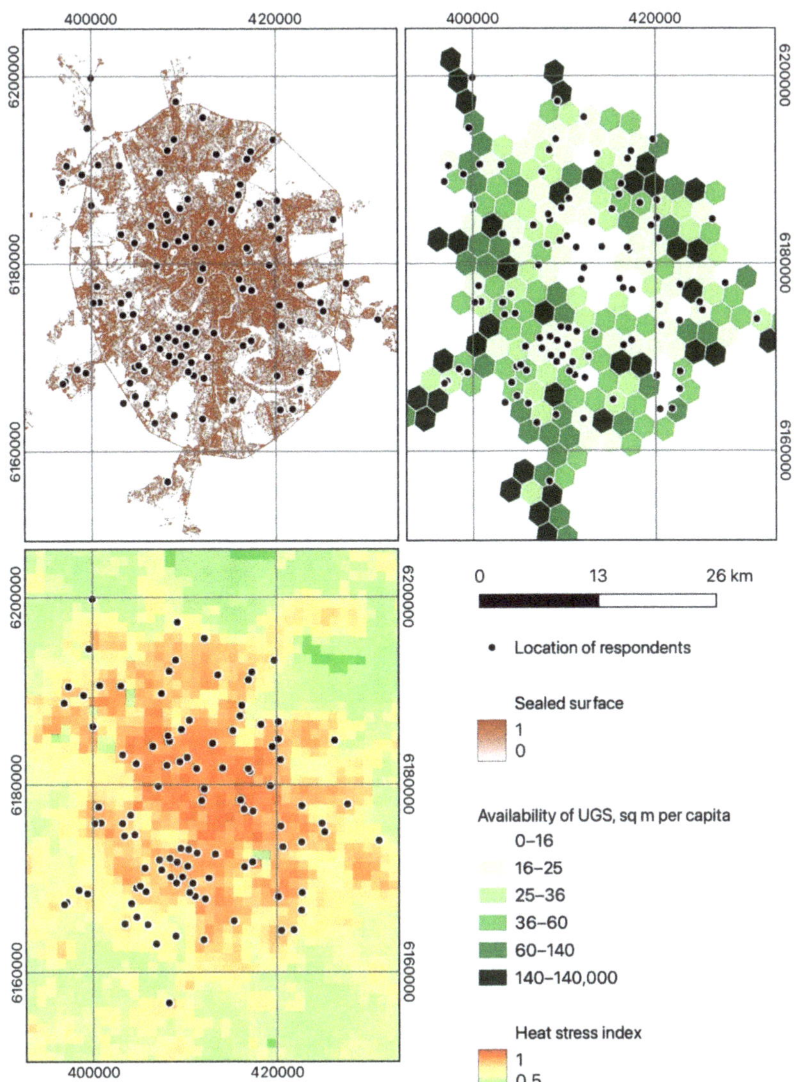

Figure 16. Location of survey respondents (**top left**), UGS availability (**top right**), and the heat stress index (**bottom left**) in Moscow.

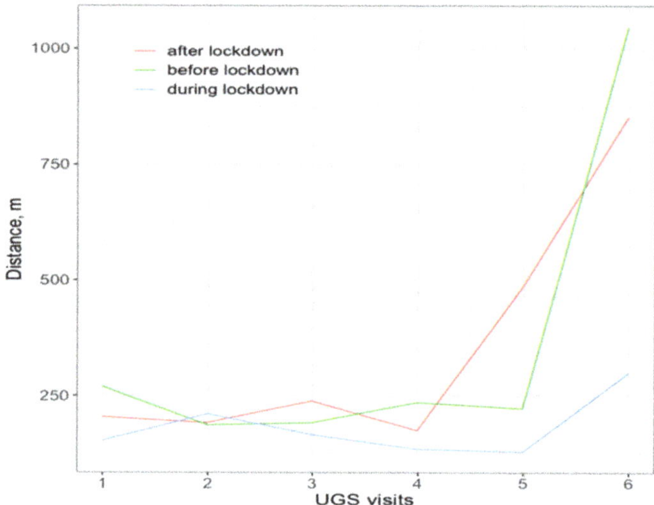

Figure 17. UGS visitation by survey respondents (1—every day, 2—> once a week, 3—several times in a month, 4—once a month, 5—< once a month, 6—never).

3.3. Needs for UGS and COVID-Related Adaptation of Urban Planning and Design in General and in Regard to UGS Based on Expert Interviews

All of the interviewees stated that the COVID-19 pandemic has become a kind of the point of no return for previous unsuccessful urban development and urban life: it has exposed all existing problems such as climate change or social inequality and has accelerated the negative transformations that started before the coronavirus. The COVID-19 pandemic challenged the system "work–home–rest" and narrowed it down to the home office and home school with limited access to outdoor facilities for rest and relaxation. This negatively impacted human health and well-being, e.g., through the increase in computer visual syndrome and stress due to social isolation as well as through a decrease in contact with nature and outdoor activities.

Like many other megacities, Moscow is a complex system that attracts many people, with well-developed economic and social infrastructure, rich cultural life, and job opportunities. At the same time, as stated by the scientists and academia, due to the high density of population, city life is characterized by close contacts, which are perceived as the main risk in the context of the spread of the COVID-19 pandemic, and pronounced inequality. With the closure of cities and quarantines, only the necessary items that support livelihood remain (food, medicine, and transportation).

Both these aspects—narrowing of spatial configuration of the system "work–home–rest" and appearance of virtual working/schooling modes along with the perceived risk of high spread of COVID-19 infection—resulted in the deurbanization process:

> *"The megacity, so attractive as a center of growth and development, with a diversity of services and movement, increasing density of population, communications and building development, is now losing its former universality. In the months of self-isolation, there was a tendency for an "escape" from big cities to nature"* (from interviews with practitioners: architects and landscape architects).

These two identified problems revealed the need to re-organize public spaces in order to ensure social distancing and provide an alternative option for contact with nature when the most parks, street alleys, playgrounds, and outdoor sport facilities in Moscow were closed (period of March–June 2020) as a measure to stop the spread of the COVID-19 pandemic.

Most of the experts highlighted that Moscow had several "preventive" measures which helped to react fast. The already existing strategy (e.g., State and City programs for improvement of streets and public spaces, back- and courtyards in housing complexes, multifunctional greening projects) helped to better adapt to the new COVID-19 circumstances.

However, deurbanization was not a new phenomenon: *"the since centuries existing tradition to go to summer cottages ("dachas") outside the city boundaries allows to escape from the city"* and *"to have contact with nature as another measure to cope with the COVID-19"* (from interviews with public authorities and decision-makers, practitioners, and scientists). Multi-story housing areas (both as a legacy from the Soviet period and newly built complexes) also played an important role in mitigation and adaptation to the COVID-19 circumstances: *" ... people in such housing complexes feel more comfortable than in the city center because they have a back- or courtyards, the area is already landscaped and green according to new standards providing three times more green space per capita than in the central districts"* (from interviews with public authorities and decision-makers and NGOs).

Below, we summarized the main lessons learned from the city's responses to COVID-19 related to urban sustainability and resilience and how the urban fabric of Moscow could be improved:

- during the pandemic, the number of people involved in outdoor sports activities has greatly increased since it was an alternative to many rest activities while other indoor spaces were closed;
- COVID-19 reveals the importance of nature in coping with the problems related to physical and mental health;
- there is a need to create more outdoor recreation spaces which support healthy lifestyles and contact with nature;
- UGS should be inclusive to consider the needs of different population groups, providing safe, attractive, and comfortable conditions by respecting the rules of the post-COVID period.

All above mentioned have defined that public spaces in the city and UGS should not only be re-organized but also apply a new ideology, which includes updating the concept and design of urban open spaces in terms of their multifunctionality, new aesthetical, architectural, and landscape architecture potentials (from interviews with practitioners, scientists and NGOs):

> *"UGS and parks should have an ideology–these are places that can promote and support a healthy lifestyle and ensure that people feel well there"* (from interviews with scientists).

Urban nature is perceived to provide two main values—social and environmental: *"different types of green support different sustainability aspects: parks and public spaces-for recreation, the natural areas for nature conservation"* (from interviews with scientists). All UGS have a big influence on the quality of life in the city. They shape urban living spaces. Here, smart and inclusive urban planning plays an essential role: *" ... any intervention, even decorative, should address environmental issues and needs of end-users"* (from interviews with practitioners, scientists and NGOs).

The interviews also unlocked the interesting aspect of city greening strategy: indoor greening as a new trend (greening homes due to limited access to parks and corporate landscaping to domesticate the workplace):

> *"We have experienced a boom of indoor greening during the COVID-19–it started with greening homes as a measure of limited access to parks and continued as a corporate landscaping in order to domesticate the workplace". "As was stated by many online markets, with the pandemic the interests of citizens in potted plants grew to 50% (comparing to the year before)"* (from interviews with practitioners).

According to the data from several landscape design studios in Moscow, the demand for planting designers to decorate the large indoor spaces such as shopping malls and business centers, offices and coworking spaces have increased.

"In Moscow which has experienced severe lockdown, balconies and terraces have become places for public meetings and also seen as private green islands" (from interviews with NGOs).

The analysis of expert interviews enables us to reveal the influence of the pandemic on the general design and development of UGS in Moscow and on the character of human-nature interactions:

- various multifunctional greening and site improvement measures for the post-pandemic time are now planned:

  "The list of greening, landscaping and site improvement measures for the post-pandemic time (from 2022 onwards) is very ambitious: there are dozens of new parks, more than 2000 courtyards, more than 320 school and kindergarten green areas" (from interviews with public authorities and decision-makers);

- further integration of UGS into the urban green infrastructure as one integrated urban network of interconnected green areas as premises for better connectiveness, accessibility and availability of UGS:

  "The parks, forests, gardens, and other green spaces in the city cannot exist autonomously; they must be competently integrated into the united interconnected urban network. An important task that Moscow now faces is to form an interconnected park framework of the city with green corridors which would allow citizens to move from one park to another" (from interviews with scientists);

- better accessibility of UGS:

  "COVID-19 pandemic underlines the need for adaptation of 15-min-city-concept in Moscow with its high population density and a special urban structure connected with geological and environmental features" (from interviews with practitioners);

  "pedestrian accessibility of public spaces is important, and this should be considered in designing urban space" (from interviews with NGOs);

- ensuring social distance in UGS while enabling social partnership:

  "We now must maintain social distancing also in outdoor public spaces such as parks which role as the social institute was highly recognized, where people not only get in contact with nature but also socialize and get close to each other" (from interviews with public authorities and decision-makers);

  "Pandemic influenced the design of the urban environment" ordering landscape planners and architects to develop *"public space interventions in preparation for post-lockdown urban life"* (from interviews with practitioners);

  "Moscow has a big number of large UGS such as urban forests and forest-parks which seem to be the safest places in the city due to the possibility of social distancing", *" . . . but also many public parks actually are redesigned to be adapted to the time of social distancing"* (from interviews with scientists).

4. Discussion

4.1. Impact of COVID-19 on Citizens' Life and Well-Being and Developed Coping Strategies

Multiple studies confirmed that the current crisis has demonstrated the decisive, potentially agile role of urban areas and especially UGS in the recovery after the pandemic [56–65]. Both the survey findings and results of expert interviews revealed several impacts of the COVID-19 pandemic on urban development, urban life, and how the pandemic has accelerated the transformations that began before the coronavirus. First of all, in Moscow like in many other cities in the world, the COVID-19 pandemic affected ideas about spatial structures and the types of urban life; a spatial configuration of the system "work–home–rest" was transformed into a home office and home schooling with the limited access to outdoor facilities for rest. As result, several negative effects on human physical and mental health appeared, also linked to a decrease in social contacts, access to nature, and outdoor activities. Secondly, the idea of a megacity as a center of economic growth, well-developed social infrastructure, safety and prosperity started to lose its universality and attractiveness during the pandemic. It was replaced by the tendency for an "escape" to nature and to the countryside. Due to the high density of population, the city life is characterized by close human physical contact, which is perceived as the main risk in the context of the spread of the COVID-19 pandemic, and pronounced inequality. Thus, the transformation of spatial configuration of the system "work–home- rest" and introducing of virtual working/schooling modes along with the perceived risk of high spread of COVID-19 infection resulted in the deurbanization process when a big part of the urban population actively used suburban and rural areas of Moscow for safe living conditions. Both identified problems revealed the need to re-organize public spaces to ensure social distancing and provide an alternative option for contact with nature when the most parks, street alleys, playgrounds and outdoor sport facilities in Moscow were closed (period of March–June 2020). The main implications of COVID-19 for current and future urban planning, landscaping, and development of resilient solutions in Moscow can be summarized as follows:

- COVID-19 reveals the importance of urban green not only for its ecological value (air purification, creating comfortable living conditions) but also for providing social benefits (contact with nature as a coping strategy to address the issues of physical and mental health resulting from isolation and lockdown);
- It underlines the need to create more outdoor recreation spaces that support healthy lifestyles and human-nature interactions (e.g., during the pandemic, the number of people involved in outdoor sports activities has greatly increased since it was an alternative to many rest activities while other indoor spaces were closed);
- UGS in the post-pandemic city should not only be re-organized but also adopt a new philosophy, which includes updating the concept and design of urban open spaces in terms of their multifunctionality, new aesthetic, architectural and landscape potentials, and opportunities for inclusive urban planning and governance (active community involvement). The recent studies [32,66,67] also underline that the involvement of citizens with different backgrounds is needed in the process of policymaking and practices of green area management in the post-COVID time. Some publications confirmed that in many cities of the world community participation arises in the time of crisis [66–68].
- New city greening strategy: indoor greening (greening homes due to limited access to parks and corporate landscaping to domesticate the workplace).

Several studies highlighted the ecological value of urban nature and the role of UGS not only for mitigating climate change but also for creating healthy living conditions and post-COVID resilience. The cooling effect of UGS has been thoroughly described and modeled [69,70], whereas the effect of UGS on air humidity is more complex. In impervious surfaces (build-up areas, roads, etc.), infiltration is minimal, and a major part of water outflow is performed by intensive evaporation and surface run-off following rainfall events. Therefore, in a long-term perspective, the impervious surfaces do not increase air humidity and provide an ecosystem disservice by strengthening urban heat islands [71]. In contrast, UGS can mitigate urban heat island by decreasing air temperature and increasing air humidity [72,73]. Based on the UGS spatial analysis, we can conclude that climate regulation services are better supplied by the larger green areas in the suburbs, whereas small green spaces in the central parts did not have a considerable effect on climate comfort what also confirmed by [74]. Nevertheless, the social value of UGS of all sizes and types was highlighted by both respondents and experts. They highlighted a range of benefits such as contact with nature, outdoor activities, and socializing.

4.2. UGS Availability and Use during and after COVID-19 Pandemic: Main Pathways and Evidence of Changed Human-Nature Interactions

The research results show that citizens highly value urban nature. They underlined a need for accessible UGS, most notably for breathing fresh air, reducing stress, relaxing, and observing and enjoying nature. The survey also revealed the particular health effects resulting from the reduction in UGS visitation due to COVID-19 restrictions and highlighted the healing value of nature in coping with a crisis such as COVID-19. These findings are also supported by several studies in different countries [56–58]. The reason is in the multifunctionality and capacity of UGS to mitigate some of the negative effects of the COVID-19 pandemic on human health and wellbeing.

Changes in human-nature interactions were also further investigated, analysed, and interpreted. In general, our research results correspond with the main findings on pandemic influence on the human-nature interactions revealed by Soga et al. [59]: (1) changes in opportunity, (2) changes in capability, and (3) changes in motivation. These three tendencies were adopted from the model of behavior developed by Michie et al. [75] where *opportunity* concerns the factors that facilitate or make interaction with nature possible; *capability* refers to individual's psychological and physical capacity to engage in interactions with nature; and *motivation* is understood as a person's brain processes that energize and direct behavior (motivation to interact with nature). Our identified changes in human-environmental interactions include:

(1) *in regard to opportunity*: decreased availability of UGS. Many respondents during the pandemic/closure of parks and outdoor facilities especially missed spending time outdoors and meeting other people. That highlights the fact that while UGS normally provides places for social integration and socializing, during the COVID-19 social isolation UGS was especially valued in regard to physical health and well-being (self-recovery);

(2) *in regard to capability*: the findings indicate that the capacity to engage with nature is strongly related to opportunity. For example, the closure of many UGS in Moscow reduced the use of nature. That is linked to stress due to social isolation and lack of socializing because public green spaces are highly valued as a place for social interaction;

(3) *in regard to motivation*: both the survey and expert interviews confirmed increased positive attitudes of citizens towards nature (increase in recreational activity after the lockdown was ended), increased interest in outdoor physical activity and being in nature as a measure for stress reduction, and decreased symptoms of depression linked to promoting the healthy lifestyle.

Such increased motivation for physical activity may be a key driver for increasing UGS use especially in the neighborhood with the walking distance to nearest UGS [29,31,34]. Such UGS often offer the only available alternative for recreation or socializing. Increased use of neighborhood UGS contributes to improving human health and well-being which is in line with other studies [26,35,59,61]. These studies reported that although people worried about COVID-19 and its impacts, increased nature exposure was associated with greater happiness, higher subjective wellbeing and life satisfaction. These conclusions are in line with the findings of our survey.

The positive side of the pandemic isolation was the adoption of remote working practice that has released free time for other activities. That promoted positive interactions with nature during visiting natural environments in their neighborhood. This conclusion is also in line with the findings of Soga et al. [59].

Our results also confirmed the idea of Uchiyama and Kohsaka [32] that environmental contexts might influence the motivation of visiting UGS by providing scenery and an opportunity to visit UGS next to the citizens' residential spaces (availability of UGS).

The revealed changes in the frequency of UGS during COVID-19 lockdowns also correspond with the findings of Berdejo-Espinola et al. [61] and Venter et al. [62], showing that UGS were discovered by people during the pandemic (new park users) and rediscovered by other individuals (previous users, re-engaged and increased their use). Study of Korpilo et al. [63] confirmed that patterns of UGS use were associated with the quality of residential green areas. However, there is also evidence that some regular users decreased their visits to green spaces during the pandemic, which can be also explained by the uncertainty of some UGS (outdoor safety) as proposed by Freeman and Eykelbosh [35]. According to their findings, the closure of parks, amenities, and green spaces restricts opportunities for healthy outdoor activity and stress relief and may drive individuals to access less suitable and more congested spaces.

However, along with these three pathways, our research also identified the additional one—a response to COVID-19 that resulted in an impact on human-nature interaction which can be defined as an *adaptation measure*. This pathway extends all three previous ones (opportunity, capability, and motivation). The adaptation/coping strategy with restrictions for use of outdoor UGS, lockdown and risk of infection included an increase in deurbanization process (active use of summer cottages in suburban and rural areas of Moscow) and indoor greening measures.

4.3. Main Lessons Learned from the City Responses to the COVID-19 and Adaptation of Urban Development Strategy in Regard to Urban (Green) Planning and Design in Post-Pandemic Time

The analysis of the questionnaire survey and expert interviews as well as participatory observations revealed the influence of COVID-19 on further urban development strategy in Moscow, as well as design and ideology of urban open public space, including UGS. Survey claimed that respondents have in general good accessibility and availability of UGS, but the structure and quality of UGS do not fully correspond to their preferences and needs. The site improvement measures were revealed from the interviews and non-participatory observations which mostly correspond to the needs claimed by the survey respondents. The plans are to create dozens of parks, 2000 courtyards and 320 school green areas. It also corresponds with the current needs and additional types of UGS expressed by most respondents (e.g., greening in their neighborhood or district: green front- and backyards, more green spaces near home, more pocket parks and small gardens/squares for silent recreation, more street greening and site improvement, e.g., vacant lots and brownfields).

The aspects of UGS provisions (better accessibility and availability of UGS) were mentioned as one of the key directions in the urban green planning and design policies (e.g., challenges for adaptation of 15-min-city-concept in Moscow with its high population density and a special urban structure connected with geological and environmental features and pedestrian accessibility of public spaces for different categories of citizens). Several studies [76,77] have observed that the 15-20-min city is in line with concepts promoting dimensions of proximity, emphasizing walkability and social interactions within cities. This concept is becoming more acceptable and appropriate for the agenda of different global western cities. However further research is needed to demonstrate how new concepts can be replicated in other cities [78]. In the case of Moscow, one of the main tasks is to transform and renovate existing infrastructures (without major restructuring) and to benefit from bicycle lanes, parks, and greener spaces.

Moreover, the issue of social inclusion refers not only to the UGS use but also to socially inclusive governance and management, which can be achieved through stronger citizens' involvement in urban planning and greening. Several studies confirmed that public participation in urban and regional planning can increase the chance of the successful implementation of new solutions [66–68,79,80]. The revealed difference between the post-COVID and pre-COVID greening projects in Moscow has shown that stakeholders actively applied a participatory approach and strong community involvement in the discussion of the greening projects through different online tools (e.g., newly appeared program "Active Citizen" to encourage citizens to participate in decision-making process).

As a response to climate change and other societal challenges reinforced by COVID-19, eco-efficient urban infrastructure and an environmentally friendly approach to urban development has been applied (e.g., replacement of asphalt in UGS, ecologically friendly pavements, using more environmentally friendly surfaces for sidewalks and roads; using native species in flowerbeds and lawns to support wildlife habitats, reconstruction of UGS in line with the original historical plans to preserve the identity of the place and historical heritage, etc.). As in many other cities worldwide [80,81], in Moscow, several strategies are developed and applied to further integrate UGS into the united urban green infrastructure consisted of interconnected green areas, which may allow citizens to move from one park to another. UGS nearby and contact with nature were crucial for mitigating the negative impacts of COVID-19 on quality of life by providing safe space for activities together with emotional and mental health benefits [81,82]. This issue underlines the need for further development and maintenance of UGS. A big challenge of ensuring social distance in UGS while enabling social partnership is addressed through implementing new design elements and additional walking routes and opportunities for interactions, underlining the value of UGS not only for supporting human health and well-being but also as a place for socializing.

5. Limitations and Future Research

Some limitations need to be addressed regarding this study's development in future research. The online survey method could not always provide the verified data on the respondents. Sometimes surveys do not include enough residents from older age groups and from lower educational levels [83]. That could misrepresent the sample. The sample size in our study is relatively low for such a huge city like Moscow and is biased in relation to gender, age and education level. It could be explained by the sampling method using social network websites and targeting recruitment as well as the low level of activeness in survey participation in Russia. Although the number of social media users is increasing every year, the average user profile is biased in terms of age, education, and other demographic criteria [84]. Skewedness of data toward younger users has been also observed in our study. We tried to obtain the data from people older than 50, but the sample was skewed to people 20–40 years old.

Although we have found some statistical significance and dependence on some parameters of the sample such as age and presence of children in the family. Thus, we tried not to draw conclusions according to these groups and presented the recommendations in a more general way. We used a combination of methods (e.g., GIS mapping, expert interviews, non-participatory observation, and photo-documentation) to limit possible errors of the survey method, to verify and supplement the results obtained from the questionnaire.

Limitations of data collection/availability during COVID-19 restriction have meant that the traditional way of non-participatory observation of UGS use in that time has been limited to small-scale sampling (lower number due to sporadic observation) since many of the public UGS were closed or had limited access. This cannot provide city-scale panoramic analysis and delivers only sporadic results. However, it underlines the need for further research to provide a comprehensive perspective for the public and decision-makers in the field of urban greening and design.

Another limitation is the limited number of studies on human-nature interactions during and after COVID-19 in megacities which, on one hand, does not allow us to provide a comprehensive comparative analysis, but on the other hand only reiterates the relevance of the current study and necessity for further research. There is an increasing number of studies presenting the results of surveys on outdoor recreation/UGS use and nature's contribution to human health and well-being in a pandemic situation, e.g., conducted in Oslo/Norway [62], Helsinki/Finland [63], Turku/Finland [82], Burlington, Vermont/USA [31], cities of Belgium [57], Birjand, Iran [60]. They confirm the increase in UGS usage and highlight that the perceived importance of UGS has also increased during the pandemic. Still, very little is published about the situation in densely populated cities/megacities such as Moscow, among them are insights from Nagoya City/Japan [32], a comparative study between Moscow/Russia and Perth/Australia [33], Brisbane/Australia [56], a survey from Asian cities of Hong Kong, Singapore, Tokyo, and

Seoul [58], New York/USA [85]. A European survey by Ugolini et al. [2] conducted in Croatia, Israel, Italy, Lithuania, Slovenia, and Spain, showed that in big cities, where outdoor recreation was restricted by the governments, the majority of those who previously visited urban green spaces regularly stopped doing so but expressed the need for UGS. Another international study by Pouso et al. [26] revealed that contact with nature helped people to cope with the impacts of COVID-19, especially for those under strict lockdown; and emotions were more positive among individuals with accessible UGS and blue-green elements in their views. Different from the questionnaire survey approach, the study of Barton et al. [29] conducted in New York, Barcelona, Berlin/Halle, Oslo, and Stockholm provides local perspectives on the importance of access to green space using Google community mobility statistics.

6. Conclusions

The study confirmed the role of urban nature as a critical urban infrastructure during the COVID-19 pandemic providing opportunities for recreation, restoration, and escape during the pandemic. The results of this study provide key insights for future resilient urban planning and policy that can fulfill a wide range of physical and psychological needs of citizens during a time of crisis and beyond. It was stated that urban planning and management in general and in regard to urban greening should take a specific and adaptive approach. The study in Moscow confirmed the value of implementing an approach that addresses the diverse needs, activities, and preferences of citizens (and especially UGS users). Our research demonstrated the impacts of COVID-19 on city development and revealed the major lessons for urban planning, design, and management of UGS. Moreover, the results presented and discussed in this paper may help to understand the importance of urban nature and human-nature interactions to implement measures to enhance the quality and quantity of UGS in Moscow. Given the complexity of UGS network, public space, mobility, and access of citizens to basic services and UGS, it is now a crucial task for planners, designers, and decision-makers to prioritize urban morphologies that are better integrated with modern technology, public health, and nature in the city. Finally, existing literature shows that the COVID-19 crisis provides an excellent opportunity for urban planners and policymakers to address the challenge of transformative actions toward creating cities that are more just, resilient, and sustainable [27–29,34,80,81].

Supplementary Materials: The following supporting information can be downloaded at: https://www.mdpi.com/article/10.3390/land11060822/s1, Questionnaire S1: The survey template.

Author Contributions: Conceptualization, D.D., M.I. and A.K.; methodology, D.D., A.K. and V.V.; software, A.K. and V.V.; validation, D.D. and A.K.; formal analysis, D.D., A.K. and Y.D.; investigation, D.D. and A.K.; resources, D.D., A.K. and V.V.; data curation, D.D., A.K. and Y.D.; writing—original draft preparation, D.D., M.I., A.K., Y.D., E.D. and V.V.; writing—review and editing, D.D., M.I., A.K., Y.D., E.D. and V.V.; visualization, D.D., A.K., V.V. and Y.D.; supervision, D.D.; project administration, D.D. and E.D.; funding acquisition, D.D. and E.D. All authors have read and agreed to the published version of the manuscript.

Funding: The research was supported in part by the German Academic Exchange Service (DAAD East-partnership) within the project on "Urban ecosystem services and their assessment: exchange of experiences between Germany and Russia" (2018–2021) and the project on Russian-German cooperation on university-practical research studies in the field of geography with a focus on "Landscape ecology", "Green Infrastructure", "Social-ecological urban Research" and "Sustainable Development" (2022–2024). This research was also supported by the Russian Science Foundation project # 19-77-30012 (recipients Yury Dvornikov and Anastasia Konstantinova, data analysis). The publication has been prepared with the support of the RUDN University Strategic Academic Leadership Program (recipients Anastasia Konstantinova, Yury Dvornikov and Elvira Dovletyarova, original draft writing, visualization).

Institutional Review Board Statement: Not applicable.

Informed Consent Statement: Not applicable.

Data Availability Statement: The data presented in this study are available on request from the corresponding author. The data are not publicly available due to privacy restrictions relating to anonymous participation in the survey.

Acknowledgments: We thank the experts participating in the interviews for their time, expertise and advice.

Conflicts of Interest: The authors declare no conflict of interest.

References

1. Kumar, P.; Druckman, A.; Gallagher, J.; Gatersleben, B.; Allison, S.; Eisenman, T.S.; Hoang, U.; Hama, S.; Tiwari, A.; Sharma, A.; et al. The nexus between air pollution, green infrastructure and human health. *Environ. Int.* **2019**, *133*, 105181. [CrossRef] [PubMed]
2. Ugolini, F.; Massetti, L.; Calaza-Martinez, P.; Cariñanos, P.; Dobbs, C.; Ostoic, S.K.; Marin, A.M.; Pearlmutter, D.; Saaroni, H.; Šauliené, I.; et al. Effects of the COVID-19 pandemic on the use and perceptions of urban green space: An international exploratory study. *Urban For. Urban Green.* **2020**, *56*, 126888. [CrossRef] [PubMed]
3. Braubach, M.; Egorov, A.; Mudu, P.; Wolf, T.; Ward Thompson, C.; Martuzzi, M. Effects of Urban Green Space on Environmental Health, Equity and Resilience. In *Nature-Based Solutions to Climate Change Adaptation in Urban Areas. Theory and Practice of Urban Sustainability Transitions*; Kabisch, N., Korn, H., Stadler, J., Bonn, A., Eds.; Springer: Cham, Switzerland, 2017.
4. WHO Regional Office for Europe. Urban Green Spaces: A Brief for Action. 2017. Available online: https://www.euro.who.int/en/health-topics/environmentand-health/urban-health/publications/2017/urban-green-spaces-a-brief-foraction-2017 (accessed on 29 November 2021).
5. Van der Berg, A.E.; Maas, J.; Verheij, R.A.; Groenewegen, P.P. Green space as a buffer between stressful life events and health. *Soc. Sci. Med.* **2010**, *70*, 1203–1210. [CrossRef] [PubMed]
6. Frumkin, H.; Bratman, G.N.; Breslow, S.J.; Cochran, B.; Kahn, P.H.J.; Lawler, J.J.; Levin, P.S.; Tandon, P.S.; Varanasi, U.; Wolf, K.L.; et al. Nature contact and human health: A research agenda. *Environ. Health Perspect.* **2017**, *125*, 75001. [CrossRef] [PubMed]
7. Hillsdon, M.; Panter, J.; Foster, C.; Jones, A. The relationship between access and quality of urban green space with population physical activity. *Public Health* **2006**, *120*, 1127–1132. [CrossRef] [PubMed]
8. Hunter, R.F.; Christian, H.; Veitch, J.; Astell-Burt, T.; Hipp, J.A.; Schipperijn, J. The impact of interventions to promote physical activity in urban green space: A systematic review and recommendations for future research. *Soc. Sci. Med.* **2015**, *124*, 246–256. [CrossRef]
9. Ekkel, D.E.; de Vries, S. Nearby green space and human health: Evaluating accessibility metrics. *Landsc. Urban Plan.* **2017**, *157*, 214–220. [CrossRef]
10. Hartig, T.; Mitchell, R.; de Vries, S.; Frumkin, H. Nature and health. *Annu. Rev. Public Health* **2014**, *35*, 207–208. [CrossRef]
11. Kabisch, N.; Haase, D. Urban nature benefits—Opportunities for improvement of health and well-being in times of global change. *WHO Newsl. Hous. Health* **2018**, *29*, 1–11.
12. Ma, B.; Zhou, T.; Lei, S.; Wen, Y.; Htun, T.T. Effects of urban green spaces on residents' well-being. *Environ. Dev. Sustain.* **2019**, *21*, 2793–2809. [CrossRef]
13. Nath, T.K.; Zhe Han, S.S.; Lechner, A.M. Urban green space and well-being in Kuala Lumpur. Malaysia. *Urban For. Urban Green.* **2018**, *36*, 34–41. [CrossRef]
14. Tsai, W.-L.; McHale, M.; Jennings, V.; Marquet, O.; Hipp, A.J.; Leung, Y.-F.; Floyd, M.F. Relationships between characteristics of urban green land cover and mental health in U.S. Metropolitan areas. *Int. J. Environ. Res. Public Health* **2018**, *15*, 340. [CrossRef]
15. Van den Bosch, M.; Sang, O. Urban natural environments as nature-based solutions for improved public health—A systematic review of reviews. *Environ. Res.* **2017**, *158*, 373–384. [CrossRef]
16. Grunewald, K.; Richter, B.; Meinel, G.; Herold, H.; Syrbe, R.-U. Proposal of indicators regarding the provision and accessibility of green spaces for assessing the ecosystem service recreation in the city" in Germany. *Int. J. Biodivers. Sci. Ecosyst. Serv. Manag.* **2017**, *13*, 26–39. [CrossRef]
17. Larson, L.R.; Keith, S.J.; Fernandez, M.; Hallo, J.C.; Shager, C.S.; Jennings, V. Ecosystem services and urban greenways: What's the public's perspective. *Ecosyst. Serv.* **2016**, *22*, 111–116. [CrossRef]
18. Zijlema, W.L.; Triguero-Mas, M.; Smith, G.; Cirach, M.; Martinez, D.; Dadvand, P.; Gascon, M.; Jones, M.; Gidlow, C.; Hurst, G.; et al. The relationship between natural outdoor environments and cognitive functioning and its mediators. *Environ. Res.* **2017**, *155*, 268–275. [CrossRef]
19. Bowler, D.E.; Lisette, M.B.-A.; Knight, T.M.; Pullin, A.S. A Systematic Review of Evidence for the Added Benefits to Health of Exposure to Natural Environments. *BMC Public Health* **2010**, *10*, 456. [CrossRef]
20. Mao, Q.; Wang, L.; Guo, Q.; Li, Y.; Liu, M.; Xu, G. Evaluating Cultural Ecosystem Services of Urban Residential Green Spaces From the Perspective of Residents' Satisfaction With Green Space. *Front. Public Health* **2020**, *8*, 226. [CrossRef]
21. Tzoulas, K.; James, P. Peoples' use of, and concerns about, green space networks: A case study of Birchwood, Warrington New Town, UK. *Urban For. Urban Green.* **2010**, *9*, 121–128. [CrossRef]

22. Andersson, E.; Langemeyer, J.; Borgström, S.; McPhearson, T.; Haase, D.; Kronenberg, J.; Barton, D.N.; Davis, M.; Naumann, S.; Röschel, N.; et al. Enabling Green and Blue Infrastructure to Improve Contributions to Human Well-Being and Equity in Urban Systems. *BioScience* **2019**, *69*, 566–574. [CrossRef]

23. Elmqvist, T.; Setälä, H.; Handel, S.; van der Ploeg, S.; Aronson, J.; Blignaut, J.; Gómez-Baggethun, E.; Nowak, D.; Kronenberg, J.; de Groot, R. Benefits of restoring ecosystem services in urban areas. *Curr. Opin. Environ. Sustain.* **2015**, *14*, 101–108. [CrossRef]

24. Lopez, G.A.P.; de Souza, L.C.L. Urban green spaces and the influence on vehicular traffic noise control. *Ambiente Construído* **2018**, *18*, 161–175. [CrossRef]

25. Vidal, D.G.; Dias, R.C.; Teixeira, C.P.; Fernandes, C.O.; Filho, W.L.; Barros, N.; Maia, R.L. Clustering public urban green spaces through ecosystem services potential: A typology proposal for place-based interventions. *Environ. Sci. Policy* **2022**, *132*, 262–272. [CrossRef]

26. Pouso, S.; Borja, A.; Fleming, L.E.; Gómez-Baggethun, E.; White, M.P.; Uyarra, M.C. Contact with blue-green spaces during the COVID-19 pandemic lockdown beneficial for mental health. *Sci. Total Environ.* **2021**, *756*, 143984. [CrossRef]

27. Slater, S.J.; Christiana, R.W.; Gustat, J. Recommendations for Keeping Parks and Green Space Accessible for Mental and Physical Health During COVID-19 and Other Pandemics. *Prev. Chronic Dis.* **2020**, *17*, 200204. [CrossRef]

28. Ahmadpoor, N.; Shahab, S. Urban form: Realising the value of green space: A planners' perspective on the COVID-19 pandemic. *Town Plan. Rev.* **2021**, *92*, 49–55. [CrossRef]

29. Barton, D.; Haase, D.; Mascarenhas, A.; Langemeyer, J.; Baro, F.; Kennedy, C.; Grabowski, Z.; McPhearson, T.; Krog, N.H.; Venter, Z.; et al. Enabling Access to Greenspace during the COVID-19 Pandemic–Perspectives from Five Cities. The Nature of Cities. 4 May 2020. Available online: https://www.thenatureofcities.com/2020/05/04/enabling-access-to-greenspace-during-the-covid-19-pandemic-perspectives-from-five-cities/ (accessed on 8 February 2022).

30. Finnsson, P.T. COVID-19 Crisis Highlights the Need for Accessible and Productive Urban Green Spaces. Nordregio Magazine 2020. Issue "Postpandemic Regional Development". 2020. Available online: https://nordregio.org/nordregio-magazine/issues/post-pandemic-regional-development/covid19-crisis-highlights-the-need-for-accessible-and-productive-urban-green-spaces/ (accessed on 23 January 2022).

31. Grima, N.; Corcoran, W.; Hill-James, C.; Langton, B.; Sommer, H.; Fisher, B. The importance of urban natural areas and urban ecosystem services during the COVID-19 pandemic. *PLoS ONE* **2020**, *15*, e0243344. [CrossRef]

32. Uchiyama, Y.; Kohsaka, R. Access and Use of Green Areas during the COVID-19 Pandemic: Green Infrastructure Management in the "New Normal". *Sustainability* **2020**, *12*, 9842. [CrossRef]

33. Dushkova, D.; Ignatieva, M.; Hughes, M.; Konstantinova, A.; Vasenev, V.; Dovletyarova, E. Human Dimensions of Urban Blue and Green Infrastructure during a Pandemic. Case Study of Moscow (Russia) and Perth (Australia). *Sustainability* **2021**, *13*, 4148. [CrossRef]

34. Kleinschroth, F.; Kowarik, I. COVID-19 crisis demonstrates the urgent need for urban greenspaces. *Front. Ecol. Environ.* **2020**, *18*, 318–319. [CrossRef] [PubMed]

35. Freeman, S.; Eykelbosh, A. *COVID-19 and Outdoor Safety: Considerations for Use of Outdoor Recreational Spaces*; National Collaborating Centre for Environmental Health: Vancouver, BC, Canada, 2020; Available online: https://ncceh.ca/documents/guide/covid-19-and-outdoor-safety-considerations-use-outdoor-recreational-spaces (accessed on 15 March 2022).

36. Crilley, G.; Weber, D.; Taplin, R. Predicting visitor satisfaction in parks: Comparing the value of personal benefit attainment and service levels in Kakadu National Park, Australia. *Visit. Stud.* **2012**, *15*, 217–237. [CrossRef]

37. Moyle, B.D.; Weiler, B. Revisiting the importance of visitation: Public perceptions of park benefits. *Tour. Hosp. Res.* **2017**, *17*, 91–105. [CrossRef]

38. Madureira, H.; Nunes, F.; Oliveira, J.; Madureira, T. Preferences for urban green space characteristics: A comparative study in three Portuguese cities. *Environments* **2018**, *5*, 23. [CrossRef]

39. Zhang, Y.; Van den Berg, A.E.; Van Dijk, T.; Weitkamp, G. Quality over quantity: Contribution of urban green space to neighborhood satisfaction. *Int. J. Environ. Res. Public Health* **2017**, *14*, 535. [CrossRef]

40. Clark, A.; Holland, C.; Katz, J.; Peace, S. Learning to see: Lessons from a participatory observation research project in public spaces. *Int. J. Soc. Res. Methodol.* **2009**, *12*, 345–360. [CrossRef]

41. Robson, C. *Real World Research: A Resource for Social Scientists and Practitioners-Researchers*; Blackwell Pushers Inc.: Oxford, UK, 1993.

42. Whyte, W.H. *The Social Life of Small Urban Spaces*; The Conservation Foundation: Washington, DC, USA, 1980.

43. Gorelick, N.; Hancher, M.; Dixon, M.; Ilyushchenko, S.; Thau, D.; Moore, R. Google Earth Engine: Planetary-scale geospatial analysis for everyone. *Remote Sens. Environ.* **2017**, *202*, 18–27. [CrossRef]

44. Thorsson, S.; Rayner, D.; Palm, G.; Lindberg, F.; Carlström, E.; Börjesson, M.; Nilson, F.; Khorram-Manesh, A.; Holmer, B. Is Physiological Equivalent Temperature (PET) a superior screening tool for heat stress risk than Wet-Bulb Globe Temperature (WBGT) index? Eight years of data from the Gothenburg half marathon. *Br. J. Sports Med.* **2021**, *55*, 825–830. [CrossRef]

45. Desai, M.; Navale, A.; Dhorde, A.G. Evolution of heat index (Hi) and physiological equivalent temperature (pet) index at Mumbai and Pune cities, India. *Mausam* **2021**, *72*, 915–934. [CrossRef]

46. Varentsov, M.L.; Samsonov, T.E.; Kislov, A.V.; Konstantinov, P.I. Simulations of Moscow agglomeration heat island within the framework of the regional climate model COSMO-CLM. *Vestn. Mosk. Univ. Seriya 5 Geogr.* **2017**, *6*, 25–37.

47. Kislov, A.V.; Alekseeva, L.I.; Varentsov, M.I.; Konstantinov, P.I. Climate Change and Extreme Weather Events in the Moscow Agglomeration. *Russ. Meteorol. Hydrol.* **2020**, *45*, 498–507. [CrossRef]
48. Zemtsov, S.; Shartova, N.; Varentsov, M.; Konstantinov, P.; Kidyaeva, V.; Shchur, A.; Timonin, S.; Grischchenko, M. Intraurban social risk and mortality patterns during extreme heat events: A case study of Moscow, 2010–2017. *Health Place* **2020**, *66*, 102429. [CrossRef] [PubMed]
49. Meuser, M.; Nagel, U. The expert interview and changes in knowledge production. In *Interviewing Experts*; Bogner, A., Littig, B., Menz, W., Eds.; Palgrave Macmillan: London, UK, 2009; pp. 17–42.
50. Dvornikov, Y.A.; Vasenev, V.I.; Romzaykina, O.N.; Grigorieva, V.E.; Litvinov, Y.A.; Gorbov, S.N.; Dolgikh, A.V.; Korneykova, M.V.; Gosse, D.D. Projecting the urbanization effect on soil organic carbon stocks in polar and steppe areas of European Russia by remote sensing. *Geoderma* **2021**, *399*, 115039. [CrossRef]
51. Morabito, M.; Crisci, A.; Messeri, A.; Orlandini, S.; Raschi, A.; Maracchi, G.; Munafò, M. The impact of built-up surfaces on land surface temperatures in Italian urban areas. *Sci. Total Environ.* **2016**, *551–552*, 317–326. [CrossRef] [PubMed]
52. García, P.; Pérez, E. Mapping of soil sealing by vegetation indexes and built-up index: A case study in Madrid (Spain). *Geoderma* **2016**, *268*, 100–107. [CrossRef]
53. Richter, S.; Haase, D.; Thestorf, K.; Makki, M. Carbon Pools of Berlin, Germany: Organic Carbon in Soils and Aboveground in Trees. *Urban For. Urban Green.* **2020**, *54*, 126777. [CrossRef]
54. Klimanova, O.; Kolbowsky, E.; Illarionova, O. Impacts of urbanization on green infrastructure ecosystem services: The case study of post-soviet Moscow. *BELGEO* **2018**, *4*. [CrossRef]
55. Klimanova, O.; Illarionova, O.; Grunewald, K.; Bukvareva, E. Green infrastructure, urbanization, and ecosystem services: The main challenges for Russia's largest cities. *Land* **2021**, *10*, 1292. [CrossRef]
56. Berdejo-Espinola, V.; Suárez-Castro, A.F.; Amano, T.; Oh, R.R.Y.; Fielding, K.S.; Fuller, R.A. Urban green space use during a time of stress: A case study during the COVID-19 pandemic in Brisbane, Australia. *People Nat.* **2021**, *3*, 10218. [CrossRef]
57. Da Schio, N.; Philips, A.; Fransen, K. The impact of the COVID-19 pandemic on the use of and attitudes towards urban forests and green spaces: Exploring the instigators of change in Belgium. *Urban For. Urban Green.* **2021**, *65*, 127305. [CrossRef]
58. Lu, Y.; Zhao, J.; Xueying, W.; Lo, S.M. Escaping to nature during a pandemic: A natural experiment in Asian cities during the COVID-19 pandemic with big social media data. *Sci. Total Environ.* **2021**, *777*, 146092. [CrossRef]
59. Soga, M.; Evans, M.J.; Cox, D.T.C.; Gaston, K.J. Impacts of the COVID-19 pandemic on human–nature interactions: Pathways, evidence and implications. *People Nat.* **2021**, *3*, 518–527. [CrossRef] [PubMed]
60. Khalilnezhad, M.R.; Ugolini, F.; Massetti, L. Attitudes and Behaviors toward the Use of Public and Private Green Space during the COVID-19 Pandemic in Iran. *Land* **2021**, *10*, 1085. [CrossRef]
61. Berdejo-Espinola, V.; Zahnow, R.; Suárez-Castro, A.F.; Rhodes, J.R.; Fuller, R.A. Changes in Green Space Use During a COVID-19 Lockdown Are Associated with Both Individual and Green Space Characteristics. *Front. Ecol. Evol.* **2022**, *10*, 804443. [CrossRef]
62. Venter, Z.; Barton, D.; Gundersen, V.; Figari, H.; Nowell, M. Urban nature in a time of crisis: Recreational use of green space increases during the COVID-19 outbreak in Oslo, Norway. *Environ. Res. Lett.* **2020**, *15*, 104075. [CrossRef]
63. Korpilo, S.; Kajosaari, A.; Rinne, T.; Hasanzadeh, K.; Raymond, C.M.; Kyttä, M. Coping with Crisis: Green Space Use in Helsinki Before and During the COVID-19 Pandemic. *Front. Sustain. Cities* **2021**, *3*, 713977. [CrossRef]
64. UN Habitat. *Cities and Pandemics: Towards a More Just, Green and Healthy Future*; Tuts, R., Knudsen, C., Moreno, E., Williams, C., Khor, N., Eds.; United Nations Human Settlements Programme (UN-Habitat): Nairobi, Kenya, 2021.
65. Graziano, T. Smart Technologies, Back-to-the-Village Rhetoric, and Tactical Urbanism: Post-COVID Planning Scenarios in Italy. *Int. J. E-Plan. Res. (IJEPR)* **2021**, *10*, 80–93. [CrossRef]
66. Falanga, R. *Citizen Participation during the COVID-19 Pandemic: Insights from Local Practices in European Cities*; Friedrich Ebert Stiftung: Lisboa, Portugal, 2020; Available online: http://library.fes.de/pdf-files/bueros/lissabon/17148.pdf (accessed on 19 January 2022).
67. Jaeweon, Y.; Sae-Woom, J.; Dongoh, H.; Seung-Won, K.; Ju-Chul, J. Post COVID-19 Visioning of Urban Comprehensive Plan through Citizen Participation: Focusing on the Citizen Participation of Busan Metropolitan City. *J. Korea Plan. Assoc.* **2021**, *56*, 156–168.
68. Pantic, M.; Cilliers, J.; Cimadomo, G.; Montaño, F.; Olufemi, O.; Torres Mallma, S.; van den Berg, J. Challenges and Opportunities for Public Participation in Urban and Regional Planning during the COVID-19 Pandemic—Lessons Learned for the Future. *Land* **2021**, *10*, 1379. [CrossRef]
69. Su, W.; Chang, Q.; Liu, X.; Zhang, L. Cooling effect of urban green and blue infrastructure: A systematic review of empirical evidence. *Shengtai Xuebao* **2021**, *41*, 2902–2917. [CrossRef]
70. Jamali, F.S.; Khaledi, S.; Razavian, M.T. Evaluating the Effect of Urban Green Infrastructure on Mitigating Temperature: A Case Study of Tehran. In *Resilient and Responsible Smart Cities*; Advances in Science, Technology & Innovation; Springer: Cham, Switzerland, 2022; pp. 229–236. [CrossRef]
71. Dugord, P.A.; Lauf, S.; Schuster, C.; Kleinschmit, B. Land use patterns, temperature distribution, and potential heat stress risk—The case study Berlin, Germany. *Comput. Environ. Urban Syst.* **2014**, *48*, 86–98. [CrossRef]
72. Senosiain, J.L. Urban regreeneration: Green urban infrastructure as a response to climate change mitigation and adaptation. *Int. J. Des. Nat. Ecodyn.* **2020**, *15*, 33–38. [CrossRef]

73. Belčáková, I.; Świader, M.; Bartyna-Zielińska, M. The green infrastructure in cities as a tool for climate change adaptation and mitigation: Slovakian and polish experiences. *Atmosphere* **2019**, *10*, 552. [CrossRef]

74. Varentsov, M.I.; Konstantinov, P.I.; Shartova, N.V.; Samsonov, T.E.; Kargashin, P.E.; Varentsov, A.I.; Fenner, D.; Meier, F. Urban heat island of the Moscow megacity: The long-term trends and new approaches for monitoring and research based on crowdsourcing data. In Proceedings of the IOP Conference Series: Earth and Environmental Science (2020), Moscow, Russia, 26–28 November 2019; Volume 606. No. 012063. [CrossRef]

75. Michie, S.; Van Stralen, M.M.; West, R. The behaviour change wheel: A new method for characterising and designing behaviour change interventions. *Implement. Sci.* **2011**, *6*, 42. [CrossRef]

76. Weng, M.; Ding, N.; Li, J.; Jin, X.; Xiao, H.; He, Z.; Su, S. The 15-min walkable neighborhoods: Measurement, social inequalities and implications for building healthy communities in urban China. *J. Transp. Health* **2019**, *13*, 259–273. [CrossRef]

77. Capasso Da Silva, D.; King, D.A.; Lemar, S. Accessibility in practice: 20-min city as a sustainability planning goal. *Sustainability* **2020**, *12*, 129. [CrossRef]

78. Moreno, C.; Allam, Z.; Chabaud, D.; Gall, C.; Pratlong, F. Introducing the "15-Minute City": Sustainability, Resilience and Place Identity in Future Post-Pandemic Cities. *Smart Cities* **2021**, *4*, 93–111. [CrossRef]

79. Gohari, S.; Baer, D.; Nielsen, B.F.; Glicher, E.; Situmorang, W.Z. Prevailing approaches and practices of citizen participation in smart city projects: Lessons from Trondheim, Norway. *Infrastructures* **2020**, *5*, 36. [CrossRef]

80. Mouratidis, K. How COVID-19 reshaped quality of life in cities: A synthesis and implications for urban planning. *Land Pol.* **2021**, *2021*, 105772. [CrossRef]

81. Sharifi, A.; Khavarian-Garmsir, A.R. The COVID-19 pandemic: Impacts on cities and major lessons for urban planning, design, and management. *Sci. Total Environ.* **2020**, *749*, 142391. [CrossRef]

82. Fagerholm, N.; Eilola, S.; Arki, V. Outdoor recreation and nature's contribution to well-being in a pandemic—Case Turku, Finland. *Urban For. Urban Green.* **2021**, *64*, 127257. [CrossRef]

83. Biasio, L.R.; Bonaccorsi, G.; Lorini, C.; Pecorelli, S. Assessing COVID-19 vaccine literacy: A preliminary online survey. *Hum. Vaccines Immunother.* **2021**, *17*, 1304–1312. [CrossRef] [PubMed]

84. Tenerelli, P.; Demšar, U.; Luque, S. Crowdsourcing indicators for cultural ecosystem services: A geographically weighted approach for mountain landscapes. *Ecol. Indic.* **2016**, *64*, 237–248. [CrossRef]

85. Lopez, B.; Kennedy, C.; McPhearson, T. Parks Are Critical Urban Infrastructure: Perception and Use of Urban Green Spaces in NYC During COVID-19. *Preprints* **2020**. [CrossRef]

 *land*

Article

Impact of COVID-19 Lock-Downs on Nature Connection in Southern and Eastern Africa

Ju-hyoung Lee [1,*], Madalitso Mkandawire [2], Patrick Niyigena [3], Abonisiwe Xotyeni [4], Edwin Itamba [5] and Sylvester Siame [6]

[1] Department of Forest Resources, Yeungnam University, 280 Daehak-ro, Gyeongsan 38541, Korea
[2] Kasungu District Agriculture Office, Kasungu P.O. Box 17, Malawi; madalitso.mkandawire@agriculture.gov.mw
[3] Rwanda Youth in Agribusiness Forum, KG 569 Street, Kigali P.O. Box 7202, Rwanda; pani2101@student.miun.se
[4] Park Chung Hee School of Policy and Saemaul (PSPS), Yeungnam University, Gyeongsan 38541, Korea; aboxotyeni@ynu.ac.kr
[5] Statistics and Logistic Officer, Sengerema District Council, Mwanza P.O. Box 175, Tanzania; elimu.msingi@sengeremadc.go.tz
[6] Forestry Department, Ministry of Green Economy and Environment, Lusaka P.O. Box 50042, Zambia; sylvester.siame@mgee.gov.zm
* Correspondence: jhlee9@yu.ac.kr; Tel.: +82-53-810-2927

Citation: Lee, J.-h.; Mkandawire, M.; Niyigena, P.; Xotyeni, A.; Itamba, E.; Siame, S. Impact of COVID-19 Lock-Downs on Nature Connection in Southern and Eastern Africa. *Land* 2022, 11, 872. https://doi.org/10.3390/land11060872

Academic Editors: Francesca Ugolini and David Pearlmutter

Received: 11 May 2022
Accepted: 6 June 2022
Published: 8 June 2022

Abstract: The response of African countries immediately after the COVID-19 pandemic declaration was rapid and appropriate, with low infections and mortality rates until June 2020. Severe lock-down measures were effective in Africa; however, the reduction in the amount of natural experience influences the quality of life in modern society. This study is conducted as an international comparative study in five African countries on changes in the perception of health recovery and outdoor activities in urban forests during the COVID-19 pandemic. An online survey was conducted with 430 respondents to investigate the relationships between COVID-19 stress, indoor activity, appreciation of greenspaces, perception of health recovery, and use of greenspaces. A structural equation model was used for analysis. The visit frequency and staying time in urban forests after lock-down dramatically decreased, raising concerns about nature-deficit disorder across the target countries after the end of the pandemic. This study confirmed urban dwellers' desire for natural experiences and health recovery during the pandemic and predicts an explosive increase in urban forest utilization after the pandemic has ended.

Keywords: COVID-19 in Africa; lock-down; urban forests; health recovery; outdoor recreation

1. Introduction

As COVID-19 swept through the world, many observers predicted that African countries would experience a much more deadly epidemic than those in Europe and the United States [1–3]. Because of insufficient hospital beds and fragile health systems, comorbidities such as Ebola, human immunodeficiency virus (HIV), tuberculosis, malnutrition of the young population, overcrowded living environments, etc., were expected to affect infection and mortality negatively [4–6]. However, many African countries have unexpectedly maintained low morbidity and mortality rates through strong lock-down measures in the early stages [1,5,7,8].

There were around 160,000 confirmed COVID-19 cases in Italy and 560,000 cases in the United States in April 2020, compared with approximately 14,000 cases in the African countries. [7].

Domestic and international travel restrictions, curfews, and the closing of offices, schools, and jobs in response to infectious diseases in Africa have been in effect since the days of Lassa Fever and Ebola before COVID-19. In particular, the Ebola virus, which started in Sierra Leone in March 2014 in Africa and was declared a pandemic in June by the World Health Organization (WHO) is suspected of 28,616 probable and 11,310 deaths in Guinea, Sierra Leone, and Liberia [9]. While an early response to COVID-19 prevented foreign inflow and domestic transmission, the government had enough time to implement strategies such as preparing medical systems for diagnosis, the quarantine of goods, contact tracing, and social distancing [10]. However, a weak health care system and insufficient medical equipment are still threats to combatting diseases in underdeveloped and developing countries in Africa [4,10,11].

The African continent's healthcare and hygiene sectors are weak, including inadequate or outdated healthcare facilities and systems, low public investment in healthcare costs per capita (Europe and the US: over \$2500/capita; most west African countries: below \$50/capita), and a dense population living in poverty and unsanitary living conditions [10]. In the case of the United States, there are 34 intensive care unit beds per 100,000 people compared with only four intensive beds in Kenya [7].

The medical system in Africa has been strengthened through the cooperation and support of international organizations to cope with HIV, malaria, and Ebola treatments, and the African Center for Disease Control was opened and began operating in 2017 [12]. Despite this, most African governments focused on the most effective containment measures in the initial COVID-19 response.

As soon as the COVID-19 outbreak began in Africa, the African Union quickly responded with tiered public health and social measures (PHSM) systems, the most effective tool to combat the fast-spreading epidemic [13]. A clear awareness of the realities of the medical and social health systems enabled governments in Africa to quickly decide on strict responses in the first step of the COVID-19 outbreak [7,13]. At the start of the pandemic, the majority of governments put measures in place such as closing borders, shutting down markets, suspending internal and international flights, and instituting bans on people gatherings [7,11,13].

As a result of the rapid and thorough lock-down in African countries, it was possible to lower the initial infection rate. However, there were side effects such as restrictions on outdoor recreational activities, the interruption of nature experiences, and reductions in the amount of nature experience.

The reward hypothesis is one of the oldest hypotheses in outdoor recreation and leisure research, stating that leisure plays a rewarding role after a hard time [14]. In this study, appreciation of the natural environment and increased outdoor activity in urban forests as a reward for extreme freedom restrictions were established as a research hypothesis.

During the pandemic, a target area survey was conducted on urban populations in the capitals of Malawi, Rwanda, South Africa, Tanzania, and Zambia. This study was conducted to investigate and analyze changes in stress level caused by COVID-19, appreciation of the natural environment, recreational behavior in urban forests, and perceptions of health recovery in a natural setting.

In an online survey on outdoor recreation conducted on American adults in 2020 right after the outbreak of COVID-19, most of the respondents did not want to risk infection due to outdoor recreation during the pandemic, and this perception was based on decreasing overall outdoor recreation [15]. A survey of Canadian adults showed a significant decrease in physical activity after COVID-19 was reported [16], and a study of Vermont residents in the US also found that outdoor recreational activities declined during the pandemic, except hiking and gardening [17]. Urban parks, community centers, recreational centers, and public places of social activity were closed worldwide during COVID-19, and people's work, learning, and leisure activity patterns mainly changed [18].

Recently, research was conducted on overcoming nature-deficit disorder through frequent exposure to the natural environment [19]. The pandemic is decreasing people's nature experience and changing their perceptions of the natural environment due to restrictions on outdoor recreation in the natural environment worldwide.

This study aims to investigate and analyze perceptions of urban forests in limited and disconnected situations and to identify the relationship between the COVID-19-stress experienced by urban residents and their recreational behaviors and perceptions of health recovery in nature. A structural equation model (SEM) research model is used to investigate the perception of nature experience in five African countries during the pandemic era [20]. The research hypotheses are as follows: (i) mental stress caused by restricted face-to-face communication and limited freedom of activity; (ii) increased screen and internet use due to increased indoor living time; (iii) appreciation of urban forest due to (i) and (ii); (iv) actual behavioral changes such as increasing visit frequency; (v) perception of health recovery in the natural environment.

2. Materials and Methods

2.1. Survey Procedure

The data were collected between April and June 2020, when governments of the five countries began to take countermeasures, including school closures, stay-at-home rules, workplace closures, and restrictions on internal movement, following the WHO pandemic declaration of March 2020 [21]. The target countries were Malawi, Rwanda, South Africa, Tanzania, and Zambia in the southeastern region of Africa, and the people living in the capitals of each country were chosen. Respondents were contacted through email due to restrictions on physical meetings during the survey period [1,5,7,8].

Because of the difficulty of finding respondents in the COVID-19 era, during which social, economic, and human life was abnormal, we secured a list of respondents with cooperative researchers (the co-authors) working in the forestry or environment departments of government agencies in each country. The respondents were selected from groups of individuals who participated in government campaigns or program meetings with the theme of community development, natural landscape, biodiversity conservation, and climate change by the ministry of environment and forestry.

All the participants were informed through consent forms of the research aim, subjects, and ethics at the beginning of the surveys. They knew their right to participate or refuse during the survey questionnaire, such as ending the survey immediately. Excluding errors and insufficient responses, a total of 430 responses were used (Table 1).

Table 1. Characteristics of participants and current status in each country.

Country, Surveyed Area	Forested Area (% of Land Surface)	Population (×1000)	GDP Growth Rates			GDP/ People ($)	Survey Respondents					
							Total (Pers.)	Female (%)	Age (%)			
			'19	'20	'21				~29	~39	~49	50~
Malawi, Lilongwe	22,417 sq.km (23.7%)	19,129	5.5	0.9	2.2	636	95	46.3	29.5	43.2	25.3	2.1
Rwanda, Kigali	2760 sq.km (11.1%)	12,952	10.1	3.5	6.7	797	65	47.6	73.8	16.9	3.1	5.2
South Africa, Pretoria	170,500 sq.km (14.0%)	59,308	0.2	−5.8	4.0	5655	94	42.5	39.4	26.6	26.6	7.4
Tanzania, Dodoma	457,450 sq.km (51.6%)	59,734	6.3	2.0	4.6	1076	74	37.8	12.2	48.6	39.2	0.0
Zambia, Lusaka	448,140 sq.km (60.2%)	18,383	1.5	−3.5	2.3	985	102	51.9	8.8	39.2	48.0	3.9

(Source: Worldbank 2020, IMF 2020, and survey results in target areas).

2.2. Survey Method and Statistical Analysis

To explore the research hypotheses, we applied developed survey questions on the perception of nature experience in five African countries during the pandemic [20]. The survey questionnaires were revised into 12 questions in 4 categories: COVID-19 Stress level, indoor activity increase, enhanced appreciation of urban forests and natural environment, and agreement with health recovery in nature (Table 2) [20]. The participants responded using a 5-point scale.

Table 2. Interview questionnaire's reliability by categories.

Category	Items (Since COVID-19, …)	Var.	Cronbach's α				
			MA	RW	SA	TA	ZA
Stress level	I am unsatisfied with the restricted daily life (activity, visitation, work, school).	ST1	0.826	0.860	0.831	0.801	0.878
	I am unsatisfied with limited communication opportunities with other people.	ST2					
	It has decreased communication with other people (except family).	ST3					
	It has increased the communication with my family. *	ST4					
Indoor activity	I don't have enough leisure activities.	ID1	0.821	0.792	0.803	0.841	0.862
	The screen time for visiting websites and watching TV has increased.	ID2					
	I prefer online activities to be offline (shopping, learning, communicating, etc.).	ID3					
Appreciation of urban forests	I get a positive feeling when I visit urban forests. *	PR1	0.793	0.774	0.836	0.810	0.775
	It became more challenging to go to the outdoor natural environment in urban forests.	PR2					
	I love to experience the outdoor nature around me, particularly during the pandemic. *	PR3					
Perception of health recovery in nature	Through nature experience in the urban forest, I felt my mental recovery. *	HE1	0.786	0.797	0.775	0.852	0.784
	Through nature experience in the urban forest, I felt my body recover. *	HE2					

Note. * reversed items (Questions were adapted from [20]).

The questionnaires were translated and used for field surveys in each country, and the 12 questions suitable for statistical significance were finally selected. The internal consistency reliability of each country was analyzed reliably with Cronbach's alpha scores greater than 0.65 [22].

Behavior change in urban forest experience was analyzed with the public health and social measures (PHSM) Index provided by the Africa Center for Disease Control and Prevention (CDC) by comparing outdoor recreation behaviors before and after COVID-19. Tiered PHSM systems are a core component of effective COVID-19 preparedness, response, and risk communication [13]. The WHO classifies the PHSM Severity Index into six indicators: (1) the earing of face masks, (2) school closures, (3) office, business, and institution closures, (4) bans on people gathering, and (5) bans on domestic and (6) international travel and movement [23].

Before and after the COVID-19 lock-downs, outdoor recreation was quantified on two five-point scales for frequency, duration of greenspaces visits, and transportation time. Visit frequency measured the average frequency of visits to urban forests and urban green areas: (5) every day; (4) 2–3 times/week; (3) 1 time/week; (2) 1 time/month; and (1) rarely. Staying time measured the average length of staying time: (5) a day; (4) 4 h; (3) 2 h; (2) 1 h; and (1) 30 min. Transport time was the average travel time to the destination: (5) less than 15 min.; (4) less than 30 min.; (3) less than 1 h; (2) 1~2 h; and (1) more than 2 h.

Appreciation of urban forest, or appreciation of a natural environment in an urban setting, was affected by exogenous variables during the pandemic. Stress level was an exogenous variable related to observed interpersonal communications and relationship variables. Indoor activity was an exogenous variable affected by observational variables, including (on- or offline) screen time and visiting websites due to increase staying time at home. Outdoor recreation included the behavior of the outdoor activity, while health recovery (in nature) was related to the attention to nature-based human health during the pandemic era. (Figure 1) [20].

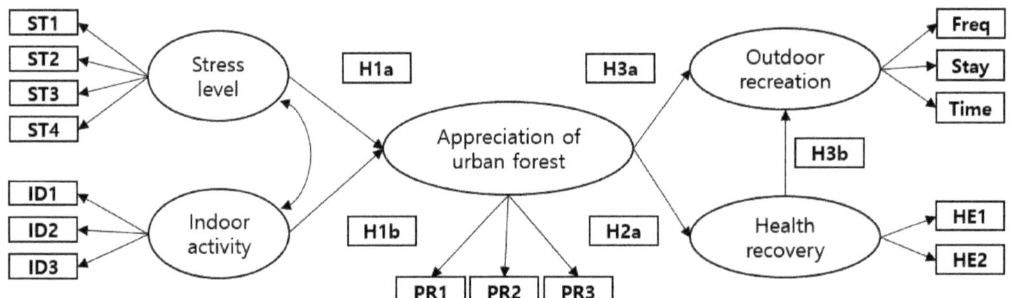

Figure 1. Structural equation modeling diagram (Adapted from [20]). Freq: visit frequency to urban forests; stay: staying time in urban forests; time: transport time to natural areas in urban settings. H1: respondent's appreciation of urban forests comes from stress level and indoor activity; H2: people's perception of health recovery during the COVID-19 pandemic comes from appreciation of urban forests; H3: behavior changes in outdoor activity come from both appreciation of urban forests and expectations for health recovery.

The collected data were analyzed using an analysis of variance (ANOVA) approach to examine the differences in each country and with structural equation modeling (SEM) to explore the relationships between variables. SPSS 25 and AMOS 25 were used for statistical analysis, and confirmatory factor analysis (CFA) was conducted to estimate the measured variables' fitness and each type's characteristics [24]. Path analysis was performed to test the influence relationships between the measured variables. Construct reliability, average variance extracted, and covariance values for the measured variables were calculated to examine the reliability and validity of the research model. To determine the intervening effect's statistical significance, we used a Sobel test.

3. Results

3.1. Comparative Changes in Perception and Behavior

The question of this study began at these points: Does increasing COVID-19 stress among urban dwellers with limited contact change their perceptions of outdoor activities and nature experiences? Do the country-specific PHSM measures identified by the Africa CDC make changes that increase or decrease the amount of natural experience of urban dwellers? This study was conducted to answer these questions.

Changes in outdoor recreation behavior before and after COVID-19 in each country were collected through the PHSM index (Table 3) [13]. The changed outdoor recreation behaviors pre- and post-COVID-19 results are presented in Table 4.

In the research SEM model, the increases in stress and indoor activities led to a rise in the appreciation of urban forests. The increased interest in health under a pandemic led to a perception of health recovery in the natural environment and outdoor activity (Figure 1). However, due to the governments' lock-down measures (Table 3), the total amount of natural experiences, such as frequency of nature experience and length of stay, of respondents in all the countries decreased significantly during the COVID-19 pandemic (Table 4). The difference in the frequency of visits before and after was verified at a high significance level.

These results were consistent with the indicators of schools closing, canceled public events, restrictions on gatherings, international travel controls, and facial coverings and mask-wearing that were common in PHSM systems during the period (Table 3). The staying time in the visited urban forests decreased in all except Malawi and was statistically significant by t-test. In other words, the visit frequency to urban forests and the time spent in urban forests both decreased during COVID-19, indicating that the total amount of city residents' natural experiences reduced significantly (Table 4). The transport time for urban forest visits increased in Malawi and Tanzania and decreased in Rwanda, South Africa, and Zambia. However, since statistical significance was not secured (South Africa and Zambia) or there were mixed cases of increase (Malawi) or decrease (Tanzania), the circumstances of each country were different, so an integrated interpretation was difficult.

Table 3. Comparison PHSM index between pre-and post-COVID-19 outbreak (Resource: Africa CDC).

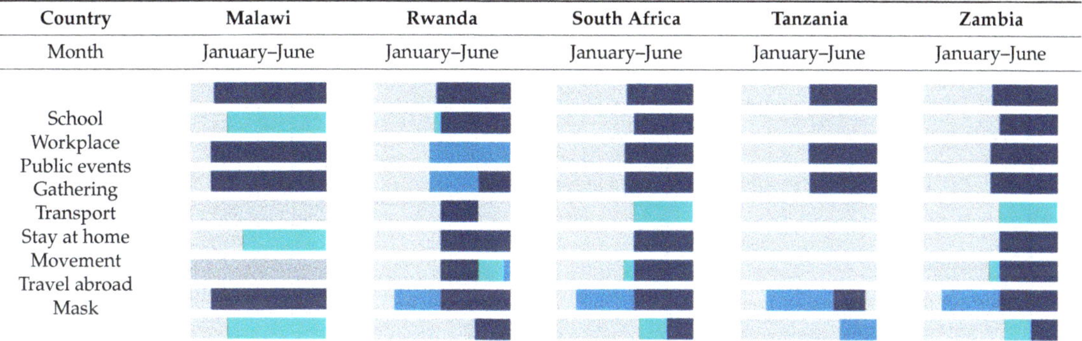

Country	Malawi	Rwanda	South Africa	Tanzania	Zambia
Month	January–June	January–June	January–June	January–June	January–June

Notes. School: school closings; Workplace: workplace closings; Public events: canceled public events; Gathering: restrictions on gatherings; Transport: closed public transport; Stay at home: stay at home requirements; Movement: restrictions on internal movement; Travel abroad: international travel controls; Mask: facial coverings or mask-wearing requirements (Resource: Africa CDC). PHSM index: no data; recommended; subnational only; national.

Table 4. Changed outdoor recreation behavior during COVID-19 pandemic.

Month	Visit Frequency			Staying Time			Transport Time		
	January	June	t-value	January	June	t-value	January	June	t-value
Malawi	2.61	2.15	3.009 **	3.07	3.51	−2.993 **	3.66	3.96	−3.243 **
Rwanda	3.66	2.38	6.022 ***	3.48	2.48	4.309 ***	3.49	3.00	2.075 *
South Africa	2.77	2.34	4.535 ***	2.71	1.96	5.647 ***	3.06	2.98	1.051
Tanzania	3.44	2.00	8.021 ***	2.97	2.00	7.083 ***	3.16	3.49	−2.042 *
Zambia	3.32	2.59	4.630 ***	3.25	2.67	4.338 ***	3.48	3.27	1.902
Average	3.11	2.30	11.110 ***	3.08	2.55	7.214 ***	3.37	3.35	0.350

Notes. *p*-value: * $p < 0.05$, ** $p < 0.01$, *** $p < 0.001$. All questionnaires were measured on a 5-point scale from 5 (every day) to 1 (rarely) for visit frequency, from 5 (a day) to 1 (less than 30 min) for staying time, and from 5 (less than 15 min) to 1 (more than 1 h) for transport time.

The COVID-19 stress was highest in Tanzania, and for the indoor activity category, Zambia and Malawi responded highly, reflecting an increase in indoor staying time due to the earliest restrictions implemented in Malawi (Table 5). South Africa showed the highest appreciation of urban forests during the pandemic.

Most of the respondents answered that it was newly recognized that the natural environment helped people recover their psychological and physical health during the pandemic, and Tanzania showed the highest number. Respondents from all the countries answered that the effect of physical health recovery was greater than that of psychological health recovery, in contrast to an Asian study that showed high responses to psychological health to the same question [20]. The visit frequency (T = 11.110; p = 0.000) and duration of visits (T = 7.214; p = 0.000) to urban forests significantly decreased due to lock-down measures; these data did not match the research hypothesis. Hypothesis H3 was rejected and dropped from SEM diagram.

Table 5. Survey results overview: appreciation and perception of nature in five countries.

	Stress Level	Indoor Activity	Appreciation of Urban Forests	Perception of Health Recovery
Malawi	3.74 ± 0.39	4.10 ± 0.28 [a,b]	3.51 ± 0.36	3.81 ± 0.40 [b]
Rwanda	3.70 ± 0.42	3.89 ± 0.46 [b]	3.62 ± 0.45	3.80 ± 0.47 [b]
South Africa	3.82 ± 0.48	3.80 ± 0.46 [b]	3.74 ± 0.43	3.93 ± 0.42 [a,b]
Tanzania	3.93 ± 0.38	3.99 ± 0.40 [a,b]	3.61 ± 0.48	4.23 ± 0.40 [a]
Zambia	3.57 ± 0.44	4.29 ± 0.39 [a]	3.54 ± 0.48	4.05 ± 0.48 [a,b]
F-value	2.042	4.895	0.968	3.450
p-value	0.088	0.001	0.425	0.009

Notes. [a,b] same letter not significant ($\alpha = 0.05$) Scheffe's post-hoc test; MANOVA Wilks' lambda = 4.033; $p < 0.000$; partial eta squared = 0.04.

3.2. Confirmatory Factor Analysis

A CFA was performed to test the validities of the variables (Table 6). Before confirming the CFA model, the variables' redundancies were tested. All the items were calculated with high R^2 values significantly by good concentration validity.

Table 6. Confirmatory factor analysis results.

Category		Variables	Malawi	Rwanda	South Africa	Tanzania	Zambia
Stress level	ß-coeff.	ST1	0.708	0.912	0.675	0.875	0.852
		ST2	0.695	0.704	0.762	0.958	0.967
		ST3	0.683	0.834	0.828	0.509	0.788
		ST4	0.890	0.716	0.718	0.486	0.616
		CR	0.840	0.855	0.813	0.829	0.879
		AVE	0.571	0.599	0.522	0.569	0.651
Indoor activity	ß-coeff.	ID1	0.753	0.573	0.554	0.720	0.762
		ID2	0.787	0.716	0.913	0.993	0.884
		ID3	0.792	0.910	0.831	0.753	0.827
		CR	0.890	0.735	0.797	0.879	0.879
		AVE	0.729	0.509	0.577	0.713	0.709
Appreciation of urban forests	ß-coeff.	PR1	0.631	0.628	0.864	0.674	0.756
		PR2	0.611	0.849	0.779	0.678	0.770
		PR3	0.998	0.735	0.757	0.976	0.673
		CR	0.833	0.760	0.839	0.800	0.731
		AVE	0.635	0.519	0.635	0.580	0.506
Health recovery perception	ß-coeff.	HE1	0.729	0.894	0.772	0.942	0.742
		HE2	0.937	0.743	0.830	0.794	0.869
		CR	0.843	0.787	0.732	0.890	0.851
		AVE	0.732	0.651	0.578	0.803	0.742
Model fit summary		Chi2	103.768	60.609	92.394	137.957	136.846
		P-level	0.000	0.105	0.000	0.000	0.000
		GFI	0.860	0.865	0.877	0.791	0.822
		AGFI	0.773	0.781	0.801	0.740	0.790
		NFI	0.810	0.858	0.840	0.846	0.815
		IFI	0.888	0.967	0.916	0.918	0.872
		TLI	0.840	0.952	0.910	0.894	0.881
		CFI	0.884	0.965	0.913	0.912	0.916
		RMSEA	0.071	0.054	0.056	0.051	0.063

Notes. ß-coeff.: standardized coefficient, CR: construct reliability, AVE: average variance extracted, P-level: probability level.

The exogenous variables were composed of four variables of stress level and three of indoor activity. The endogenous variables consisted of two variables of health recovery and three of appreciation of urban forests.

The construct reliability (CR) and average variance extracted (AVE) were calculated to test discriminant validity. The validity of the AVE range value was: very good (>0.7), acceptable (0.7~0.5), and not acceptable (<0.5). If the CR value was above 0.7 and the AVE was above 0.5, the data could be determined to have convergent validity [25,26]. The research CFA model was confirmed with high CR and AVE values for internal consistency and convergence validity, which indicated a good data fit. The model fit test by each country is shown in Table 6. The model was suitable when the chi-square value was small and the probability value was considerable ($p > 0.10$). The indices of model fit by chi square were not acceptable for Malawi (103.768, $p = 0.000$), South Africa (92.394, $p = 0.000$), Tanzania (137.957, $p = 0.000$), and Zambia (136.846, $p = 0.0000$), but they were acceptable for Rwanda at 60.609 ($p = 0.105$).

The minimum and maximum indices of the five countries were goodness-of-fit index (GFI) = 0.791~0.877 and adjusted goodness-of-fit (AGFI) = 0.773~0.801. The GFI and AGFI statistics ranged between 0 and 1, and the recommended values, a good fit, were more than 0.90. The GFI value of 0.791~0.877 was less than 0.9 due to the relatively small sample size [27,28].

The normed fit index (NFI) was 0.810~0.858. The normed fit index (NFI) was used as an alternative to CFI, but one did not require the chi-square test, and the range was from 0 to 1, with 1 = perfect fit [29]. The incremental fit index (IFI) = 0.872~0.967, the Tucker–Lewis index (TLI) = 0.840~0.952, and the comparative fit index (CFI) = 0.884~0.965. The CFI value was close to 0.9, which showed a relatively good fit [30]. The TLI is relatively independent of sample size, and a TLI value approaching 1 indicates a good fit [28,29,31,32].

The root mean square error of approximation (RMSEA) = 0.051~0.071. RMSEA is an index of the difference between the observed covariance matrix per degree of freedom and the hypothesized covariance matrix, which denotes the model [25,33]. A good model fit by RMSEA was recommended as smaller than 0.06, in general [34], from 0.05 to 0.07 are acceptable [35], 0.08 to 0.1 are marginal, and values above 0.1 are poor [36]. The CFA model fit was evaluated to be acceptable (Table 6).

3.3. Structural Equation Modeling for the Research Hypotheses

SEM was conducted to examine the hypothesis of the relationship between COVID-19 stress and changed perceptions of health in the urban forests of five African countries' people (Table 7, Appendix A (Table A1)). The model fit indices by chi square were not acceptable for Malawi (χ^2 = 119.963, $p = 0.000$), South Africa (χ^2 = 130.390, $p = 0.000$), Tanzania (χ^2 = 151.270, $p = 0.000$), and Zambia (χ^2 = 140.689, $p = 0.000$), but they were acceptable for Rwanda at χ^2 = 64.672 ($p = 0.079$). The different model fit test results are given in Table 7 and Appendix A (Table A1). The first hypothesis of the relationship of stress level to appreciation of urban forests as deterministic variables was examined. The hypothesis of a positive effect of stress level on urban forest appreciation was accepted in Rwanda, South Africa, and Zambia with standardized coefficients of 0.419, 0.363, and 0.375, respectively (t = 3.081, 3.175, and 3.281, respectively; $p < 0.01$). The level of the symbolic meaning of increasing stress in the COVID-19 pandemic was significantly related to the appreciation of urban forests.

However, the hypothesis of the significant effect of indoor activity, which represented the changed indoor-oriented lifestyle of increased screen-watching time (internet-accessed digital device tablet-PC, smartphone, etc.) to appreciation of urban forests, was supported in two countries of Rwanda and Zambia, with standardized coefficients of 0.447 (t = 2.787, $p < 0.01$) and 0.285 (t = 2.006, $p < 0.05$), respectively.

The second hypothesis of the relationship of health recovery perception to appreciation of urban forests was examined. Except in Tanzania, the appreciation of urban forests and nature experiences was significantly related to recovering health in nature ($p < 0.01$~0.001). The results supported the hypothesis that the appreciation of urban forests and nature experiences caused nature-related health recovery perceptions in Rwanda, South Africa, and Zambia.

Table 7. Research hypotheses tested by SEM model fit.

Hypothesis: Direction		Malawi	Rwanda	South Africa	Tanzania	Zambia
H1a: Stress level →	Appreciation of urban forests	Reject	**Accept**	**Accept**	Reject	**Accept**
H1b: Indoor activity →	Appreciation of urban forests	Reject	**Accept**	Reject	Reject	**Accept**
H2: Appreciation of urban forests →	Perception of health recovery	**Accept**	**Accept**	**Accept**	Reject	**Accept**

Because of bans on direct interpersonal contact and increased indoor time, in relation to COVID-19 stress, it was observed that the appreciation of urban forests was enhanced in Rwanda and Zambia. The increased appreciation of nature experiences in urban forests during the COVID-19 pandemic was related to the perception of nature-based health recovery in Malawi, Rwanda, South Africa, and Zambia.

The appreciation of urban forests intervened between stress level, indoor activity, and health recovery. A Sobel test was conducted on the research model, assuming that either stress level or indoor activity affected health recovery perception (Table 8) [37,38].

Table 8. The indirect effect of explanatory variables on health outcomes evaluated with the Sobel test.

Indirect Effect		Malawi	Rwanda	South Africa	Tanzania	Zambia
Stress level → Urban forest → Health	Z-value	0.850	**2.363**	1.909	0.152	**2.576**
	p	0.197	**0.009**	**0.028**	0.439	**0.004**
Indoor activity → Urban forest → Health	Z-value	0.657	**2.226**	0.667	0.713	**1.806**
	p	0.255	**0.013**	0.252	0.237	**0.035**

Notes. Urban forest: appreciation of urban forests, Health: perception of health recovery.

Appreciation of urban forests acted as an intervening variable between stress level and health recovery perception in Rwanda ($p = 0.009$), South Africa ($p = 0.028$), and Zambia ($p = 0.004$). The second test showed that appreciation of urban forests intervened between indoor activity and health recovery perception in Rwanda ($p = 0.013$) and Zambia ($p = 0.035$). The Sobel test results showed that the appreciation of urban forests in Rwanda and Zambia played a mediating role between stress level during the COVID-19 pandemic and the perception of health recovery.

We found that the stress level and indoor activities affected the appreciation of urban forests associated with the perception of health recovery through research model examination. Thus, the research hypotheses of H1a, H1b, and H2a were fully supported in Rwanda and Zambia.

4. Discussion

4.1. Distancing from Nature during Lock-Down Measures

The tendency of a decreasing amount of nature experience during the COVID-19 pandemic and increasing stress levels was observed in this study. We studied the bases of these phenomena from limited interpersonal communication during lock-down. In the world after the pandemic declaration, people's stress increased due to a series of government regulations on face-to-face contact, and residents' daily lives in densely populated cities meant disconnection and isolation. In particular, underdeveloped countries imposed strong lock-down measures to prevent the collapse of the health system due to large-scale infection, and the majority of people with weak financial status faced difficulties due to income reduction, food insecurity, and threats to their livelihoods [5,9,10,20,39].

In Africa, where experience in coping with diseases such as tuberculosis, HIV, Ebola, and H1N1 influenza has been accumulated, immediate and strong lock-down policies were implemented after the pandemic. In the five countries of this study (Malawi, Rwanda, South Africa, Tanzania, and Zambia), after the pandemic declaration, strong containment policies were implemented against schools, public gatherings, group contact, travel, and the internal movement of people [7,8,10,11,13].

We found increased stress levels due to limited interpersonal communication during lock-down. In the United States, Poland, Canada, and Australia, compared to the pre-COVID-19 period people complained of psychological disorders such as isolation, depression, helplessness, and stress due to significantly increased indoor dwelling time, limited communication, and insufficient outdoor activities [40–43]. The increase in indoor life has been indicated by passive and low-movement habits such as reading, pc games, tablets, and smartphones [44–49]. Despite interest in various indoor exercises, including yoga, the total amount of activity tended to decrease [50].

Quantitative changes in outdoor recreation activities during the COVID-19 pandemic have been reported worldwide, including in the US [15–17,51]. In this study, a decrease in the frequency of visits to urban forests was observed in five countries.

In a long-term survey of 64,000 people in the United States, despite national and state parks closing and recreation program stoppage due to the federal government's COVID-19 prevention policy, visits to the forest increased [51]. It was reported that the frequency of outdoor recreation among American citizens increased by 43% during the COVID-19 pandemic [52]. In Oslo's urban green areas, pedestrian activity increased during the pandemic, and it was identified that urban nature acted as an escape from various aspects of restriction stress [53].

The frequency of visits to the natural environment has been identified as an essential factor influencing nature experience and outdoor recreation [54–56]. In a comparative study of urban residents in three Asian countries with pro-environmental attitudes [57], the perception and behavior of urban forest visitors in three western European countries for health recovery [58], and a comparative analysis of visitors to urban forests in Korea and Germany with the theme of nature experience aversion, visit frequency was always determined as an important influencing factor [19]. The decrease in the frequency of visits to urban forests after lock-down observed in this study was predicted as a factor that could increase COVID-19 stress and lower the quality of life under the pandemic.

4.2. Perception of Recovery in Human Health in Nature during the COVID-19 Pandemic

We predicted that the stress caused by COVID-19 was predicted to express an increase in appreciation of urban forest experiences and affect behavioral changes, such as increased outdoor activities and the perception of health recovery in urban forests. The research hypothesis assumed that the flow of increased interest in health after the pandemic would lead to health recovery perception through natural experiences. All five countries showed very high responses to the idea of the recovery of human health in nature.

The H2 hypothesis of the SEM model assumed a relationship between the appreciation strengthened during the pandemic and the perception of health recovery in nature, and this was supported in four countries, all except Tanzania. As a result, factors such as indoor lifestyle, communication disconnection, and stress in Rwanda and Zambia influenced the perception of health recovery in nature with appreciation of urban forest experiences as a parameter.

Many previous studies have examined health promotion in forests and nature [40–43,59–68]. In a study on the perception of psychological and physical health promotion effects in urban forests conducted in Berlin (Germany), Vienna (Austria), and Zurich (Switzerland), the health recovery function in urban forests was affirmed [58]. Electoencephalography (EEG) analyses have suggested that the alpha wave (the conscious and relaxed brain status with a frequency of 8 to 13 hertz), which is increased during relaxation, has also been observed to increase when the people are exposed to a forest environment, thereby affirming that forest experiences have a therapeutic effect and a significant impact on stress reduction [41,42,69,70]. In studies on human physiological substances, the cortisol hormone, detected at a high concentration when people are in a stressed state, has shown a statistically significant decrease when subjects are exposed to a forest environment, thus supporting forest experience programs as a physiological treatment therapy [71–73].

The effectiveness of trekking in forests to promote people's physical health, such as muscular endurance, bone density, and cardiorespiratory function, was studied by planning an appropriate exercise load considering the slope and distance of forest trails in mountainous areas. This has been named "terrain therapy" as a treatment therapy [71,74–77]. The recovery of human health in a forest environment has been shown through research on its effects on mental health treatment, such as alcoholism, gambling addiction, and gaming addiction, and on the improvement of various psychological disorders, such as interpersonal stress, depression, and obsession disorder [41,61,78–82]. In the pandemic, urban green spaces were also newly defined as areas important for life that maximize natural human resilience [83–85].

The results of this study, which demonstrated the relationship between the appreciation of urban forests during the pandemic and the perception of health recovery, are in line with the results of studies in the United States and Europe that reported an increase in the amount of outdoor activity to maintain health [51,86].

4.3. Study Limitations

This study had several research limitations that need to be improved in further studies. First, it is necessary to verify whether the sample is representative demographically, geographically, historically, and culturally because the sample collected during the international comparative study was assumed to represent the research target country and region [87–89]. Of course, this study surveyed the citizens of the capital of each country.

Still, in a country where the capital and the provinces have economic, social, and cultural differences, residents of the capital may not represent residents of the entire country. This is another typical problem in a cross-national study of the sample homogeneity issue at the same time. This includes whether the sample group of this study, which was smaller under the pandemic situation, could represent the population. To compensate for these weaknesses, it is necessary to deepen the group composition and interpretation of research results by including sociological or anthropological factors in future studies.

Despite briefly mentioning a decrease in GDP growth rates in five countries before and after the COVID-19, the interpretation of the post-COVID-19 socioeconomic and industrial impact was insufficient. COVID-19 caused severe food insecurity, as well as the extreme decline and partial collapse of tourism, the financial sector, and education systems [90–94].

Considering that macro-and micro-socioeconomic impacts affect even the political arena [9,95], an integrated interpretation of the economic situation and the pandemic is necessary for future research.

After the lock-down, there was an apparent decrease in citizens' urban forest use rates, so we had to revise the SEM model. Although the high appreciation of urban forests and the will to recover health in nature were analyzed, it was impossible to integrate and interpret observed behavioral changes, such as visit frequency and stay time. An analysis of the trend of behavioral change after the pandemic is lifted is necessary as a pre-post-follow-up survey.

5. Conclusions

This study aimed to investigate the relationship between urban forest appreciation and health recovery perception with stress, outdoor recreation, and lock-down measures in most but not all of the investigated countries during the COVID-19 pandemic through an SEM model. The study results showed that, in Rwanda, South Africa, and Zambia, COVID-19 stress led to an appreciation of urban forests and was related to health recovery perception. However, at the same time, social distancing from nature due to COVID-19 lock-down rules appeared to decrease the amount of natural experience, such as the visit frequency and stay time.

As the research results showed, appreciation of the natural environment in urban forests developed into high expectations for health recovery in nature. The role and function of the urban forest that has worked so far should be strengthened as a health recovery place that can overcome the stress of COVID-19.

The COVID-19 lock-down measures were a response to a crisis of humanity, and they reinforced the desire to experience nature and restore nature-based health. The results of this study can serve as a basis for predicting the expectation for health recovery and healing activities in urban forests after the end of the pandemic.

Author Contributions: Conceptualization, J.-h.L.; methodology, J.-h.L.; software, J.-h.L.; validation, J.-h.L.; formal analysis, J.-h.L.; investigation, M.M., P.N., A.X., E.I., and S.S.; resources, M.M., P.N., A.X., E.I., and S.S.; data curation, M.M., P.N., A.X., E.I., and S.S.; writing—original draft preparation, J.-h.L.; writing—review and editing, J.-h.L.; visualization, J.-h.L.; project administration, J.-h.L. All authors have read and agreed to the published version of the manuscript.

Funding: This research received no external funding.

Institutional Review Board Statement: Not applicable.

Informed Consent Statement: Not applicable.

Data Availability Statement: Not applicable.

Acknowledgments: The authors would like to thank the Park Chung Hee School of Policy and Saemaul (PSPS), Yeungnam University, Korea, for building the research cooperation relationship with five governments of African countries.

Conflicts of Interest: The authors declare no conflict of interest.

Appendix A

Table A1. Research hypotheses tested by SEM model fit.

Hypothesis: Direction			Malawi	Rwanda	South Africa	Tanzania	Zambia
H1a: Stress level →	Appreciation of urban forest	Estimate	0.106	0.419	0.363	0.027	**0.375**
		S.E.	0.118	0.136	0.152	0.174	0.114
		CR	0.898	3.081	3.175	0.002	3.281
		p	0.497	0.002	0.001	0.998	0.001
		Result	Reject	Accept	Accept	Reject	Accept
H1b: Indoor activity →	Appreciation of urban forest	Estimate	0.091	**0.447**	0.114	0.314	**0.285**
		S.E.	0.134	0.160	0.167	0.215	0.142
		CR	0.679	2.787	0.680	1.463	2.006
		p	0.497	0.005	0.496	0.143	0.045
		Result	Reject	Accept	Reject	Reject	Accept
H2: Appreciation of urban forest →	Perception of health recovery	Estimate	**0.454**	**0.571**	0.563	0.067	**0.402**
		S.E.	0.172	0.155	0.177	0.082	0.097
		CR	2.647	3.695	3.175	0.812	4.123
		p	0.008	0.000	0.001	0.417	0.000
		Result	Accept	Accept	Accept	Reject	Accept
Model fit test		Chi2	119.963	64.672	130.390	151.270	140.689
		p-value	0.000	0.079	0.000	0.000	0.000
		GFI	0.838	0.859	0.838	0.769	0.812
		AGFI	0.747	0.781	0.747	0.639	0.707
		NFI	0.780	0.848	0.774	0.722	0.810
		IFI	0.859	0.961	0.847	0.795	0.869
		TLI	0.807	0.946	0.792	0.720	0.822
		CFI	0.854	0.959	0.843	0.788	0.865
		RMSEA	0.122	0.048	0.052	0.167	0.057

References

1. Wadvalla, B.A. How Africa has tackled COVID-19. *BMJ* **2020**, *370*, m2830. [CrossRef] [PubMed]
2. Berhan, Y. Will Africa be devastated by COVID-19 as many predicted? Perspective and Prospective. *Ethiop. J. Health Sci.* **2020**, *30*, 459. [CrossRef] [PubMed]
3. Massinga Loembé, M.; Tshangela, A.; Salyer, S.J.; Varma, J.K.; Ouma, A.E.O.; Nkengasong, J.N. COVID-19 in Africa: The spread and response. *Nat. Med.* **2020**, *26*, 999–1003. [CrossRef] [PubMed]
4. Divala, T.; Burke, R.M.; Ndeketa, L.; Corbett, E.L.; MacPherson, P. Africa faces difficult choices in responding to COVID-19. *Lancet* **2020**, *395*, 1611. [CrossRef]
5. Musanabaganwa, C.; Cubaka, V.; Mpabuka, E.; Semakula, M.; Nahayo, E.; Hedt-Gauthier, B.L.; NG, K.C.S.; Murray, M.B.; Kateera, F.; Mutesa, L.; et al. One hundred thirty-three observed COVID-19 deaths in 10 months: Unpacking lower than predicted mortality in Rwanda. *BMJ Glob. Health* **2021**, *6*, e004547. [CrossRef] [PubMed]
6. International Organization for Migration. Regional Strategic Preparedness and Response Plan COVID-19. IOM East and Horn of Africa. April–December 2020. Available online: https://www.iom.int/news/iom-regional-office-launches-covid-19-strategic-preparedness-and-response-plan-east-and-horn-africa (accessed on 7 May 2022).
7. El-Sadr, W.M.; Justman, J. Africa in the path of COVID-19. *N. Engl. J. Med.* **2020**, *383*, e11. [CrossRef]
8. Tembo, J.; Maluzi, K.; Egbe, F.; Bates, M. COVID-19 in Africa. *BMJ* **2021**, *372*, n457. [CrossRef]
9. Saleh, M. Impact of COVID-19 on Tanzania political economy. *Int. J. Adv. Stud. Soc. Sci. Innov.* **2020**, *4*, 24–36. [CrossRef]
10. Mbow, M.; Lell, B.; Jochems, S.P.; Cisse, B.; Mboup, S.; Dewals, B.G.; Jaye, A.; Dieye, A.; Yazdanbakhsh, M. COVID-19 in Africa: Dampening the storm? *Science* **2020**, *369*, 624–626. [CrossRef]
11. Ogbolosingha, A.J.; Singh, A. COVID-19 pandemic: Review of impediments to public health measures in Sub-Saharan Africa. *Am. J. Prev. Med.* **2020**, *6*, 68–75. [CrossRef]
12. Mwisongo, A.; Nabyonga-Orem, J. Global health initiatives in Africa–governance, priorities, harmonisation and alignment. *BMC Health Serv. Res.* **2016**, *16*, 245–254. [CrossRef] [PubMed]
13. Africa CDC. COVID-19 Africa Hotspot Dashboard. Available online: https://africacdccovid.org/ (accessed on 27 April 2022).
14. Bammel, G.; Burrus-Bammel, L.L. *Leisure and Human Behaviour*; Wm. C. Brown Company Publishers: Dubuque, IA, USA, 1982; 361p.
15. Landry, C.E.; Bergstrom, J.; Salazar, J.; Turner, D. How Has the COVID-19 Pandemic Affected Outdoor Recreation in the US? A Revealed Preference Approach. *Appl. Econ. Perspect. Policy* **2021**, *43*, 443–457. [CrossRef]
16. Rhodes, R.E.; Liu, S.; Lithopoulos, A.; Zhang, C.Q.; Garcia-Barrera, M.A. Correlates of perceived physical activity transitions during the COVID-19 pandemic among Canadian adults. *Appl. Psychol. Health Well-Being* **2020**, *12*, 1157–1182. [CrossRef]
17. Morse, J.W.; Gladkikh, T.M.; Hackenburg, D.M.; Gould, R.K. COVID-19 and human-nature relationships: Vermonters' activities in nature and associated nonmaterial values during the pandemic. *PLoS ONE* **2020**, *15*, e0243697. [CrossRef] [PubMed]
18. Ainsworth, B.E.; Li, F. Physical activity during the coronavirus disease-2019 global pandemic. *J. Sport Health Sci.* **2020**, *9*, 291. [CrossRef] [PubMed]
19. Lee, J.H.; Lee, S.J. Nature experience influences nature aversion: Comparison of South Korea and Germany. *Soc. Behav. Personal.* **2018**, *46*, 161–176. [CrossRef]
20. Lee, J.H.; Cheng, M.; Syamsi, M.N.; Lee, K.H.; Aung, T.R.; Burns, R.C. Accelerating the Nature Deficit or Enhancing the Nature-Based Human Health during the Pandemic Era: An International Study in Cambodia, Indonesia, Japan, South Korea, and Myanmar, following the Start of the COVID-19 Pandemic. *Forests* **2022**, *13*, 57. [CrossRef]
21. World Health Organization. Coronavirus Disease (COVID-19) Pandemic: WHO Characterizes COVID-19 as a Pandemic. Available online: https://www.who.int/emergencies/diseases/novel-coronavirus-2019/events-as-they-happen (accessed on 27 April 2022).
22. Vaske, J.J. *Survey Research and Analysis: Applications in Parks, Recreation and Human Dimensions*; Venture Publishing: State College, PA, USA, 2008; 635p.
23. World Health Organization. *A Systematic Approach to Monitoring and Analysing Public Health and Social Measures (PHSM) in the Context of the COVID-19 Pandemic: Underlying Methodology and Application of the PHSM Database and PHSM Severity Index (No. WHO/EURO: 2020-1610-41361-56329)*; World Health Organization Regional Office for Europe: Copenhagen, Denmark, 2020; 45p.
24. Marsh, H.W.; Dowson, M.; Pietsch, J.; Walker, R. Why multicollinearity matters: A reexamination of relations between self-efficacy, self-concept, and achievement. *J. Educ. Psychol.* **2004**, *96*, 518. [CrossRef]
25. Chen, F.F. Sensitivity of goodness of fit indexes to lack of measurement invariance. *Struct. Equ. Modeling A Multidiscip. J.* **2007**, *14*, 464–504. [CrossRef]
26. Hair, J.F.; Ringle, C.M.; Sarstedt, M. PLS-SEM: Indeed a silver bullet. *J. Mark. Theory Pract.* **2011**, *19*, 139–152. [CrossRef]
27. Mulaik, S.A.; James, L.R.; Van Alstine, J.; Bennett, N.; Lind, S.; Stilwell, C.D. Evaluation of goodness-of-fit indices for structural equation models. *Psychol. Bull.* **1989**, *105*, 430–445. [CrossRef]
28. Kim, H.; Ku, B.; Kim, J.Y.; Park, Y.J.; Park, Y.B. Confirmatory and exploratory factor analysis for validating the phlegm pattern questionnaire for healthy subjects. *Evid-Based Complementary Altern. Med.* **2016**, *2016*, 2696019. [CrossRef] [PubMed]
29. Shadfar, S.; Malekmohammadi, I. Application of Structural Equation Modeling (SEM) in restructuring state intervention strategies toward paddy production development. *Int. J. Acad. Res. Bus. Soc. Sci.* **2013**, *3*, 576. [CrossRef]

30. Bentler, P.M. Comparative fit indexes in structural models. *Psychol. Bull.* **1990**, *107*, 238–246. [CrossRef] [PubMed]
31. Marsh, H.W.; Balla, J.R.; McDonald, R.P. Goodness-of-fit indexes in confirmatory factor analysis: The effect of sample size. *Psychol. Bull.* **1988**, *103*, 391. [CrossRef]
32. Marsh, H.W.; Balla, J.R.; Hau, K.T. An evaluation of incremental fit indices: A clarification of mathematical and empirical properties. In *Advanced Structural Equation Modeling Techniques*; Marcoulides, G.A., Schumacker, R.E., Eds.; Psychology Press: Mahwah, NJ, USA, 1996; pp. 315–353.
33. Cangur, S.; Ercan, I. Comparison of model fit indices used in structural equation modeling under multivariate normality. *J. Mod. Appl. Stat. Methods* **2015**, *14*, 14. [CrossRef]
34. Hu, L.T.; Bentler, P.M. Cutoff criteria for fit indexes in covariance structure analysis: Conventional criteria versus new alternatives. *Struct. Equ. Modeling A Multidiscip. J.* **1999**, *6*, 1–55. [CrossRef]
35. Steiger, J.H. Understanding the limitations of global fit assessment in structural equation modeling. *Personal. Individ. Differ.* **2007**, *42*, 893–898. [CrossRef]
36. Fabrigar, L.R.; MacCallum, R.C.; Wegener, D.T.; Strahan, E.J. Evaluating the use of exploratory factor analysis in psychological research. *Psychol. Methods* **1999**, *4*, 272–299. [CrossRef]
37. Abu-Bader, S.; Jones, T.V. Statistical mediation analysis using the sobel test and hayes SPSS process macro. *Int. J. Quant. Qual. Res. Methods* **2021**, *9*, 42–61.
38. Sobel, M.E. Asymptotic intervals for indirect effects in structural equations models. In *Sociological Methodology*; Leinhart, S., Ed.; Jossey-Bass: San Francisco, CA, USA, 1982; pp. 290–312.
39. Arndt, C.; Davies, R.; Gabriel, S.; Harris, L.; Makrelov, K.; Robinson, S.; Levy, S.; Simbanegavi, W.; van Seventer, D.; Anderson, L. COVID-19 lockdowns, income distribution, and food security: An analysis for South Africa. *Glob. Food Secur.* **2020**, *26*, 100410. [CrossRef] [PubMed]
40. Zeng, C.; Lin, W.; Li, N.; Wen, Y.; Wang, Y.; Jiang, W.; Zhang, J.; Zhong, H.; Chen, X.; Luo, W.; et al. Electroencephalography (EEG)-Based Neural Emotional Response to the Vegetation Density and Integrated Sound Environment in a Green Space. *Forests* **2021**, *12*, 1380. [CrossRef]
41. Wang, Y.; Xu, M. Electroencephalogram Application for the Analysis of Stress Relief in the Seasonal Landscape. *Int. J. Environ. Res. Public Health* **2021**, *18*, 8522. [CrossRef] [PubMed]
42. Olszewska-Guizzo, A.; Escoffier, N.; Chan, J.; Puay Yok, T. Window view and the brain: Effects of floor level and green cover on the alpha and beta rhythms in a passive exposure EEG experiment. *Int. J. Environ. Res. Public Health* **2018**, *15*, 2358. [CrossRef]
43. Bae, Y.M.; Lee, Y.; Kim, S.M.; Piao, Y.H. A comparative study on the forest therapy policies of Japan and Korea. *J. Korean Soc. For. Sci.* **2014**, *103*, 299–306. [CrossRef]
44. Chopdar, P.K.; Paul, J.; Prodanova, J. Mobile shoppers' response to COVID-19 phobia, pessimism and smartphone addiction: Does social influence matter? *Technol. Forecast. Soc. Change* **2022**, *174*, 121249. [CrossRef]
45. Elhai, J.D.; Yang, H.; McKay, D.; Asmundson, G.J. COVID-19 anxiety symptoms associated with problematic smartphone use severity in Chinese adults. *J. Affect. Disord.* **2020**, *274*, 576–582. [CrossRef]
46. Limone, P.; Toto, G.A. Psychological and emotional effects of Digital Technology on Children in COVID-19 Pandemic. *Brain Sci.* **2021**, *11*, 1126. [CrossRef]
47. Daglis, T. The Increase in Addiction during COVID-19. *Encyclopedia* **2021**, *1*, 1257–1266. [CrossRef]
48. David, M.E.; Roberts, J.A. Smartphone Use during the COVID-19 Pandemic: Social Versus Physical Distancing. *Int. J. Environ. Res. Public Health* **2021**, *18*, 1034. [CrossRef]
49. Nathan, A.; George, P.; Ng, M.; Wenden, E.; Bai, P.; Phiri, Z.; Christian, H. Impact of COVID-19 Restrictions on Western Australian Children's Physical Activity and Screen Time. *Int. J. Environ. Res. Public Health* **2021**, *18*, 2583. [CrossRef] [PubMed]
50. García-Tascón, M.; Sahelices-Pinto, C.; Mendaña-Cuervo, C.; Magaz-González, A.M. The impact of the COVID-19 confinement on the habits of PA practice according to gender (male/female): Spanish case. *Int. J. Environ. Res. Public Health* **2020**, *17*, 6961. [CrossRef] [PubMed]
51. Rice, W.L.; Mateer, T.; Taff, B.D.; Lawhon, B.; Reigner, N.; Newman, P. Longitudinal Changes in the Outdoor Recreation Community's Reaction to the COVID-19 Pandemic: Final Report on a Three-Phase National Survey of Outdoor Enthusiasts. *SocArXiv Gnjcy Cent. Open Sci.* **2020**, 1–10. [CrossRef]
52. Yamori, K.; Goltz, J.D. Disasters without Borders: The Coronavirus Pandemic, Global Climate Change and the Ascendancy of Gradual Onset Disasters. *Int. J. Environ. Res. Public Health* **2021**, *18*, 3299. [CrossRef]
53. Venter, Z.S.; Barton, D.N.; Gundersen, V.; Figari, H.; Nowell, M. Urban nature in a time of crisis: Recreational use of green space increases during the COVID-19 outbreak in Oslo, Norway. *Environ. Res. Lett.* **2020**, *15*, 104075. [CrossRef]
54. Newhouse, N. Implications of attitude and behavior research for environmental conservation. *J. Environ. Educ.* **1990**, *22*, 26–32. [CrossRef]
55. Soga, M.; Gaston, K.J. Extinction of experience: The loss of human–nature interactions. *Front. Ecol. Environ.* **2016**, *14*, 94–101. [CrossRef]
56. Wolsko, C.; Lindberg, K. Experiencing connection with nature: The matrix of psychological well-being, mindfulness, and outdoor recreation. *Ecopsychology* **2013**, *5*, 80–91. [CrossRef]
57. Kim, D.; Avenzora, R.; Lee, J.H. Exploring the Outdoor Recreational Behavior and New Environmental Paradigm among Urban Forest Visitors in Korea, Taiwan, and Indonesia. *Forests* **2021**, *12*, 1651. [CrossRef]

58. Lee, J.H.; Lee, D.J. Nature experience, recreation activity and health benefits of visitors in mountain and urban forests in Vienna, Zurich and Freiburg. *J. Mt. Sci.* **2015**, *12*, 1551–1561. [CrossRef]

59. Kotera, Y.; Richardson, M.; Sheffield, D. Effects of Shinrin-Yoku (Forest Bathing) and Nature Therapy on Mental Health: A Systematic Review and Meta-analysis. *Int. J. Ment. Health Addict.* **2020**, *20*, 337–361. [CrossRef]

60. Shin, W.S.; Yeoun, P.S.; Yoo, R.W.; Shin, C.S. Forest experience and psychological health benefits: The state of the art and future prospect in Korea. *Environ. Health Prev. Med.* **2010**, *15*, 38. [CrossRef] [PubMed]

61. Oh, K.H.; Shin, W.S.; Khil, T.G.; Kim, D.J. Six-step model of nature-based therapy process. *Int. J. Environ. Res. Public Health* **2020**, *17*, 685. [CrossRef] [PubMed]

62. Ebenberger, M.; Arnberger, A. Exploring visual preferences for structural attributes of urban forest stands for restoration and heat relief. *Urban For. Urban Green.* **2019**, *41*, 272–282. [CrossRef]

63. Nilsson, K.; Sangster, M.; Gallis, C.; Hartig, T.; De Vries, S.; Seeland, K.; Schipperijn, J. (Eds.) *Forests, Trees and Human Health*; Springer Science & Business Media: New York, NY, USA; Dordrecht, The Netherlands; Heidelberg, Germany; London, UK, 2010; 427p.

64. Park, B.J.; Shin, C.S.; Shin, W.S.; Chung, C.Y.; Lee, S.H.; Kim, D.J.; Kim, Y.H.; Park, C.E. Effects of forest therapy on health promotion among middle-aged women: Focusing on physiological indicators. *Int. J. Environ. Res. Public Health* **2020**, *17*, 4348. [CrossRef]

65. Arnberger, A.; Eder, R.; Allex, B.; Ebenberger, M.; Hutter, H.P.; Wallner, P.; Bauer, N.; Zaller, J.G.; Frank, T. Health-related effects of short stays at mountain meadows, a river and an urban site—Results from a field experiment. *Int. J. Environ. Res. Public Health* **2018**, *15*, 2647. [CrossRef]

66. Kim, J.; Park, D.B.; Seo, J.I. Exploring the Relationship between Forest Structure and Health. *Forests* **2020**, *11*, 1264. [CrossRef]

67. Sacchelli, S.; Grilli, G.; Capecchi, I.; Bambi, L.; Barbierato, E.; Borghini, T. Neuroscience application for the analysis of cultural ecosystem services related to stress relief in forest. *Forests* **2020**, *11*, 190. [CrossRef]

68. Ohe, Y.; Ikei, H.; Song, C.; Miyazaki, Y. Evaluating the relaxation effects of emerging forest-therapy tourism: A multidisciplinary approach. *Tour. Manag.* **2017**, *62*, 322–334. [CrossRef]

69. Hong, J.Y.; Lee, J.H. Analysis of Electroencephalogram and Electrocardiogram Changes in Adults in National Healing Forests Environment. *J. People Plants Environ.* **2018**, *21*, 575–589. [CrossRef]

70. Lim, Y.S.; Kim, J.; Khil, T.; Yi, J.; Kim, D.J. Effects of the Forest Healing Program on Depression, Cognition, and the Autonomic Nervous System in the Elderly with Cognitive Decline. *J. People Plants Environ.* **2021**, *24*, 107–117. [CrossRef]

71. Yu, Y.M.; Lee, Y.J.; Kim, J.Y.; Yoon, S.B.; Shin, C.S. Effects of forest therapy camp on quality of life and stress in postmenopausal women. *For. Sci. Technol.* **2016**, *12*, 125–129. [CrossRef]

72. Jung, W.H.; Woo, J.M.; Ryu, J.S. Effect of a forest therapy program and the forest environment on female workers' stress. *Urban For. Urban Green.* **2015**, *14*, 274–281. [CrossRef]

73. Lee, M.M.; Park, B.J. Effects of Forest Healing Program on Depression, Stress and Cortisol Changes of Cancer Patients. *J. People Plants Environ.* **2020**, *23*, 245–254. [CrossRef]

74. Li, Q.; Kobayashi, M.; Kumeda, S.; Ochiai, T.; Miura, T.; Kagawa, T.; Imai, M.; Wang, Z.; Otsuka, T.; Kawada, T. Effects of forest bathing on cardiovascular and metabolic parameters in middle-aged males. *Evid-Based Complementary Altern. Med.* **2016**, *2016*, 2587381. [CrossRef] [PubMed]

75. Ideno, Y.; Hayashi, K.; Abe, Y.; Ueda, K.; Iso, H.; Noda, M.; Lee, J.S.; Suzuki, S. Blood pressure-lowering effect of Shinrin-yoku (Forest bathing): A systematic review and meta-analysis. *BMC Complementary Altern. Med.* **2017**, *17*, 409. [CrossRef]

76. Baek, J.E.; Shin, H.J.; Kim, S.H.; Kim, J.Y.; Park, S.; Sung, S.Y.; Cho, H.-Y.; Hahm, S.C.; Lee, M.G. The Effects of Forest Healing Anti-aging Program on Physical Health of the Elderly: A Pilot Study. *Korean Soc. Phys. Med.* **2021**, *16*, 81–90. [CrossRef]

77. Kim, T.; Song, B.; Cho, K.S.; Lee, I.S. Therapeutic potential of volatile terpenes and terpenoids from forests for inflammatory diseases. *Int. J. Mol. Sci.* **2020**, *21*, 2187. [CrossRef]

78. Park, S.; Kim, S.; Kim, G.; Choi, Y.; Kim, E.; Paek, D. Evidence-Based Status of Forest Healing Program in South Korea. *Int. J. Environ. Res. Public Health* **2021**, *18*, 10368. [CrossRef]

79. Dolling, A.; Nilsson, H.; Lundell, Y. Stress recovery in forest or handicraft environments–An intervention study. *Urban For. Urban Green.* **2017**, *27*, 162–172. [CrossRef]

80. Kim, J.G.; Khil, T.G.; Lim, Y.; Park, K.; Shin, M.; Shin, W.S. The psychological effects of a campus forest therapy program. *Int. J. Environ. Res. Public Health* **2020**, *17*, 3409. [CrossRef] [PubMed]

81. Bielinis, E.; Jaroszewska, A.; Łukowski, A.; Takayama, N. The effects of a forest therapy programme on mental hospital patients with affective and psychotic disorders. *Int. J. Environ. Res. Public Health* **2020**, *17*, 118. [CrossRef] [PubMed]

82. Doimo, I.; Masiero, M.; Gatto, P. Forest and wellbeing: Bridging medical and forest research for effective forest-based initiatives. *Forests* **2020**, *11*, 791. [CrossRef]

83. Naomi, A.S. Access to Nature Has Always Been Important; with COVID-19, it Is Essential. *HERD Health Environ. Res. Des. J.* **2020**, *13*, 242–244. [CrossRef]

84. Tomita, A.; Vandormael, A.M.; Cuadros, D.; Di Minin, E.; Heikinheimo, V.; Tanser, F.; Slotow, R.; Burns, J.K. Green environment and incident depression in South Africa: A geospatial analysis and mental health implications in a resource-limited setting. *Lancet Planet. Health* **2017**, *1*, e152–e162. [CrossRef]

85. Nawrath, M.; Guenat, S.; Elsey, H.; Dallimer, M. Exploring uncharted territory: Do urban greenspaces support mental health in low-and middle-income countries? *Environ. Res.* **2021**, *194*, 110625. [CrossRef]
86. Brode, N. 15% of Americans Plan to Hike More than Usual Due to COVID-19. *Civic Science*. Available online: https://civicscience.com/how-americans-are-fighting-cabin-fever/ (accessed on 4 February 2022).
87. Smith, T.D.; Yu, Z.; Le, A.B. An examination of work interference with family using data from a representative sample of workers participating in the general social survey and NIOSH quality of worklife module. *J. Fam. Issues* **2020**, *41*, 2356–2376. [CrossRef]
88. Amichai-Hamburger, Y.; Hayat, Z. The impact of the Internet on the social lives of users: A representative sample from 13 ountries. *Comput. Hum. Behav.* **2011**, *27*, 585–589. [CrossRef]
89. Banerjee, S.; Galizzi, M.M.; Hortala-Vallve, R. Trusting the trust game: An external validity analysis with a UK representative sample. *Games* **2021**, *12*, 66. [CrossRef]
90. Gamil, Y.; Alhagar, A. The impact of pandemic crisis on the survival of construction industry: A case of COVID-19. *Mediterr. J. Soc. Sci.* **2020**, *11*, 122. [CrossRef]
91. Gursoy, D.; Chi, C.G. Effects of COVID-19 pandemic on hospitality industry: Review of the current situations and a research agenda. *J. Hosp. Mark. Manag.* **2020**, *29*, 527–529. [CrossRef]
92. Laing, T. The economic impact of the Coronavirus 2019 (COVID-2019): Implications for the mining industry. *Extr. Ind. Soc.* **2020**, *7*, 580–582. [CrossRef] [PubMed]
93. Gupta, M.; Abdelmaksoud, A.; Jafferany, M.; Lotti, T.; Sadoughifar, R.; Goldust, M. COVID-19 and economy. *Dermatol. Ther.* **2020**, *33*, e13329. [CrossRef] [PubMed]
94. Shankar, K. The impact of COVID-19 on IT services industry-expected transformations. *Br. J. Manag.* **2020**, *31*, 450. [CrossRef]
95. Bizoza, A.; Sibomana, S. Indicative socio-economic impacts of the novel coronavirus (COVID-19) outbreak in Eastern Africa: Case of Rwanda. *SSRN* **2020**, *3586622*, 1–23. [CrossRef]

Article

Perceived Qualities, Visitation and Felt Benefits of Preferred Nature Spaces during the COVID-19 Pandemic in Australia: A Nationally-Representative Cross-Sectional Study of 2940 Adults

Xiaoqi Feng [1,2,3,*] **and Thomas Astell-Burt** [2,3]

[1] School of Population Health, University of New South Wales, Sydney, NSW 2052, Australia
[2] Population Wellbeing and Environment Research Lab (PowerLab), NSW, Australia; thomasab@uow.edu.au
[3] School of Health and Society, University of Wollongong, Wollongong, NSW 2522, Australia
* Correspondence: xiaoqi.feng@unsw.edu.au

Abstract: We investigated how the perceived quality of natural spaces influenced levels of visitation and felt benefits during the COVID-19 pandemic in Australia via a nationally representative online and telephone survey conducted on 12–26 October (Social Research Centre's Life in Australia[TM] panel aged > 18 years, 78.8% response, n = 3043). Our sample was restricted to those with complete information (n = 2940). Likert scale responses to 18 statements regarding the quality of local natural spaces that participants preferred to visit were classified into eight quality domains: access; aesthetics; amenities; facilities; incivilities; potential usage; safety; and social. These domains were then summed into an overall nature quality score (mean = 5.8, range = 0–16). Associations between these quality variables and a range of nature visitation and felt benefits were tested using weighted multilevel models, adjusted for demographic and socioeconomic confounders. Compared with participants in the lowest perceived nature quality quintile, those in the highest quality quintile had higher odds of spending at least 2 h in their preferred local nature space in the past week (Odds Ratio [OR] = 3.40; 95% Confidence Interval [95%CI] = 2.38–4.86), of visiting their preferred nature space almost every day in the past four weeks (OR = 3.90; 2.77–5.47), and of reporting increased levels of nature visitation in comparison with before the COVID-19 pandemic (OR = 3.90; 2.54–6.00). Participants in the highest versus lowest perceived nature quality quintile also reported higher odds of feeling their visits to nature enabled them to take solace and respite during the pandemic (OR = 9.49; 6.73–13.39), to keep connected with their communities (OR = 5.30; 3.46–8.11), and to exercise more often than they did before the pandemic (OR = 3.88; 2.57–5.86). Further analyses of each quality domain indicated time in and frequency of visiting nature spaces were most affected by potential usage and safety (time in nature was also influenced by the level of amenity). Feelings of connection and solace were most affected by potential usage and social domains. Exercise was most influenced by potential usage, social and access domains. In conclusion, evidence reported in this study indicates that visits to nature and various health-related benefits associated with it during the COVID-19 pandemic were highly contingent upon numerous qualities of green and blue spaces.

Keywords: nature; green space; blue space; quality; visit; benefit; solace; physical activity; social connection; safety

Citation: Feng, X.; Astell-Burt, T. Perceived Qualities, Visitation and Felt Benefits of Preferred Nature Spaces during the COVID-19 Pandemic in Australia: A Nationally-Representative Cross-Sectional Study of 2940 Adults. *Land* **2022**, *11*, 904. https://doi.org/10.3390/land11060904

Academic Editors: Francesca Ugolini and David Pearlmutter

Received: 11 May 2022
Accepted: 10 June 2022
Published: 14 June 2022

Publisher's Note: MDPI stays neutral with regard to jurisdictional claims in published maps and institutional affiliations.

1. Introduction

Epidemiological, experimental and ethnographic research accumulated over many decades indicates that contact with nature (i.e., green and blue spaces) can provide relief from stress and restore depleted cognitive resources, enrich lives with more socially and physically active recreational pursuits, and endow communities with cleaner and cooler microclimates [1–3]. The net-health benefits of these entwined pathways include lower risks of cardiovascular diseases [4–11], diabetes [12–14] and death [15], while a smaller number of studies also identify lower risks of loneliness [16,17], cognitive decline and dementia [18–27].

Numerous studies around the world have shone a light on how contact with nature has played an important role in coping with protracted socioeconomic upheaval and significant emotional distress during the COVID-19 pandemic [28]. Many have zeroed in on the mental health benefits of nature during this difficult period [29]. Some work has focused on the inequalities that were exacerbated by lockdowns [30] and the compensation of private gardens for a lack of access to public green spaces [31,32]. Others have shown how the rapid transition to 'remote work' in some countries enabled more people to benefit from spending time in nature [33]. Notable for its absence from current research, however, was the critical issue of green and blue space *quality*, which was not mentioned once in a comprehensive narrative review of studies published [28]. To our knowledge, only one study has examined the role of green space quality during the pandemic [29].

Green and blue space quality is a critical issue for health and mental health in particular, as many previous studies conducted before the COVID-19 pandemic demonstrate [34–37]. Qualitative evidence bears this out in ways that are largely absent from the epidemiological literature [38,39]. For example, many people will visit natural settings that they identify as providing sources of non-judgmental, ego-free and dependable support [40]; it would be erroneous to consider all green and blue spaces as affording such precious experiences. Meanwhile, some natural settings can, as a result of regular visitation over time and over generations, become invested with individual and shared meanings that (re)generate feelings of connection and belonging within and between nearby communities [41]. Clearly, not all green and blue spaces share equal value. If a nearby green or blue space has a deficit of what people might need or desire (i.e., 'good' qualities) and/or an abundance of things they might seek to avoid (i.e., 'bad' qualities), a situation may transpire where there are lots of nature spots nearby but this does not necessarily translate into better health.

Accordingly, we sought to look beyond the sole issues of availability and accessibility to investigate more holistically the degree to which the nature–health association during the COVID-19 pandemic was dependent upon the various qualities of the nearby natural settings that people prefer to visit. We examined the extent to which different aspects of visitation and contrasting health benefits were contingent upon the presence of a range of specific qualities, both good and bad. We hypothesized that people tended to visit and benefit more from green and/or blue spaces that they felt had more of the qualities they value.

2. Materials and Methods

2.1. Data

A nationally-representative survey of the Australian adult population aged $\geq$ 18 years was conducted between 12 and 26 October 2020 via the Social Research Centre's Life in Australia^{TM} panel. Participation was in the English language only. Panel members were recruited in 2016 via their landline or cell phone using a dual-frame random digit dialling design (RDD) with a 30:70 split between landline and cell phone sample frames. Respondents in households were selected using an alternating next or last birthday approach via the landline method wherein households were occupied by at least two residents in scope. The phone answerer was the selected respondent via the cell phone method. In each case, invitation to join the panel was for one member per household only. The panel was refreshed in 2018 using mobile phone RDD only and again in 2019 with online-only participants using a G-NAF (Geocoded National Address File) sample frame and push-to-web methodology. In each of these cases, refreshment was required to balance demographics of the panel with respect to the Australian population. Online panel members were invited to participate in the survey via email and SMS, followed by emails, telephone calls, voicemails and SMS in week 2 of the survey period to encourage completion. The Social Research Centre's interviewers and supervisors had received training in the Life in Australia^{TM} panel, survey procedures and sample management protocols, respondent liaison procedures, strategies to maintain cooperation, and detailed examination of the survey questionnaire developed by the researchers. An incentive of a supermarket or department store gift card, direct payment into a PayPal account, or donation to a designated charity was offered to all panel members to the value of AUD 10.00 each. Ethical approval for the survey was granted by the University of Wollongong HREC.

Approximately 78.8% (n = 3043) of the Life in Australia^{TM} panel participated in our survey (95.0% completing online, 5.0% completing via telephone). A total of 19.9% of panel members could not be contacted during the survey period and 1.3% of invited members declined to participate. Response propensity weights were constructed by the Social Research Centre using logistic regression to limit the impact of non-participation on sample representativeness, taking into account geography, age group, gender, annual household income, citizenship status, language(s) spoken other than English, country of birth, Aboriginal or Torres Strait Islander status, number of adults and children in the household, employment status, marital status, highest education, television viewing and internet browsing habits, smoking and drinking status, general health, life satisfaction, early adopter status, caregiving, disability status, volunteer status, concession card status, and telephony status. Our sample focussed on people with full outcome data (n = 2940; 96.6%).

2.2. Outcomes: Nature Space Visitation and Felt Benefits

Six binary outcomes describing participants' visitation of, and felt benefits from nearby natural settings during the COVID-19 pandemic were examined. Given evidence indicating at least 120 min of time in nature may support general health, we used this as a cut-point for responses to the question: *"Approximately how many hours did you spend in green spaces and/or blue spaces in total over the last 7 days?"* We allowed for the possibility that for some people, the natural setting(s) they would prefer to visit might not be accessible for multiple reasons and so responses to the following question were also examined: *"In the past four weeks (including the weekends), how often have you visited your preferred local green space and/or blue space?"* Responses were classified as 'at least once a week' (combining 'almost daily' and '1–4 times weekly') or 'less than once a week' (combining '2–3 times in the past month', 'once or less in the past month', and 'never'). The third visitation-focussed outcome was used to determine if participants had increased their visit frequency since the pandemic began with the following question: *"Since the COVID-19 pandemic and social distancing began in Australia, to what extent, if at all, do you agree or disagree with each of the following statements? (A) I now visit green spaces and/or blue spaces more often than before the COVID-19 pandemic"*. This question set also included three statements pertaining to felt benefits, as follows: *"(B) Green spaces and/or blue spaces have helped me to stay connected with my neighbours during the COVID-19 pandemic. (C) Green spaces and/or blue spaces have brought me solace and respite in these challenging times. (D) I now walk and/or exercise in green spaces and/or blue spaces more frequently than before the COVID-19 pandemic"*.

2.3. Qualities of Nearby Natural Spaces

Eighteen different quality indicators on the nearby natural setting participants preferred to visit were measured in the survey (Table 1). These indicators were used to construct eight different quality domains adopted from a published green space quality auditing tool [42]. The survey question was *"Thinking of the green space and/or blue space you prefer to visit most often and your experiences in it, how much do you agree or disagree with the following statements?"* The first domain was "access", for which participants were asked whether their preferred nature spaces were well-connected by public transport, footpaths and road crossing points. The second domain, "aesthetics", examined views concerning potential for exploration and interesting discoveries, as well as pleasant natural vistas and biodiverse soundscapes. "Amenities" were the third domain to permit acknowledgement of adjacent reasons people might visit the natural setting for, such as before or after shopping and dining out. The amenities domain also included a separate question on provision of shade along footpaths from tree canopy cover. The fourth domain described "facilities", including those used for physical activity, public bathrooms, and seating. "Incivilities" were the fifth domain and were measured by a single item pertaining to perception of quality and maintenance of the natural space. The sixth domain described various types of "usage" including breaks from day-to-day routines, opportunities to feel some distance from cognitive demands, and spaces that children can play outdoors and/or that a participant feels they can walk and/or exercise in. The seventh domain attended to "safety", with specific focus on safety during the evening/night. The eighth and final domain considered "social" factors, such as whether the green/blue space was viewed as a shared setting for neighbours, friends and/or family to meet. Where possible, these questions were derived from existing literature. For instance, *"there is much to explore and discover there"* in the aesthetics domain was drawn from Hartig's perceived restorativeness scale [43]. The answer set to all eighteen quality indicators was a five-point Likert scale.

Table 1. Eighteen nature space quality indicators nested within eight quality domains.

"Thinking of the green space and/or blue space you prefer to visit most often and your experiences in it, how much do you agree or disagree with the following statements?"
[strongly disagree, disagree, neither agree nor disagree, agree, strongly agree]

Domain 1: Access
"It is well connected by footpaths and safe road crossing points"
"Public transport is available nearby"

Domain 2: Aesthetics
"There is much to explore and discover there"
"My attention is drawn to many interesting things there"
"It is a place I can enjoy watching and/or listening to wildlife (e.g., birds)"

Domain 3: Amenities
"There are cafes, and/or shops, and/or supermarkets and/or restaurants nearby"
"There is lots of tree canopy along footpaths that provide shade from heat and direct sunlight"

Domain 4: Facilities
"There are free or low cost recreation facilities, such as outdoor gyms, sports grounds and/or swimming pools in it or nearby"
"There are public toilets available in it or nearby"
"There are benches in it or nearby on which I can sit and relax"

Domain 5: Incivilities
"I consider it to be high quality and well maintained"

Domain 6: Potential usage
"Spending time there gives me a break from my day-to-day routine"
"This is a place to get away from the things that usually demand my attention"
"It is a good place for children to play outdoors"
"I go there for walks and/or to exercise"

Domain 7: Safety
"This is a place I feel safe to visit during the evening/night"

Domain 8: Social
"It is a social hub for the local community"
"This is a place to spend time with friends and/or family"

We classified each indicator to "disagree/ambivalent" (scoring zero, combining "strongly disagree", "disagree", and "neither agree nor disagree"), "agree" (scoring 1) and "strongly agree" (scoring 2). Participants' mean scores across all quality indicators within each domain were calculated. Degree of correlation between domain means was assessed using Pearson's correlation coefficients (Table 2), from which it was evident that most domains were weak-to-moderately correlated (e.g., the only correlation >0.6 was for the aesthetics and usage domains: coefficient = 0.619, *p*-value < 0.001). Those domain means were then summed across all domains to give a total quality score for nearby green and blue spaces. This total quality score was normally distributed with an overall mean of 5.84 (standard deviation = 3.05) and ranged from zero to 16. We classified it into quintiles, for which the interval means and other parameters are reported in Table 3.

Table 2. Pearson's correlation coefficients for quality domains of natural settings.

	Access	Aesthetics	Amenities	Facilities	Incivilities	Potential Usage	Safety	Social
Access	1							
Aesthetics	0.164	1.000						
p-value	<0.001							
Amenities	0.462	0.354	1.000					
p-value	<0.001	<0.001						
Facilities	0.520	0.282	0.507	1.000				
p-value	<0.001	<0.001	<0.001					
Incivilities	0.425	0.338	0.427	0.520	1.000			
p-value	<0.001	<0.001	<0.001	<0.001				
Potential Usage	0.375	0.619	0.441	0.424	0.483	1.000		
p-value	<0.001	<0.001	<0.001	<0.001	<0.001			
Safety	0.124	0.257	0.147	0.162	0.215	0.273	1.000	
p-value	<0.001	<0.001	<0.001	<0.001	<0.001	<0.001		
Social	0.396	0.396	0.450	0.559	0.477	0.494	0.216	1
p-value	<0.001	<0.001	<0.001	<0.001	<0.001	<0.001	<0.001	

Table 3. Description of the nature space total quality score quintile intervals.

Quintiles	Mean	Standard Deviation	Quintile Bounds	
1 (low)	2.02	1.00	0.00	3.30
2	4.11	0.45	3.33	4.83
3	5.53	0.39	4.87	6.17
4	7.08	0.57	6.20	8.03
5 (high)	10.53	1.83	8.07	16.00

2.4. Confounders

Variables that denote factors known to influence both human behaviour and psychological and social wellbeing, and also where people live and access to green and/or blue space, were measured using survey responses. These included gender, age, country of birth, language spoken at home, relationship status, highest educational qualification, annual household income, economic status (e.g., employed, retired, unemployed), perceived financial difficulties, housing type (e.g., house, flat), and urban/rural. The urban/rural variable was extended to 15 categories to account for substantial geographical variations across the states and territories of Australia, differentiating between participants living in major cities (e.g., Sydney, Melbourne) from those living in regional and rural areas of the same states (e.g., Rest of New South Wales, Rest of Victoria).

2.5. Statistical Analysis

Cross-tabulations, percentages and means were used to describe the study sample and the patterning of the total nature space quality scores across participants' characteristics. Weighted linear regressions were used to assess associations between the total nature space quality scores and participants' characteristics. Separate weighted logistic regressions were then used to examine associations between each of the nature space visitation and felt benefit outcomes with the quality scores, adjusting for confounding variables. All analyses were conducted in Stata V.14 (StataCorp., College Station, TX, USA).

3. Results

3.1. Sample Description and Differences in Nature Space Total Quality Scores

Weighted descriptive statistics of the study sample are reported in Table 4, as well as unadjusted mean nature space total quality scores and adjusted coefficients from a weighted multiple linear regression. The ratio of females to males is almost equal. About 53% of the sample was aged between 25 years and 54 years. About two-thirds of the sample were born in Australia and nearly four-fifths did not speak a language other than English at home. Approximately 71% of the sample were in a couple with or without children, whereas just over 15% were living in single-person households. University degrees were held by 27.6% of participants whereas the highest qualification for 13.2% was fewer than 12 years of education. In total, 59% of the sample had annual household incomes up to AUD 100 k, whereas 34.6% had incomes greater than or equal to AUD 101 k per year. Unemployment was at 9.5%, retirement at 20.7%, employment at 61.1% and those living with disability at 2.6%. Employment varied with respect to remote work, with 29.7% having no remote work option whereas 13.9% working remotely full-time. Nearly 10% of the sample reported financial difficulty relative to 26.5% who were comfortable. Approximately 37.6% of the sample was resident in the cities of Sydney or Melbourne, with 66.7% living in major cities.

Table 4. Sample description, mean quality of preferred nearby natural setting, and adjusted differences, weighted for national representativeness.

Total Sample *n* = 2940	*n* (%)	Mean (SE)	Coef (95%CI) [*p*-Value]
Gender (ref: Female)	49.4%	5.6 (0.1)	
Male	50.5%	6.0 (0.1)	0.432 (0.138, 0.726) [0.004]
Other	0.2%	6.9 (1.2)	0.950 (−1.091, 2.990) [0.361]
Age group (ref: 18–24 years)	10.0%	5.2 (0.3)	
25–34 years	18.5%	6.2 (0.2)	0.977 (0.225, 1.729) [0.011]
35–44 years	18.1%	6.2 (0.2)	0.834 (0.105, 1.563) [0.025]
45–54 years	16.3%	5.9 (0.2)	0.764 (0.036, 1.491) [0.040]
55–64 years	15.1%	5.5 (0.1)	0.515 (−0.192, 1.222) [0.153]
65–74 years	14.0%	5.4 (0.2)	0.604 (−0.231, 1.440) [0.156]
≥75 years	6.9%	5.4 (0.2)	0.588 (−0.321, 1.496) [0.205]
Undetermined	1.1%	6.9 (0.9)	1.792 (−0.046, 3.630) [0.056]
Country of birth (ref: Australia)	66.1%	5.7 (0.1)	
Overseas, not English-speaking	19.1%	5.8 (0.2)	−0.392 (−0.946, 0.162) [0.165]
Overseas, English-speaking	14.7%	5.9 (0.2)	0.161 (−0.296, 0.618) [0.489]
Undetermined	0.2%	7.6 (1.5)	0.638 (−2.987, 4.263) [0.730]
Language other than English at home (ref: Yes)	20.1%	6.1 (0.2)	
No	79.9%	5.7 (0.1)	−0.375 (−0.946, 0.196) [0.198]
Relationship status (ref: Living alone)	15.2%	5.5 (0.2)	
Alone with kids	6.9%	5.6 (0.3)	0.232 (−0.426, 0.890) [0.489]
Couple without kids	27.7%	5.6 (0.1)	0.139 (−0.287, 0.565) [0.521]
Couple with kids	43.4%	6.0 (0.1)	0.413 (−0.037, 0.864) [0.072]
Cohabiting, unrelated	2.7%	6.3 (0.4)	0.592 (−0.212, 1.397) [0.149]
Other	4.1%	4.9 (0.4)	−0.576 (−1.380, 0.228) [0.160]
Highest educational qualification (ref: <Year 12)	13.2%	5.5 (0.2)	
Year 12	19.0%	5.6 (0.2)	−0.020 (−0.574, 0.533) [0.943]
Advanced diploma/certificate	37.2%	5.7 (0.1)	0.069 (−0.385, 0.522) [0.767]
Bachelor degree	18.8%	6.2 (0.1)	0.334 (−0.182, 0.850) [0.204]
Postgraduate degree	8.8%	6.1 (0.1)	0.195 (−0.360, 0.751) [0.491]
Undetermined	3.0%	5.0 (0.3)	−0.310 (−1.008, 0.388) [0.384]

Table 4. *Cont.*

Total Sample *n* = 2940	*n* (%)	Mean (SE)	Coef (95%CI) [*p*-Value]
Annual household income (ref: ≤50 K)	27.1%	5.5 (0.1)	
AUD 51 K–AUD 100 K	31.9%	5.7 (0.1)	−0.135 (−0.530, 0.259) [0.501]
AUD 101 K–AUD 150 K	18.0%	6.1 (0.2)	−0.115 (−0.627, 0.396) [0.658]
≥AUD 151 K	16.6%	6.1 (0.2)	−0.136 (−0.776, 0.503) [0.675]
Undetermined	6.4%	5.4 (0.3)	−0.337 (−0.996, 0.322) [0.316]
Economic status (ref: Employed, never remotely)	29.7%	5.8 (0.2)	
Employed, work remotely sometimes	11.0%	6.0 (0.2)	0.091 (−0.469, 0.650) [0.751]
Employed, work remotely often	6.5%	6.2 (0.3)	0.228 (−0.386, 0.842) [0.466]
Employed, work remotely always	13.9%	6.3 (0.2)	0.210 (−0.281, 0.701) [0.402]
Unemployed	9.5%	5.3 (0.3)	−0.377 (−0.993, 0.239) [0.231]
Retired	20.7%	5.4 (0.1)	−0.183 (−0.758, 0.393) [0.534]
Disabled	2.6%	4.9 (0.3)	−0.636 (−1.325, 0.053) [0.070]
Other	5.3%	5.9 (0.3)	0.121 (−0.505, 0.747) [0.705]
Undetermined	0.9%	6.6 (0.7)	0.812 (−0.615, 2.238) [0.265]
Economic difficulty (ref: Comfortable)	26.5%	6.2 (0.1)	
Doing ok	44.2%	5.7 (0.1)	−0.460 (−0.809, −0.111) [0.010]
Getting by	19.3%	5.6 (0.2)	−0.560 (−0.990, −0.131) [0.011]
Difficult	9.9%	5.3 (0.2)	−0.801 (−1.398, −0.205) [0.009]
Undetermined	0.2%	3.8 (0.5)	−2.308 (−3.715, −0.902) [0.001]
Housing (ref: House)	75.0%	5.8 (0.1)	
Flat	17.7%	6.0 (0.2)	0.335 (−0.089, 0.760) [0.122]
Farmhouse	5.2%	4.8 (0.3)	−0.936 (−1.470, −0.403) [0.001]
Retirement village	0.9%	5.4 (0.5)	−0.202 (−1.300, 0.895) [0.718]
Other	1.2%	5.3 (0.7)	−0.044 (−1.305, 1.217) [0.945]
Geographic area (ref: Greater Sydney)	18.8%	5.8 (0.2)	
Rest of New South Wales	13.2%	5.4 (0.2)	−0.072 (−0.603, 0.459) [0.790]
Greater Melbourne	18.8%	6.2 (0.2)	0.352 (−0.141, 0.845) [0.162]
Rest of Victoria	7.4%	5.5 (0.3)	0.006 (−0.610, 0.623) [0.984]
Greater Brisbane	11.1%	5.6 (0.2)	−0.106 (−0.699, 0.487) [0.726]
Rest of Queensland	8.9%	5.8 (0.2)	0.182 (−0.387, 0.752) [0.530]
Greater Adelaide	5.7%	5.6 (0.3)	−0.016 (−0.649, 0.617) [0.960]
Rest of South Australia	1.3%	6.1 (0.5)	0.508 (−0.588, 1.604) [0.363]
Greater Perth	10.3%	6.0 (0.2)	0.173 (−0.407, 0.752) [0.559]
Rest of Western Australia	1.5%	5.1 (0.5)	−0.319 (−1.362, 0.724) [0.549]
Greater Hobart	0.7%	6.6 (0.6)	1.027 (−0.191, 2.245) [0.098]
Rest of Tasmania	1.0%	5.8 (0.5)	0.372 (−0.632, 1.377) [0.467]
Greater Darwin	0.2%	5.0 (1.0)	−0.763 (−2.610, 1.083) [0.418]
Rest of Northern Territory	0.0%	5.4 (1.1)	−0.533 (−2.139, 1.072) [0.515]
Australian Capital Territory	1.1%	5.3 (0.4)	−0.140 (−1.060, 0.779) [0.765]
Constant			5.330 (4.178, 6.482) [<0.001]

SE: Standard Error; 95%CI: 95% Confidence Interval; Note: all means, standard errors, regression coefficients and 95% confidence intervals are weighted for national representativeness.

The mean nature quality scores tended to be higher among males in comparison with females (Table 4). Mean nature quality scores also tended to be higher among participants aged 25–54 years in comparison to those aged 18–24 years, participants who felt their financial circumstances were comfortable relative to those who were not, and those in houses or flats relative to a farmhouse. Variations in mean nature space quality scores between other demographic and socioeconomic groups were small and not statistically significant.

3.2. Associations between Nature Space Quality Scores and Visitation and Felt Benefits

Higher quintiles of nature space total quality scores were consistently and positively associated with each of the visitation and felt benefit outcome variables, after adjustment for confounding variables and weighted for national representativeness (Figure 1).

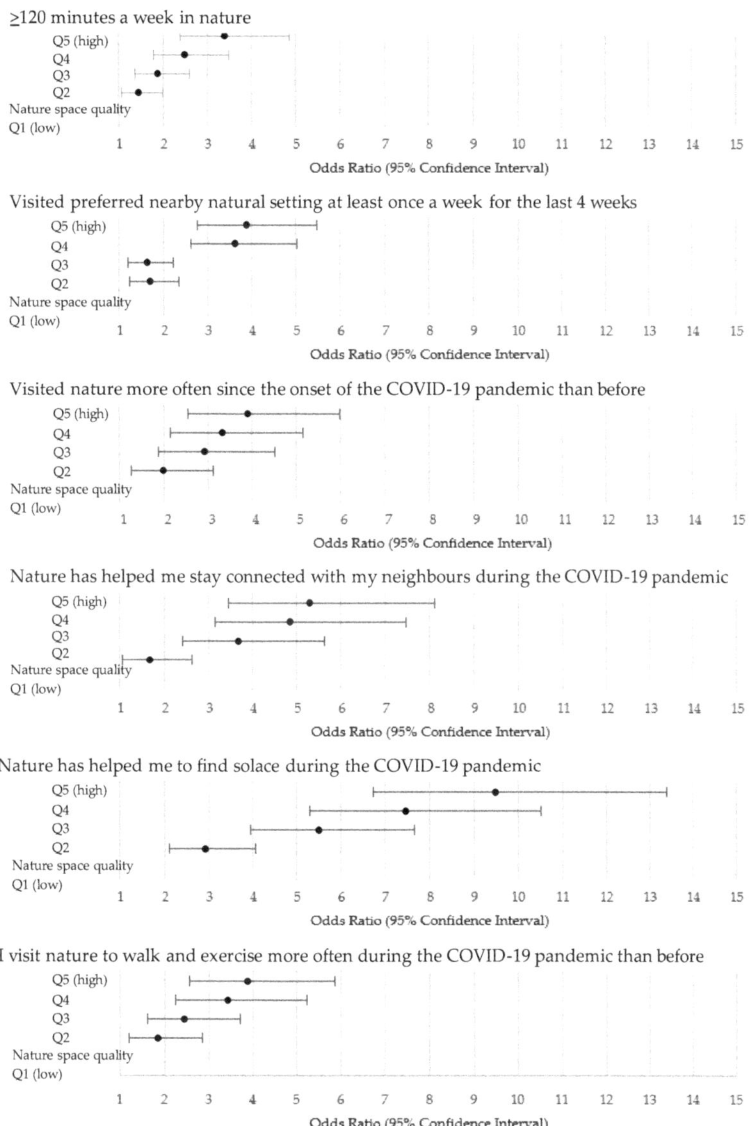

Figure 1. Adjusted associations between nature space total quality scores and visitation and felt benefits, weighted for national representativeness.

In comparison with quintile 1 (i.e., lowest quality), participants in quintile 5 (i.e., highest quality) had 3.4 times greater odds of spending two hours or more in nature a week (odds ratio [OR] 3.40, 95% confidence interval [95%CI] 2.38–4.86), 3.9 times greater odds of visiting their nearby preferred nature setting at least once a week for the last four weeks (OR 3.90, 95%CI 2.77–5.47), and 3.9 times greater odds of reporting that they visit green and blue spaces more often now than before the COVID-19 pandemic (OR 3.90, 95%CI 2.54–5.99). In order of magnitude, participants in quintile 5 compared with quintile 1 for nature space total quality scores had 9.5 times greater odds of reporting that time in nature had been a source of solace during the pandemic (OR 9.49, 95%CI 6.73–13.39; wider confidence intervals are in part indicative of smaller sample sizes), 5.3 times greater odds of reporting that green and blue spaces had enabled them to stay connected with their neighbors (OR 5.30, 95%CI 3.46–8.11), and 3.9 times greater odds of reporting more engagement in physical activity now than before the pandemic (OR 3.88, 95%CI 2.57–5.86).

Associations between each nature space quality domain and the six visitation and felt benefit outcome variables are reported in Table 5. Models were initially fitted for each outcome and quality domain score separately, followed by models that adjusted for all quality domain scores simultaneously. Every quality domain score was associated with more favorable outcomes when analyzed in isolation and adjusted for confounding. Many domains were no longer statistically significantly associated with the outcomes after adjusting for other domains. Potential usage was the only domain that was consistently associated with increased odds of all outcomes. The magnitude of odds ratios for potential usage were also consistently larger than in other domains. The access domain was only associated with taking exercise in nature more often than before the COVID-19 pandemic. Aesthetics, facilities and incivilities were not associated with any of the outcomes, while amenities were only associated—negatively—with the odds of spending at least two hours a week in nature. Higher levels of safety were important for both achieving at least two hours a week in nature and for visiting the preferred nearby natural setting at least once a week for the last four weeks, but not for any of the other outcomes. The social domain was positively associated with staying connected, finding solace, and taking more physical activity in nature, but none of the visitation-focused outcomes.

Table 5. Adjusted associations between qualities of nearby natural settings and visitation and felt benefits, weighted for national representativeness.

	Single Quality Domain Model	Multi Quality Domain Model
Visitation and felt benefit	Odds Ratio (95% Confidence Interval) [*p*-value]	
≥120 min a week in nature		
Access	1.408 (1.145, 1.732) [0.001]	0.864 (0.652, 1.145) [0.310]
Aesthetics	1.921 (1.594, 2.314) [<0.001]	1.062 (0.838, 1.345) [0.619]
Amenities	1.391 (1.121, 1.727) [0.003]	0.721 (0.540, 0.962) [0.026]
Facilities	1.610 (1.299, 1.997) [<0.001]	0.926 (0.670, 1.278) [0.639]
Incivilities	1.524 (1.302, 1.785) [<0.001]	1.029 (0.825, 1.285) [0.797]
Potential usage	3.631 (2.803, 4.704) [<0.001]	3.510 (2.434, 5.062) [<0.001]
Safety	1.759 (1.464, 2.114) [<0.001]	1.402 (1.143, 1.719) [0.001]
Social	1.823 (1.491, 2.230) [<0.001]	1.242 (0.943, 1.634) [0.122]
Visited preferred nearby natural setting at least once a week for the last 4 weeks		
Access	2.067 (1.491, 2.864) [<0.001]	1.235 (0.814, 1.876) [0.321]
Aesthetics	2.156 (1.600, 2.904) [<0.001]	0.974 (0.679, 1.397) [0.887]
Amenities	1.730 (1.203, 2.489) [0.003]	0.748 (0.493, 1.135) [0.172]
Facilities	1.913 (1.387, 2.638) [<0.001]	0.930 (0.601, 1.437) [0.742]
Incivilities	1.499 (1.192, 1.886) [0.001]	0.798 (0.599, 1.064) [0.125]
Potential usage	5.399 (3.408, 8.553) [<0.001]	5.521 (3.020, 10.094) [<0.001]
Safety	2.255 (1.649, 3.082) [<0.001]	1.782 (1.268, 2.506) [0.001]
Social	1.988 (1.425, 2.773) [<0.001]	1.108 (0.726, 1.693) [0.634]
Visited nature more often since the onset of the COVID-19 pandemic than before		

Table 5. *Cont.*

	Single Quality Domain Model	Multi Quality Domain Model
Access	1.683 (1.352, 2.095) [<0.001]	1.148 (0.876, 1.505) [0.317]
Aesthetics	1.685 (1.398, 2.031) [<0.001]	1.039 (0.811, 1.330) [0.764]
Amenities	1.599 (1.274, 2.008) [<0.001]	0.964 (0.723, 1.285) [0.802]
Facilities	1.729 (1.378, 2.171) [<0.001]	1.043 (0.759, 1.433) [0.795]
Incivilities	1.373 (1.159, 1.627) [<0.001]	0.868 (0.699, 1.079) [0.203]
Potential usage	2.821 (2.205, 3.611) [<0.001]	2.405 (1.695, 3.412) [<0.001]
Safety	1.359 (1.137, 1.624) [0.001]	1.071 (0.886, 1.294) [0.480]
Social	1.792 (1.476, 2.176) [<0.001]	1.248 (0.964, 1.615) [0.092]
Nature has helped me stay connected with my neighbours during the COVID-19 pandemic		
Access	1.819 (1.492, 2.216) [<0.001]	1.191 (0.921, 1.539) [0.182]
Aesthetics	1.913 (1.599, 2.290) [<0.001]	1.188 (0.944, 1.495) [0.142]
Amenities	1.697 (1.370, 2.102) [<0.001]	0.907 (0.694, 1.186) [0.476]
Facilities	1.840 (1.486, 2.279) [<0.001]	0.877 (0.656, 1.173) [0.377]
Incivilities	1.688 (1.434, 1.986) [<0.001]	1.084 (0.882, 1.330) [0.444]
Potential usage	2.863 (2.279, 3.598) [<0.001]	1.699 (1.227, 2.351) [0.001]
Safety	1.515 (1.280, 1.793) [<0.001]	1.147 (0.956, 1.377) [0.139]
Social	2.304 (1.897, 2.798) [<0.001]	1.670 (1.285, 2.171) [<0.001]
Nature has helped me to find solace during the COVID-19 pandemic		
Access	2.200 (1.804, 2.682) [<0.001]	1.134 (0.864, 1.489) [0.365]
Aesthetics	2.698 (2.258, 3.224) [<0.001]	1.124 (0.892, 1.416) [0.322]
Amenities	2.193 (1.784, 2.696) [<0.001]	0.885 (0.667, 1.175) [0.398]
Facilities	2.215 (1.804, 2.720) [<0.001]	0.805 (0.588, 1.101) [0.174]
Incivilities	2.039 (1.757, 2.366) [<0.001]	1.112 (0.904, 1.370) [0.315]
Potential usage	8.275 (6.330, 10.816) [<0.001]	6.358 (4.494, 8.996) [<0.001]
Safety	1.792 (1.519, 2.114) [<0.001]	1.205 (0.998, 1.456) [0.053]
Social	2.583 (2.137, 3.122) [<0.001]	1.311 (1.008, 1.705) [0.044]
I visit nature to walk and exercise more often during the COVID-19 pandemic than before		
Access	2.001 (1.605, 2.494) [<0.001]	1.418 (1.081, 1.860) [0.012]
Aesthetics	1.805 (1.502, 2.170) [<0.001]	1.120 (0.875, 1.432) [0.369]
Amenities	1.706 (1.361, 2.137) [<0.001]	0.952 (0.715, 1.267) [0.736]
Facilities	1.819 (1.452, 2.278) [<0.001]	0.948 (0.689, 1.304) [0.743]
Incivilities	1.441 (1.218, 1.706) [<0.001]	0.860 (0.693, 1.068) [0.172]
Potential usage	2.966 (2.336, 3.765) [<0.001]	2.163 (1.534, 3.051) [<0.001]
Safety	1.427 (1.193, 1.707) [<0.001]	1.100 (0.910, 1.330) [0.325]
Social	1.980 (1.629, 2.407) [<0.001]	1.363 (1.047, 1.774) [0.021]

SE: Standard Error; 95%CI: 95% Confidence Interval; All models weighted for national representativeness and adjusted for gender, age group, country of birth, language spoken at home, relationship status, highest educational qualification, annual household income, economic status, perceived financial difficulty, housing status, and geographic region.

4. Discussion

Key findings from this study affirm our hypothesis on the importance of having higher quality green and blue spaces nearby and their positive role in enabling people to keep connected with their neighbours, feel a sense of solace and maintain or increase their level of physical activity during the COVID-19 pandemic. These results present a major advance in research on nature and various aspects of mental, physical and social health experienced during the pandemic that has largely ignored the issue of quality [28]. Our results show—as many already suspected—that simply having green and/or blue space nearby is not always sufficient to elicit favorable outcomes [38,44]. This is perhaps no more vividly illustrated than the over nine-fold increase in the odds of finding solace through nature during the pandemic where those natural settings were of the highest quality quintile versus the lowest. The dose–response patterns for each of the outcomes with respect to the overall quality of nearby nature spaces demonstrate how this is not a curvilinear effect, with higher quality scores and odds of favorable outcomes following positive linearity.

Analyzing separate domains of nature space quality revealed which ones appeared to be of greater importance to specific outcomes. This is important as a common policy-relevant area for improvement with current research on perceived green space quality and various health outcomes is that data are often insufficient to discern which qualities matter most and least [45,46]. It was notable that the domain describing potential usage was consistently and positively associated with visitation and felt benefits. This is expected, as indicators in this domain emphasized nature as a setting for rest, recuperation, restoration of depleted cognitive resources, physical activity and as a play space for children. These indicators might attend more to some participants' intrinsic motivations for seeking time in nature in comparison to other domains where descriptive elements are present, but might not necessarily be of fundamental importance to the individual responding. This domain and others that were associated with some outcomes but not all, such as the safety and social domains, appeared to be more important than others, such as incivilities, facilities and aesthetics that were no longer statistically significant in multi-domain models. To some extent, this will be due to partial overlap between each of the domains, most notably for the aesthetics and potential usage domains; the presence of wildlife and other cues that prompt interest and exploration in the aesthetic domain are concomitant with the desire to feel away from the day-to-day routine of cognitive demands.

Our survey included a partial assessment of biodiversity through a single indicator. This is important as there are now several studies indicating how objective and subjectively-measured biodiversity of green spaces (e.g., presence of birds and species diversity indicators) may be an important conduit by which some mental health benefit occurs [47–50], not least through attending to people's preferences [51]. However, the present indicator likely only grazes the surface of this concept and this is an area in need of further research. Similarly, it is also plausible that some of the domains, such as incivilities, might be underpowered, given its current single-item focus on quality and maintenance, wherein specific issues might resonate strongly with some people (e.g., the presence of dog feces on grass). So too might be the access domain, with its focus more on getting to the natural space, rather than the accessibility within it, which might be especially important for people with physical limitations or disabilities [52–54]. Curiously, the amenities domain was statistically significantly and negatively associated with time in nature and non-significantly negatively associated with all other outcomes after adjusting for other domains. Caution is needed in interpreting this result. On one hand, it may be driven by multicollinearity, but on the other, negative association may be due to some people actively avoiding natural settings that are within close proximity to retail strips and similar that may attenuate their restorative experiences while in nature. For instance, while some work indicates that non-natural sounds emanating from automobiles, trucks and other elements concomitant with commercial (and industrial) landscapes may be soothed by being in nature via psychoacoustic pathways [55,56], it is unclear if such annoyances are ameliorated entirely. Furthermore, other physical cues that are either located nearby or encroach within natural settings, such as neon signage to advertise workplaces and fast-food restaurants, may also be a source of distraction (perhaps, even irritation) for many people when visiting nature for rest and escape from the day-to-day demands in life.

A third key finding was that the degree of the socioeconomic gradient in the availability of quality nature spaces was atypical, with total quality scores being only slightly higher on average for people with university-level qualifications or annual household incomes over AUD 150 k. These differences were not statistically significant in fully adjusted models. However, importantly, there was a somewhat greater gap in mean total quality scores between people whose financial situation was difficult, in comparison with those whose situation was more comfortable. These differences remained statistically significant after adjustment. How a person feels about their financial situation is important and often overlooked in epidemiological studies of person-level data that tend to rely on education, income, and in some contexts, occupational class [57,58]. While difficult financial circumstances were more common in participants with less than 12 years of education versus those with a bachelor's degree (9.9% vs. 6.7%) or among those with AUD 50 k per annum or less versus those with AUD 151 k per annum or greater (17.8% vs. 2.63%), clearly there are situations in which those with higher qualifications and income categories may also be living under major financial strain and vice versa. This variable, therefore, provides more incisive utility for accounting for socioeconomic circumstances than other more routinely-used variables. However, it is also worth noting that how people perceive the quality of their nearby green and blue spaces may be, in part, influenced by the levels of financial and psychological strain under which they are presently living. Accordingly, further work might examine to what extent changes in perceived financial circumstances among people whose socioeconomic circumstances remain consistent, as measured by income and education (etc.), may influence how they perceive the natural spaces they have nearby.

It is worthwhile noting that although our survey was able to measure time spent in nature, which is important as previous studies have indicated [59,60], we did not have information on what survey participants did in those spaces specifically beyond the outcomes already analyzed. For instance, some people may have visited specific parks on Saturdays to engage with other local community members in the Parkrun movement, which various studies indicate provides opportunities for volunteering and social connection, as well as physical activity [61–63]. For many people, especially during the COVID-19 pandemic, cemeteries may have played a key role in giving people opportunities to be outdoors and connect, if not with each other, then with loved ones no longer around [64]. Others may have visited natural settings they regard as special places, perhaps due to childhood memories [65,66] or as providing opportunities to do things they feel unable to at home, such as connecting with peers through allotment gardens [67,68]. Further qualitative research and maybe further survey analysis are needed to better understand the roles in which specific types of green space and their qualities have aided coping and restoration through the pandemic.

An additional layer to this research is the well-reported socioeconomic inequities in green space and blue space availability and qualities are likely to reflect, in part, personal preferences, financial capacities and willingness to pay to live near these health-promoting resources. The intersection between these economic issues and the epidemiological literature remains a gap in knowledge, though research was carried out on these aspects. For example, Johnson and Thomassin [69] provided a model to estimate the willingness to pay for surface water quality improvements by recreational users. Further investigation utilizing longitudinal data capable of tracking changes in the qualities of green and blue spaces that occur and impacts on both the health of local residents and on population flows in and out of the areas nearby is warranted.

Beyond the strengths and limitations already discussed, this study benefits from a large and nationally representative sample of the Australian adult population, covering all states and territories. The survey contained a large range of variables used to describe green and blue spaces, permitting the identification of nature space qualities and the development of an overall score from domains established by a published study that focused on in-person auditing of parks [42]. All descriptive statistics and models were adjusted using a comprehensive set of confounding variables and also weighted to ensure parameter estimates that can be extrapolated to the adult population of Australia. The analyses use data of cross-sectional design and so the associations reported should not be interpreted as definitively revealing cause and effect. Follow-up of the same individuals over time will enable stronger epidemiological study designs with which to minimize the potential for reverse causation, wherein individuals more prone to poorer health are socioeconomically disadvantaged in part because of their circumstances and move into areas with poorer quality green and blue spaces as a result. These data and follow-up of the same individuals will also permit opportunities to study the longer-term impacts of the COVID-19 pandemic and to identify the extent to which different types of nature, preferential elements of it and ways in which people interact with it have supported recovery and flourishing.

Author Contributions: Conceptualization, X.F. and T.A.-B.; methodology, X.F. and T.A.-B.; software, T.A.-B.; validation, T.A.-B.; formal analysis, X.F. and T.A.-B.; investigation, X.F. and T.A.-B.; resources,

X.F. and T.A.-B.; data curation, X.F. and T.A.-B.; writing—original draft preparation, X.F. and T.A.-B.; writing—review and editing, X.F. and T.A.-B.; visualization, X.F. and T.A.-B.; project administration, X.F. and T.A.-B.; funding acquisition, X.F. and T.A.-B. All authors have read and agreed to the published version of the manuscript.

Funding: This work was supported by the Hort Frontiers Green Cities Fund, part of the Hort Frontiers strategic partnership initiative developed by Hort Innovation, with co-investment from the University of Wollongong (UOW) Faculty of Arts, Social Sciences and Humanities, the UOW Global Challenges initiative and contributions from the Australian Government (project number #GC15005). T.A.-B. was supported by a National Health and Medical Research Council Boosting Dementia Research Leader Fellowship (#1140317). X.F. was supported by a National Health and Medical Research Council Career Development Fellowship (#1148792). All aspects related to the conduct of this study including the views stated and the decision to publish the findings are that of the authors only.

Institutional Review Board Statement: The study was conducted according to the guidelines of the Declaration of Helsinki, and approved by the Ethics Committee of the University of Wollongong (protocol code 2020/343, 14 September 2020).

Informed Consent Statement: Informed consent was obtained from all subjects involved in the study.

Data Availability Statement: The data are not publically available.

Acknowledgments: We thank the Social Research Centre and the Life in AustraliaTM panel members. We thank Richard Mitchell (University of Glasgow) for sharing ideas on survey questions.

Conflicts of Interest: The authors declare no conflict of interest. The funders had no role in the design of the study; in the collection, analyses, or interpretation of data; in the writing of the manuscript, or in the decision to publish the results.

References

1. Markevych, I.; Schoierer, J.; Hartig, T.; Chudnovsky, A.; Hystad, P.; Dzhambov, A.M.; de Vries, S.; Triguero-Mas, M.; Brauer, M.; Nieuwenhuijsen, M.J.; et al. Exploring pathways linking greenspace to health: Theoretical and methodological guidance. *Environ. Res.* **2017**, *158*, 301–317. [CrossRef] [PubMed]
2. Hartig, T. Restoration in nature: Beyond the conventional narrative. In *Nature and Psychology: Biological, Cognitive, Developmental, and Social Pathways to Well-Being, Proceedings of the 67th Annual Nebraska Symposium on Motivation, Lincoln, NE, USA, 21–22 April 2021*; Schutte, A.R., Torquati, J., Stevens, J.R., Eds.; Springer Nature: Cham, Switzerland, 2021; Volume 67.
3. Hartig, T.; Mitchell, R.; de Vries, S.; Frumkin, H. Nature and Health. *Annu. Rev. Public Health* **2014**, *35*, 207–228. [CrossRef] [PubMed]
4. Liu, X.-X.; Ma, X.-L.; Huang, W.-Z.; Luo, Y.-N.; He, C.-J.; Zhong, X.-M.; Dadvand, P.; Browning, M.H.; Li, L.; Zou, X.-G.; et al. Green space and cardiovascular disease: A systematic review with meta-analysis. *Environ. Pollut.* **2022**, *301*, 118990. [CrossRef] [PubMed]
5. Astell-Burt, T.; Feng, X. Urban green space, tree canopy and prevention of cardiometabolic diseases: A multilevel longitudinal study of 46 786 Australians. *Int. J. Epidemiol.* **2020**, *49*, 926–933. [CrossRef] [PubMed]
6. Kardan, O.; Gozdyra, P.; Misic, B.; Moola, F.; Palmer, L.; Paus, T.; Berman, M.G. Neighborhood greenspace and health in a large urban center. *Sci. Rep.* **2015**, *5*, 11610. [CrossRef] [PubMed]
7. Moreira, T.C.L.; Polizel, J.L.; Santos, I.S.; Filho, D.F.S.; Bensenor, I.; Lotufo, P.A.; Mauad, T. Green Spaces, Land Cover, Street Trees and Hypertension in the Megacity of São Paulo. *Int. J. Environ. Res. Public Health* **2020**, *17*, 725. [CrossRef] [PubMed]
8. Dzhambov, A.M.; Markevych, I.; Lercher, P. Greenspace seems protective of both high and low blood pressure among residents of an Alpine valley. *Environ. Int.* **2018**, *121*, 443–452. [CrossRef]
9. Tamosiunas, A.; Grazuleviciene, R.; Luksiene, D.; Dedele, A.; Reklaitiene, R.; Baceviciene, M.; Vencloviene, J.; Bernotiene, G.; Radisauskas, R.; Malinauskiene, V.; et al. Accessibility and use of urban green spaces, and cardiovascular health: Findings from a Kaunas cohort study. *Environ. Health* **2014**, *13*, 20. [CrossRef]
10. Yeager, R.; Riggs, D.W.; DeJarnett, N.; Tollerud, D.J.; Wilson, J.; Conklin, D.J.; O'Toole, T.E.; McCracken, J.; Lorkiewicz, P.; Xie, Z.; et al. Association Between Residential Greenness and Cardiovascular Disease Risk. *J. Am. Heart Assoc.* **2018**, *7*, e009117. [CrossRef]
11. Yitshak-Sade, M.; James, P.; Kloog, I.; Hart, J.E.; Schwartz, J.D.; Laden, F.; Lane, K.J.; Fabian, M.P.; Fong, K.C.; Zanobetti, A. Neighborhood Greenness Attenuates the Adverse Effect of PM2.5 on Cardiovascular Mortality in Neighborhoods of Lower Socioeconomic Status. *Int. J. Environ. Res. Public Health* **2019**, *16*, 814. [CrossRef]
12. Twohig-Bennett, C.; Jones, A. The health benefits of the great outdoors: A systematic review and meta-analysis of greenspace exposure and health outcomes. *Environ. Res.* **2018**, *166*, 628–637. [CrossRef] [PubMed]

13. Dalton, A.M.; Jones, A.P.; Sharp, S.J.; Cooper, A.J.; Griffin, S.; Wareham, N.J. Residential neighbourhood greenspace is associated with reduced risk of incident diabetes in older people: A prospective cohort study. *BMC Public Health* **2016**, *16*, 1171. [CrossRef] [PubMed]
14. Bodicoat, D.H.; O'Donovan, G.; Dalton, A.M.; Gray, L.J.; Yates, T.; Edwardson, C.; Hill, S.; Webb, D.R.; Khunti, K.; Davies, M.J.; et al. The association between neighbourhood greenspace and type 2 diabetes in a large cross-sectional study. *BMJ Open* **2014**, *4*, e006076. [CrossRef]
15. Rojas-Rueda, D.; Nieuwenhuijsen, M.J.; Gascon, M.; Perez-Leon, D.; Mudu, P. Green spaces and mortality: A systematic review and meta-analysis of cohort studies. *Lancet Planet. Health* **2019**, *3*, e469–e477. [CrossRef]
16. Astell-Burt, T.; Hartig, T.; Eckermann, S.; Nieuwenhuijsen, M.; McMunn, A.; Frumkin, H.; Feng, X. More green, less lonely? A longitudinal cohort study. *Int. J. Epidemiol.* **2022**, *51*, 99–110. [CrossRef]
17. Maas, J.; van Dillen, S.M.E.; Verheij, R.A.; Groenewegen, P.P. Social contacts as a possible mechanism behind the relation between green space and health. *Health Place* **2009**, *15*, 586–595. [CrossRef] [PubMed]
18. Astell-Burt, T.; Navakatikyan, M.A.; Feng, X. Urban green space, tree canopy and 11-year risk of dementia in a cohort of 109,688 Australians. *Environ. Int.* **2020**, *145*, 106102. [CrossRef]
19. Paul, L.A.; Hystad, P.; Burnett, R.T.; Kwong, J.C.; Crouse, D.L.; van Donkelaar, A.; Tu, K.; Lavigne, E.; Copes, R.; Martin, R.V.; et al. Urban green space and the risks of dementia and stroke. *Environ. Res. Lett.* **2020**, *186*, 109520. [CrossRef]
20. Astell-Burt, T.; Feng, X. Greener neighbourhoods, better memory? A longitudinal study. *Health Place* **2020**, *65*, 102393. [CrossRef]
21. Brown, S.C.; Perrino, T.; Lombard, J.; Wang, K.; Toro, M.; Rundek, T.; Gutierrez, C.M.; Dong, C.; Plater-Zyberk, E.; Nardi, M.I.; et al. Health Disparities in the Relationship of Neighborhood Greenness to Mental Health Outcomes in 249,405 U.S. Medicare Beneficiaries. *Int. J. Environ. Res. Public Health* **2018**, *15*, 430. [CrossRef]
22. Cherrie, M.P.; Shortt, N.K.; Mitchell, R.J.; Taylor, A.M.; Redmond, P.; Thompson, C.W.; Starr, J.M.; Deary, I.J.; Pearce, J.R. Green space and cognitive ageing: A retrospective life course analysis in the Lothian Birth Cohort. *Soc. Sci. Med.* **2018**, *196*, 56–65. [CrossRef] [PubMed]
23. Cherrie, M.P.; Shortt, N.K.; Thompson, C.W.; Deary, I.J.; Pearce, J.R. Association Between the Activity Space Exposure to Parks in Childhood and Adolescence and Cognitive Aging in Later Life. *Int. J. Environ. Res. Public Health* **2019**, *16*, 632. [CrossRef] [PubMed]
24. Crous-Bou, M.; Gascon, M.; Gispert, J.D.; Cirach, M.; Sánchez-Benavides, G.; Falcon, C.; Arenaza-Urquijo, E.M.; Gotsens, X.; Fauria, K.; Sunyer, J.; et al. Impact of urban environmental exposures on cognitive performance and brain structure of healthy individuals at risk for Alzheimer's dementia. *Environ. Int.* **2020**, *138*, 105546. [CrossRef] [PubMed]
25. De Keijzer, C.; Tonne, C.; Basagaña, X.; Valentín, A.; Singh-Manoux, A.; Alonso, J.; Antó, J.M.; Nieuwenhuijsen, M.J.; Sunyer, J.; Dadvand, P.; et al. Residential Surrounding Greenness and Cognitive Decline: A 10-Year Follow-Up of the Whitehall II Cohort. *Environ. Health Perspect.* **2018**, *126*, 077003. [CrossRef]
26. Dzhambov, A.M.; Bahchevanov, K.M.; Chompalov, K.A.; Atanassova, P.A. A feasibility study on the association between residential greenness and neurocognitive function in middle-aged Bulgarians. *Arch. Ind. Hyg. Toxicol.* **2019**, *70*, 173–185. [CrossRef]
27. Ho, H.C.; Fong, K.N.K.; Chan, T.-C.; Shi, Y. The associations between social, built and geophysical environment and age-specific dementia mortality among older adults in a high-density Asian city. *Int. J. Health Geogr.* **2020**, *19*, 1–13. [CrossRef]
28. Labib, S.; Browning, M.H.; Rigolon, A.; Helbich, M.; James, P. Nature's contributions in coping with a pandemic in the 21st century: A narrative review of evidence during COVID. *Sci. Total Environ.* **2022**, *833*, 155095. [CrossRef]
29. Reid, C.E.; Rieves, E.S.; Carlson, K. Perceptions of green space usage, abundance, and quality of green space were associated with better mental health during the COVID-19 pandemic among residents of Denver. *PLoS ONE* **2022**, *17*, e0263779. [CrossRef]
30. Burnett, H.; Olsen, J.R.; Nicholls, N.; Mitchell, R. Change in time spent visiting and experiences of green space following restrictions on movement during the COVID-19 pandemic: A nationally representative cross-sectional study of UK adults. *BMJ Open* **2021**, *11*, e044067. [CrossRef]
31. Ribeiro, A.I.; Triguero-Mas, M.; Santos, C.J.; Gómez-Nieto, A.; Cole, H.; Anguelovski, I.; Silva, F.M.; Baró, F. Exposure to nature and mental health outcomes during COVID-19 lockdown. A comparison between Portugal and Spain. *Environ. Int.* **2021**, *154*, 106664. [CrossRef]
32. Poortinga, W.; Bird, N.; Hallingberg, B.; Phillips, R.; Williams, D. The role of perceived public and private green space in subjective health and wellbeing during and after the first peak of the COVID-19 outbreak. *Landsc. Urban Plan.* **2021**, *211*, 104092. [CrossRef]
33. Astell-Burt, T.; Feng, X. Time for 'green' during COVID-19? Inequities in green and blue space access, visitation and felt benefits. *Int. J. Environ. Res. Public Health* **2021**, *18*, 2757. [CrossRef] [PubMed]
34. Feng, X.; Astell-Burt, T. Residential Green Space Quantity and Quality and Child Well-being: A Longitudinal Study. *Am. J. Prev. Med.* **2017**, *53*, 616–624. [CrossRef] [PubMed]
35. Feng, X.; Astell-Burt, T. Residential green space quantity and quality and symptoms of psychological distress: A 15-year longitudinal study of 3897 women in postpartum. *BMC Psychiatry* **2018**, *18*, 348. [CrossRef]
36. Van Dillen, S.M.; de Vries, S.; Groenewegen, P.P.; Spreeuwenberg, P. Greenspace in urban neighbourhoods and residents' health: Adding quality to quantity. *J. Epidemiol. Community Health* **2012**, *66*, e8. [CrossRef]
37. Francis, J.; Wood, L.J.; Knuiman, M.; Giles-Corti, B. Quality or quantity? Exploring the relationship between Public Open Space attributes and mental health in Perth, Western Australia. *Soc. Sci. Med.* **2012**, *74*, 1570–1577. [CrossRef]

38. Astell-Burt, T.; Feng, X. Paths through the woods. *Int. J. Epidemiol.* **2022**, *51*, 1–5. [CrossRef]
39. Holland, I.; DeVille, N.V.; Browning, M.H.; Buehler, R.M.; Hart, J.E.; Hipp, J.; Mitchell, R.; Rakow, D.; Schiff, J.; White, M.; et al. Measuring Nature Contact: A Narrative Review. *Int. J. Environ. Res. Public Health* **2021**, *18*, 4092. [CrossRef]
40. Birch, J.; Rishbeth, C.; Payne, S.R. Nature doesn't judge you—How urban nature supports young people's mental health and wellbeing in a diverse UK city. *Health Place* **2020**, *62*, 102296. [CrossRef]
41. Graham, T.M.; Glover, T.D. On the Fence: Dog Parks in the (Un)Leashing of Community and Social Capital. *Leis. Sci.* **2014**, *36*, 217–234. [CrossRef]
42. Knobel, P.; Dadvand, P.; Alonso, L.; Costa, L.; Español, M.; Maneja, R. Development of the urban green space quality assessment tool (RECITAL). *Urban For. Urban Green.* **2021**, *57*, 126895. [CrossRef]
43. Hartig, T.; Korpela, K.; Evans, G.W.; Gärling, T. A measure of restorative quality in environments. *Scand. Hous. Plan. Res.* **1997**, *14*, 175–194. [CrossRef]
44. Nguyen, P.-Y.; Astell-Burt, T.; Rahimi-Ardabili, H.; Feng, X. Green Space Quality and Health: A Systematic Review. *Int. J. Environ. Res. Public Health* **2021**, *18*, 11028. [CrossRef] [PubMed]
45. Feng, X.; Astell-Burt, T.; Standl, M.; Flexeder, C.; Heinrich, J.; Markevych, I. Green space quality and adolescent mental health: Do personality traits matter? *Environ Res.* **2022**, *206*, 112591. [CrossRef]
46. Feng, X.; Astell-Burt, T. Can green space quantity and quality help prevent postpartum weight gain? A longitudinal study. *J. Epidemiol. Community Health* **2019**, *73*, 295–302. [CrossRef]
47. Aerts, R.; Honnay, O.; Van Nieuwenhuyse, A. Biodiversity and human health: Mechanisms and evidence of the positive health effects of diversity in nature and green spaces. *Br. Med. Bull.* **2018**, *127*, 5–22. [CrossRef]
48. Marselle, M.R.; Hartig, T.; Cox, D.T.; de Bell, S.; Knapp, S.; Lindley, S.; Triguero-Mas, M.; Böhning-Gaese, K.; Braubach, M.; Cook, P.A.; et al. Pathways linking biodiversity to human health: A conceptual framework. *Environ. Int.* **2021**, *150*, 106420. [CrossRef]
49. Rook, G.A. Regulation of the immune system by biodiversity from the natural environment: An ecosystem service essential to health. *Proc. Natl. Acad. Sci. USA* **2013**, *110*, 18360–18367. [CrossRef]
50. Fuller, R.A.; Irvine, K.N.; Devine-Wright, P.; Warren, P.H.; Gaston, K.J. Psychological benefits of greenspace increase with biodiversity. *Biol. Lett.* **2007**, *3*, 390–394. [CrossRef]
51. Harris, V.; Kendal, D.; Hahs, A.K.; Threlfall, C.G. Green space context and vegetation complexity shape people's preferences for urban public parks and residential gardens. *Landsc. Res.* **2018**, *43*, 150–162. [CrossRef]
52. Perry, M.; Cotes, L.; Horton, B.; Kunac, R.; Snell, I.; Taylor, B.; Wright, A.; Devan, H. "Enticing" but Not Necessarily a "Space Designed for Me": Experiences of Urban Park Use by Older Adults with Disability. *Int. J. Environ. Res. Public Health* **2021**, *18*, 552. [CrossRef] [PubMed]
53. Corazon, S.S.; Gramkow, M.C.; Poulsen, D.V.; Lygum, V.L.; Zhang, G.; Stigsdotter, U.K.; Gramkov, M.C. I Would Really like to Visit the Forest, but it is Just Too Difficult: A Qualitative Study on Mobility Disability and Green Spaces. *Scand. J. Disabil. Res.* **2019**, *20*, 1–13. [CrossRef]
54. Wojnowska-Heciak, M.; Suchocka, M.; Błaszczyk, M.; Muszyńska, M. Urban Parks as Perceived by City Residents with Mobility Difficulties: A Qualitative Study with In-Depth Interviews. *Int. J. Environ. Res. Public Health* **2022**, *19*, 2018. [CrossRef] [PubMed]
55. Dzhambov, A.M.; Dimitrova, D.D. Green spaces and environmental noise perception. *Urban For. Urban Green.* **2015**, *14*, 1000–1008. [CrossRef]
56. Dzhambov, A.M.; Dimitrova, D.D. Urban green spaces' effectiveness as a psychological buffer for the negative health impact of noise pollution: A systematic review. *Noise Health* **2014**, *16*, 157–165. [CrossRef]
57. Galobardes, B.; Shaw, M.; Lawlor, D.; Lynch, J.; Davey Smith, G. Indicators of socioeconomic position (part 2). *J. Epidemiol. Community Health* **2006**, *60*, 95. [CrossRef]
58. Galobardes, B.; Shaw, M.; Lawlor, D.A.; Lynch, J.W.; Davey Smith, G. Indicators of socioeconomic position (part 1). *J. Epidemiol. Community Health* **2006**, *60*, 7–12. [CrossRef]
59. Shanahan, D.F.; Bush, R.; Gaston, K.J.; Lin, B.B.; Dean, J.; Barber, E.; Fuller, R.A. Health Benefits from Nature Experiences Depend on Dose. *Sci. Rep.* **2016**, *6*, 28551. [CrossRef]
60. White, M.P.; Alcock, I.; Grellier, J.; Wheeler, B.W.; Hartig, T.; Warber, S.L.; Bone, A.; Depledge, M.H.; Fleming, L.E. Spending at least 120 minutes a week in nature is associated with good health and wellbeing. *Sci. Rep.* **2019**, *9*, 7730. [CrossRef]
61. Wiltshire, G.; Stevinson, C. Exploring the role of social capital in community-based physical activity: Qualitative insights from parkrun. *Qual. Res. Sport Exerc. Health* **2018**, *10*, 47–62. [CrossRef]
62. Morris, P.; Scott, H. Not just a run in the park: A qualitative exploration of parkrun and mental health. *Adv. Ment. Health* **2019**, *17*, 110–123. [CrossRef]
63. Hindley, D. "More than just a run in the park": An exploration of parkrun as a shared leisure space. *Leis. Sci.* **2020**, *42*, 85–105. [CrossRef]
64. Swensen, G.; Skår, M. Urban cemeteries' potential as sites for cultural encounters. *Mortality* **2019**, *24*, 333–356. [CrossRef]
65. Rishbeth, C.; Powell, M. Place Attachment and Memory: Landscapes of Belonging as Experienced Post-migration. *Landsc. Res.* **2013**, *38*, 160–178. [CrossRef]
66. Sobel, D. A place in the world: Adults' memories of childhood's special places. *Child. Environ. Q.* **1990**, *7*, 5–12.
67. Kingsley, J.; Foenander, E.; Bailey, A. "It's about community": Exploring social capital in community gardens across Melbourne, Australia. *Urban For. Urban Green.* **2020**, *49*, 126640. [CrossRef]

68. Kingsley, J.; Foenander, E.; Bailey, A. "You feel like you're part of something bigger": Exploring motivations for community garden participation in Melbourne, Australia. *BMC Public Health* **2019**, *19*, 745. [CrossRef]
69. Johnston, R.J.; Thomassin, P.J. Willingness to Pay for Water Quality Improvements in the United States and Canada: Considering Possibilities for International Meta-Analysis and Benefit Transfer. *Agric. Resour. Econ. Rev.* **2010**, *39*, 114–131. [CrossRef]

 *land*

Article

Change of Residents' Attitudes and Behaviors toward Urban Green Space Pre- and Post- COVID-19 Pandemic

Luyang Chen [1], Lingbo Liu [1], Hao Wu [2], Zhenghong Peng [2,*] and Zhihao Sun [1]

[1] Department of Urban Planning, School of Urban Design, Wuhan University, Wuhan 430072, China; chenluyang@whu.edu.cn (L.C.); lingbo.liu@whu.edu.cn (L.L.); sun99xt@whu.edu.cn (Z.S.)
[2] Department of Graphics and Digital Technology, School of Urban Design, Wuhan University, Wuhan 430072, China; wh79@whu.edu.cn
* Correspondence: pengzhenghong@whu.edu.cn; Tel.: +86-27-6877-3062

Abstract: The COVID-19 pandemic has changed and influenced people's attitudes and behaviors toward visiting green spaces. This paper aims to explore the association between residents' health and urban green spaces (UGS) through an in-depth study of changes in residents' use of UGS under the influence of the COVID-19 pandemic. The Wuhan East Lake Greenway Park was selected as the location for the field survey and in-depth interviews. At the same time, an online survey was also conducted (total number = 302) regarding participants' physical and mental health and their attitude and behavior toward the UGS. A paired sample *t*-test and binary logistic regression were performed to investigate the association between participants' health and UGS during COVID-19. The results show that: (1) the COVID-19 pandemic has primarily changed the leisure patterns of parks, with potential impacts on the physical and mental health of participants; (2) the purpose, frequency, timing, and preferred areas of participants' park visits have changed to varying degrees after the pandemic, highlighting the important role and benefits of UGSs; (3) the physical and mental health of participants and urban development issues reflected by UGS use are prominent. This study reveals that awareness of the construction and protection of UGSs is an important prerequisite for ensuring the health of urban residents.

Keywords: postpandemic era; urban green space; green landscape; residents' health

Citation: Chen, L.; Liu, L.; Wu, H.; Peng, Z.; Sun, Z. Change of Residents' Attitudes and Behaviors toward Urban Green Space Pre- and Post-COVID-19 Pandemic. *Land* **2022**, *11*, 1051. https://doi.org/10.3390/land11071051

Academic Editor: Francesca Ugolini

Received: 10 June 2022
Accepted: 8 July 2022
Published: 11 July 2022

Publisher's Note: MDPI stays neutral with regard to jurisdictional claims in published maps and institutional affiliations.

1. Introduction

There is a growing body of literature that recognizes the importance of urban green spaces (UGS) for residents' lives and well-being [1–4]. The rapid outbreak of novel coronavirus (COVID-19) pandemic has fundamentally changed people's lives and greatly impacted their daily lives [5,6]. Although most studies have focused on morbidity and mortality studies related to the disease, the physical and mental health [7,8] aspects of the residents are worthy of more study [9].

UGSs provide multiple ecosystem services [10] and are of great value. Not only do they benefit human physical and mental health and social development, but these roles may be amplified in special times [4,11,12]. A lot of pieces of evidence have shown that the unavailability and absence of UGSs, social isolation, and blockades during COVID-19 have a negative impact on mental health [13], presenting varying degrees of anxiety, anger, fear, irritability, reduced well-being, and other negative emotions [14–16], and increasing the risk of mental health disorders [17]. On the other hand, exposure to blue–green space is beneficial for physical and mental health [18,19], relieves stress [3], reduces anxiety, improves attention recovery [20], and increases well-being and satisfaction. To some extent, it increases motivation to exercise and enhances physical activity, thus reducing diseases such as obesity, hypertension, and hyperlipidemia [21–23]. A large proportion of the population believes that the pandemic greatly affected physical activity levels and physical health, and tries to restore physical function and relieve psychological stress by increasing walking exercise [24–27]. The aforementioned positive and negative effects are closely associated with the resident and ultimately manifest visually as changes in resident use behaviors and attitudes.

Behaviors and patterns of UGS use changed quietly as a result of the COVID-19 pandemic. During the Tokyo Shinkansen epidemic, older adults, elementary school students, and others experienced lifestyle changes, lack of exercise, stress accumulation, and decreased well-being [15]; park use and associated mental health patterns among college student populations during the pandemic raised concerns, with some studies suggesting that young people were more likely to experience negative emotions such as stress, anxiety, and depression [13,28–30]. Some studies have shown significant increases in park visitation [4,31] and increased park use [6,32,33], while some urban parks have seen decreases in visitor numbers [34], decreases in public space engagement, and reduced risk of epidemic prevention and control with the use of public urban green spaces [35]. These conflicting and changing uses need to be studied by more in-depth investigations.

The pandemic is reflected in differences in attitudes and behaviors of different groups in UGSs [26]. Additionally, the uneven distribution of UGSs at particular times can also cause changes in user behavior. Wuhan was one of the first cities to experience a major outbreak of the COVID-19 pandemic, and it went through a three-month lockdown. To understand the actual changes in the use of UGSs before and after the pandemic, we conducted a questionnaire survey in Wuhan's largest, most environmentally friendly, and most well-known green space, called East Lake, after the announcement of Wuhan's "lifting of the city lockdown" on 18 April 2020. The COVID-19 erupted in January 2020 and Wuhan lifted the lockdown in April 2020. Therefore, in the study, the "pre" refers to the time before January 2020, while the "post" refers to the time after April 2020. This survey intends to examine the changes in the relationship between UGSs and residents' health under the COVID-19 pandemic from the perspective of changes in residents' use behavior. Survey analysis was conducted to answer the above hypotheses based on the specific performance of residents of different ages, genders, occupations, and statuses in terms of their attitudes and behavioral patterns of use. First, sociodemographic information was collected to classify the population of users; then, the specific changes in residents' use of urban parks were used to study the deeper associations during the pandemic; and finally, the changing patterns of attitudes and behaviors of use were explored to consider the construction of UGSs.

2. Research Methods and Data Collection

2.1. Questionnaire Survey

To investigate the specific changes in residents' use of UGSs before and after the COVID-19 pandemic, a combination of field and online surveys was conducted in the paper. For the field survey, the respondents were selected from the core areas of five different scenic spots in East Lake and filled in voluntarily for 6–10 min by the respondents in an informed manner. During this field survey, a total of 297 questionnaires were distributed. The online method is more effective and easier for people to visualize the various options. An online questionnaire was distributed via WeChat to groups who had visited East Lake to ensure that the groups researched had realistic feelings about the use of the UGS. As a result, a total of 38 participants responded to the online questionnaire. The choice of an online survey is based on three main considerations. (1) Due to the development of the times, the popularity of digital devices and the internet has changed the daily life of a large part of the population, and the internet is increasingly used, especially by young people, to reach a large number of people through social media (WeChat) posting and forwarding. (2) Field surveys are often influenced by many factors, while an online survey is more convenient and faster to disseminate. (3) The no-contact approach is more popular due to epidemics. However, the online survey cannot comprehensively collect information from all kinds of people, especially those who cannot use electronic devices (elderly, students, etc.), resulting in insufficient data samples. Moreover, this method also cannot conduct interviews. Therefore, we decided to conduct multiple field surveys to obtain more realistic and effective information.

The questionnaire was designed to assess three main types of information: sociodemographic, self-ratings of the residents' physical and mental health, and the use of UGSs. In the first part, sociodemographic information was collected to map the personal situation of the residents to connect with the information that follows [1,36,37]. In the second part, the self-rated questionnaire was used to capture the physical and mental health status of the participants, and then the results were classified and analyzed [36,38]. In the third part, six factors, including purpose of visit, frequency of visit, mode of transportation, mode of travel, area visited, and duration of stay, which is the most intuitive and reflective of changes in UGS use, were selected for question setting to collect changes in use [6,39–42]. Table 1 shows the questionnaire content of residents' self-rating of their physical and mental health, as well as their attitudes and behavior toward green space. For the above self-rated questionnaire, the five-point Likert Scale Method was introduced to capture the intensity of participants' feelings for a given item, which means participants would choose one of five levels (for example, 1 = worse, 2 = poor, 3 = general, 4 = good, 5 = better) for a series of statements on a symmetric agree–disagree scale.

This round of research was conducted from March 2021 to November 2021, and the questionnaires were distributed and collated during the period when the pandemic was relatively stable and the contrast between before and after the COVID-19 pandemic was more obvious. A total of 335 questionnaires were distributed during this period, and 302 valid questionnaires were returned, with an efficiency rate of 90.1%.

2.2. Field Trips

2.2.1. Location Selection

The Wuhan East Lake Greenway, which has a superior natural environment and convenient transportation, was chosen as the research object. It is located in the Wuhan East Lake Scenic Area which is situated in the eastern part of the city. It is the largest internal urban park in Wuhan, serving a large part of the city, and is a representative of UGSs with a large volume.

The planning and construction are based on "Eco-Wuhan" to create a world-class greenway around the lake. Relying on humanistic and historical resources, rich in natural resources, Bruno Deacon, an official of UN-Habitat, called it a model. With a total length of 101.98 km and a width of 6 m, the East Lake Greenway connects five scenic spots of the East Lake, which is shown in Figure 1. With its unique natural scenery, historical and cultural heritage, biological diversity, plant diversity, and functional diversity, it attracts more than 40 million people and has become a preferred place for residents to travel, relax, and interact outdoors.

2.2.2. Field Research

Considering the general context of the pandemic and the specificity of public space use, priority was given to on-site behavioral observation, questionnaire distribution, targeted in-depth interviews, green space landscape classification, local staff interviews and consultations, and photo documentation to collect on-site information. In the on-site research, the following two aspects were specifically addressed. First, the participants were classified using the observation method to determine whether they were permanent residents by their behavior. For example, if residents exhibit behaviors such as excessive photo-taking, sightseeing, and punching in, we classified these residents as nonlocals, and they were filtered out in our questionnaire. Second, after the interviewees were identified, inquiries were made and targeted interviews were conducted. Finally, the identities of all the interviewees were classified, and many effective interview results were obtained.

Table 1. Measurement of residents' physical and mental health and their attitude and behavior toward green space.

Part 1: The Use of Green Space.

		"Pre-Pandemic"					"Post- Pandemic"				
1	How do you think your physical health?	Worse	poor	general	good	better	Worse	poor	general	good	better
2	How do you think your mental health?	Worse	poor	general	good	better	Worse	poor	general	good	better
3	What is the purpose of your visit to the park?	exercise	relax	Relieve the pressure	view	entertainment	exercise	relax	Relieve the pressure	view	entertainment
4	Have you been here a lot?	hardly	occasionally	sometimes	often	almost everyday	hardly	occasionally	sometimes	often	almost everyday
5	What is your mode of transportation to the park?	public transportation	private car	walking	cycling	other	public transportation	private car	walking	cycling	other
6	How do you choose to travel?	alone	community				alone	community			
7	Which areas of the park do you usually choose?	Baima scenic area	Tingtao scenic area	Luoyan scenic area	Moshan scenic area	Chuidi scenic area	Baima scenic area	Tingtao scenic area	Luoyan scenic area	Moshan scenic area	Chuidi scenic area
8	How long do you stay?	20 min	20–40 min	1 h	1–2 h	>2 h	20 min	20–40 min	1 h	1–2 h	>2 h

Part 2: The evaluation of green space.

1. What is your evaluation of the green space around your community?
☐ very dissatisfied ☐dissatisfied ☐general ☐satisfied ☐very satisfied

2. What is your evaluation of park green space?
☐ very dissatisfied ☐dissatisfied ☐general ☐satisfied ☐very satisfied

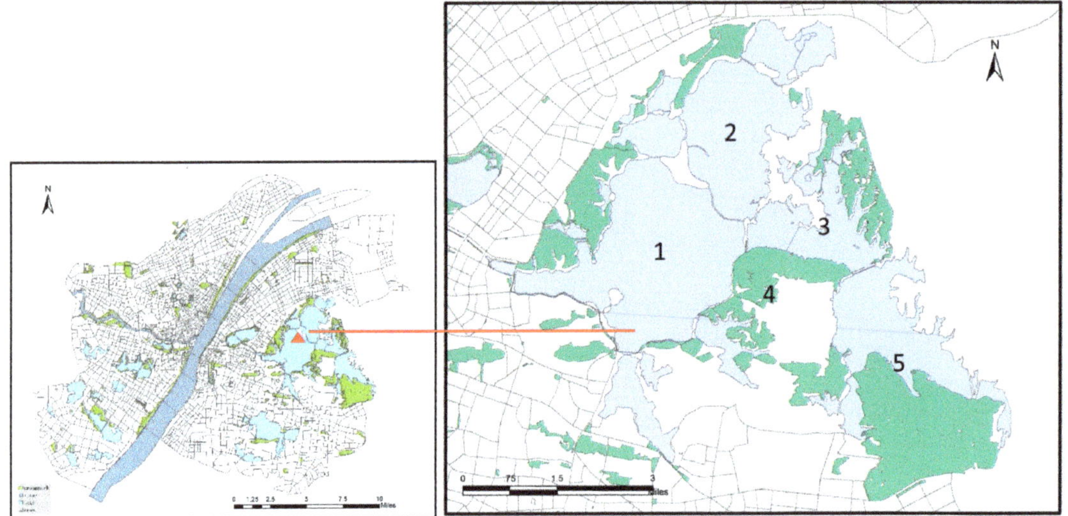

Figure 1. Scenic spot distribution map. (1 Tingtao scenic area; 2 Baima scenic area; 3 Luoyan scenic area; 4 Moshan scenic area; 5 Chuidi scenic area).

2.3. Statistical Analysis

The descriptive statistics were used to obtain and summarize the sociodemographic characteristics of the participants, their health status, and urban park use before and after the COVID-19 pandemic. For the changes in participants' health and urban park use, the paired sample *t*-test was utilized to quantitatively analyze the impact of the pandemic on them, assuming as a null hypothesis that the pandemic would not affect the participants' health and urban park use.

To analyze the relationship between the existing changes and sociodemographic characteristics of the participants, binary logistic regression was carried out for two trends of changes in residents' health, separately: the increasing trend and the decreasing trend. For the former case, the dependent variable (participants' health increased after the pandemic) was set to 1, and those that remained unchanged or decreased were set to 0. Similarly, for the latter case, the dependent variable (participants' health decreased after the pandemic) was set to 1, and those that remained unchanged or increased were set to 0. Regarding independent variables, several binary variables and categorical scale variables were included. For gender, the male was set to 1 and female was set to 0. For the income trend, the increase was set to 1 and the reduction was set to 0. In other binary variables (including identity, marital, housing ownership, change of income), 1 means yes, and 0 means no. The categorical scale variables (including age, years of residence, education, occupation, income, and evaluation of green space near home or in the park) were used as ordinal value scales.

The paired sample *t*-test and binary logistic regression were performed through the statistical analysis software SPSS 26.0.

3. Result and Analysis

3.1. Sociodemographic Characteristics

As the central city of central China, Wuhan has 11 million residents [43]. Table 2 shows some basic information about the residents of Wuhan. Among them, males accounted for 50.8%, and females for 49.2%. Excluding the elderly aged over 60, the group aged 30–39 has the highest percentage of the population, at 19.45%. Meanwhile, 33.87% of the residents have bachelor's degrees or above.

Table 2. Basic sociodemographic characteristics of Wuhan residents.

Demographic	Variable	Percentage (%)
Gender	male	50.8
	female	49.2
Age	<10	11
	10–20	7.22
	20–29	11.98
	30–39	19.45
	40–49	13.61
	50–59	15.46
	$\geq$60	21.23
Education	primary Schools	13.65
	junior High School	25.34
	high School	19.69
	bachelor's degree or above	33.87
	other	7.45

source url: The basic sociodemographic characteristics of Wuhan residents were obtained from the Hubei Provincial Statistics Bureau website. (https://tjj.hubei.gov.cn/tjsj/sjkscx/tjnj/gsztj/whs/, accessed on 27 June 2022).

Table 3 shows the sociodemographic characteristics of participants. Of the 335 questionnaires distributed, a total of 302 valid responses were collected with a final response rate of 90.1%. The participation rate of males (62.25%) was much higher than that of females (37.75%), and the respondents were mainly from 20–29 years old, accounting for 41.06% of the total. Most of them are permanent residents (78.15%), and most of them have their own houses (57.95%). Most of the participants are well educated (68.54% with a bachelor's degree or above) and have a stable job. The average monthly income is concentrated in the range of RMB 5000–8000 (30.46%) and over RMB 8000 (30.46%). Due to the impact of the COVID-19 pandemic, the salary of nearly half of the participants (43.71%) has changed, of which 77.27% were less than before the pandemic.

Table 3. The statistical results on the sociodemographic characteristics of the participants.

Demographic	Variable	N	Percentage (%)
Gender	male	188	62.25
	female	114	37.75
Age	<10	0	0
	10–20	29	9.6
	20–29	124	41.06
	30–39	70	23.18
	40–49	32	10.6
	50–59	37	12.25
	$\geq$60	10	3.31
Identity	permanent residents	236	78.15
	nonpermanent residents	66	21.85
Marital	married	126	41.72
	unmarried	176	58.28
Housing ownership	home ownership	175	57.95
	rental housing	127	42.05
Years of Residence	<1 year	29	9.6
	1–3 years	76	25.17
	>3 years	197	65.23
Education	high school and below	32	10.6
	technical college	63	20.86
	bachelor's degree or above	207	68.54
Occupation	regular occupation	154	50.99
	freelance	67	22.19
	retired	13	4.3
	current students	68	22.52

Table 3. *Cont.*

Demographic	Variable	N	Percentage (%)
	<1550	42	13.91
	1550–3500	32	10.6
Income/month	3500–5000	44	14.57
	5000–8000	92	30.46
	>8000	92	30.46
Change of income	yes	132	43.71
	no	170	56.29
Income trend	increase	30	22.73
	reduce	102	77.27

3.2. Self-Rating of Physical and Mental Health

Numerous studies have confirmed that the presence of UGSs contribute to improved quality of life in many ways. In addition to some environmental and ecological services, cities naturally provide important social and psychological benefits to human society, enriching the meaning and emotion of human life [12]. The visit to UGSs has a positive promotion effect on physical and mental health [44]; conversely, it will also produce a certain degree of fear and pain of negative emotions, and eventually lead to an increase in anxiety, depression, and other mental diseases [45].

In this study, a self-rated questionnaire was used to obtain the physical and mental health status of the participants, and the results are shown in Figure 2. Additionally, the paired sample t-test revealed that postpandemic physical and mental health levels have both decreased significantly (for physical health: $t = 5.025$, $p = 0.000$; for mental health: $t = 6.949$, $p = 0.000$). In terms of physical health, the number of participants who felt their physical and mental health status is "fair poor" and "average" has increased, and conversely, the number of participants who felt their health status is "fair good" and "good" has decreased. In addition, the changing trend in mental health also decreased significantly. The result also proved that the pandemic has a certain impact on participants' attitudes and behavior changes in park use and that it is consistent with speculation and existing research that the pandemic has a direct or indirect impact on physical and mental health [18].

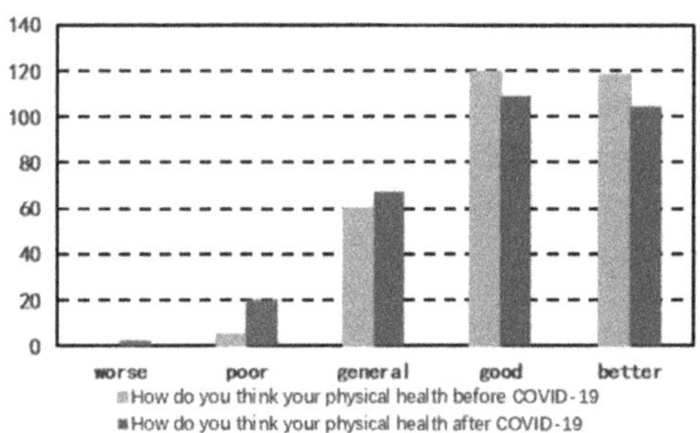

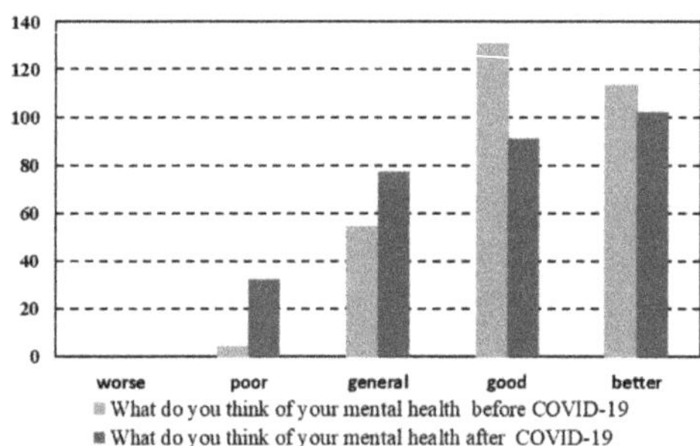

Figure 2. Self-rated comparison chart of physical health (**left**) and mental health (**right**).

3.3. Urban Park Usage

Changes in the use of urban parks were investigated from four perspectives: visit purpose, visit frequency, accompany mode, and visit duration of the parks, and the results are shown in Figure 3. Additionally, the paired sample t-test results of different perspectives are shown in Table 4, in which the visit purpose is not tested because it is a multiple-choice question. These results illustrated that the user behavior and attitude of urban parks changed under the pandemic:

(1) Compared with the changes in the purpose of participants visiting parks before and after the pandemic, the majority of participants choose to "relax", "relieve pressure", and "exercise" after the pandemic, and it is worth noting that "relax" become the first choice. Secondly, the number of participants who choose to "relieve pressure" has increased significantly, which has become an important purpose for participants to go to parks after the pandemic, reflecting the impact of the pandemic on participants' mental health and the enhanced recreational use of UGSs.

(2) In terms of the frequency of visits, it remained stable, but the number of participants who choose to visit "often" and "almost every day" showed a significant increase ($t = -4.421$, $p = 0.000$) in Table 5. Under the impact of the pandemic, the frequency of participants visiting green spaces in parks increased for health reasons, highlighting the important regulatory role of UGSs, especially urban parks, in the context of the pandemic.

(3) The change of visit mode was not significant. It indicated that the number of lone-visitors decreased slightly while the number of visits with multiple companions increased. This phenomenon may be related to the psychological changes of participants after the pandemic.

(4) According to Table 5, The change of visit duration was significant ($t = 7.052$, $p = 0.000$). The number of participants who stay for more than 2 h is significantly more than before the pandemic, which reflects that most residents prefer outdoor activities after the pandemic.

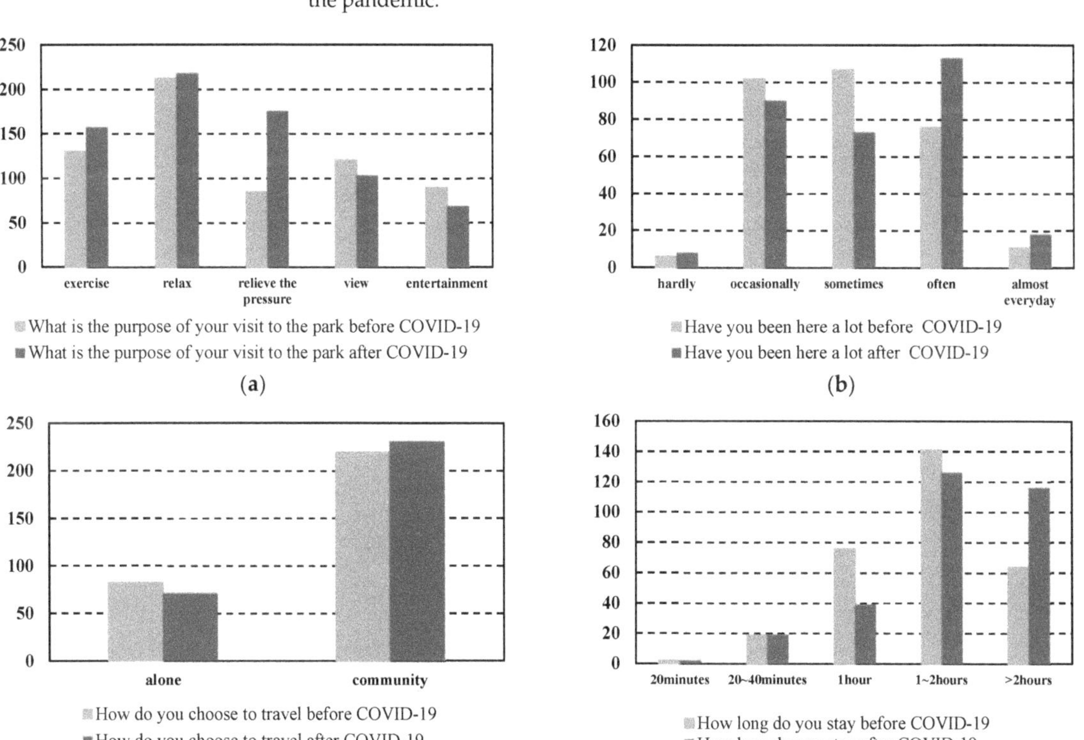

Figure 3. Urban park uses during the pandemic: (**a**) purpose, (**b**) frequency, (**c**) accompany mode, (**d**) duration.

Table 4. Paired sample *t*-test results of physical and mental health before and after the pandemic.

	Paired		Difference	*t*	*p*
	Before-	**Post-**	**(Before vs. Post)**		
Physical health	4.159 ± 0.795	3.970 ± 0.945	0.189 ± 0.653	5.025 **	0.000 **
Mental health	4.169 ± 0.761	3.871 ± 1.002	0.298 ± 0.745	6.949 **	0.000 **

$p < 0.05$; ** $p < 0.01$.

Table 5. Paired sample *t*-test results of visit frequency, accompany mode, and visit duration before and after the pandemic.

	Paired		Difference	*t*	*p*
	Before-	**Post-**	**(Before vs. Post)**		
Visit frequency	2.947 ± 0.902	3.142 ± 1.000	−0.195 ± 0.768	−4.421 **	0.000
Visit duration	2.185 ± 0.862	1.891 ± 0.903	0.295 ± 0.726	7.052 **	0.000

$p < 0.05$; ** $p < 0.01$.

The changes in the intention to visit different areas are shown in Figure 4. According to Figure 4, the number of participants choosing "Moshan Scenic Area" decreased slightly after the pandemic, while the number of participants choosing the other four areas, which are relatively far away, are all increasing. By normalizing the numbers of visits to different areas, it was found that the number of visits to the Moshan scenic area was the highest both before and after the pandemic, while the number of visits to the other four scenic areas was relatively small. Specifically, compared with "before the pandemic", the number of participants choosing the Moshan scenic area decreased "after the pandemic", while the corresponding number of participants choosing the Baima scenic area, Tingtao scenic area, and Luoyan scenic area increased significantly. In terms of geographical location, the Moshan scenic area is located in the core area of the whole park, with convenient transportation and complete infrastructure, and it is a concentrated tourist and leisure area. The Baima scenic spot and the Luoyan scenic spot are located in the northern part of the park with weak geographical advantages. These areas are dominated by ecological wetland resources and have fewer recreational facilities, which may be the main reason for the low number of visits. The Tingtao scenic spot has the largest lake in the East Lake and is dominated by several commemorative scenic spots with low attractions. The Chuidi scenic spot is located in the easternmost part of the park, close to Maanshan Forest Park. It also has rich ecological resources and is relatively secluded. Therefore, the change in visitor numbers before and after the pandemic may be caused by the differences in the location and function of the park. On the one hand, for the consideration of prevention, control, and health, it has gradually become a trend to choose areas with safe distances and safe numbers to avoid potential infection risks. On the other hand, areas with prominent advantages of ecological and natural resources become more attractive to participants.

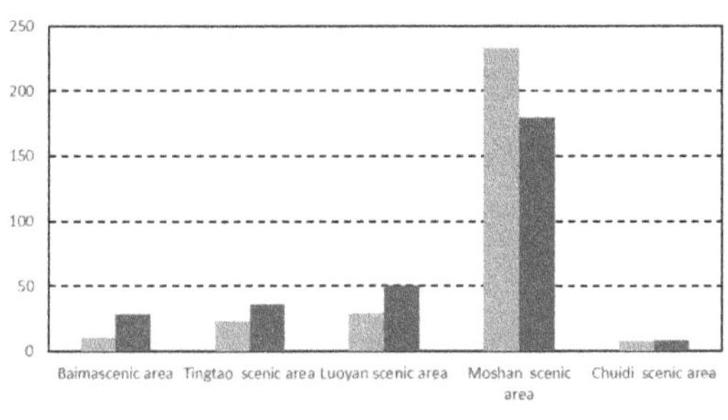

Figure 4. Comparison of urban park use areas before and after the pandemic.

3.4. The Relationship between Changes and Some Independent Variables

To investigate the influence of factors that are potentially associated with the changes in participants' health or changes in park use status, the binary logistic regression analysis was performed, and the increasing and decreasing trends in these changes were used as dependent variables, respectively, and the social characteristics of the participants were used as independent variables. Results of the binary logistic regression analysis are shown in Table 3 and reveal some distinctive features of the UGS use on participants' health before and after the pandemic. According to Table 6, on the one hand, after the outbreak: the increase in the self-rated physical health of the participants was negatively correlated with age ($p < 0.05$); the increase in the frequency of visiting UGSs was closely related to the quality of green space distribution in urban public spaces (visiting green spaces around parks and neighborhoods) ($p < 0.01$), with both poorer quality green spaces in neighborhoods and better quality green spaces in parks leading to an increase in the frequency of visiting UGS in urban parks; there is a correlation between the change in the mode of escorting more than one person to UGS and participants' income ($p < 0.05$), with higher-income participants preferring to be accompanied by more than one person to UGS; the change in the duration of visiting UGS was significantly correlated with the income level of residents ($p < 0.01$) and showed a negative correlation, and the income level of participants directly led to the change in the duration of visiting UGS.

Table 6. The relationship between increasing and decreasing trends and some independent variables.

			B	S.E.	Wald	Sig.	Exp (B)
Increase	Physical health	Age	−0.541 *	0.240	5.071	0.024	0.582
		Constant	−0.745	0.812	0.843	0.358	0.475
	Mental health						
	Visit frequency	Evaluation of green space near home	−0.843 **	0.172	24.045	0.000	0.430
		Evaluation of green space in the park	1.077 **	0.124	19.493	0.000	2.935
		Constant	−2.644 **	1.030	6.585	0.000	0.071
	Accompany mode	Income	0.528 *	0.207	6.545	0.011	1.696
		Constant	−4.463 **	0.893	24.997	0.000	0.012
	Visit duration	Income	−0.442 **	0.154	8.233	0.004	0.643
		Constant	−1.152 *	0.482	5.716	0.017	0.316
Decrease	Physical health	Gender	−0.658 *	0.300	4.817	0.028	0.518
		Evaluation of green space near home	−0.441 **	0.158	7.759	0.005	0.643
		Constant	1.290	0.720	3.212	0.073	3.633
	Mental health	Housing ownership	−0.553 *	0.272	4.139	0.042	0.575
		Evaluation of green space near home	−0.568 **	0.163	12.187	0.000	0.567
		Evaluation of green space in the park	0.708 **	0.227	9.686	0.002	2.030
		Constant	−1.135	1.021	1.237	0.266	0.321
	Visit frequency						
	Accompany mode	Gender	−2.128 *	1.045	4.415	0.042	0.119
		Constant	−0.472	1.157	0.167	0.683	0.624
	Visit duration	Period of resident	0.519 *	0.223	5.404	0.020	1.680
		Income	0.183	0.099	3.408	0.065	1.201
		Evaluation of green space near home	−0.330 *	0.157	4.431	0.035	0.719
		Evaluation of green space in the park	0.576 **	0.225	6.551	0.010	1.780
		Constant	−4.120 **	1.202	11.753	0.001	0.016

* *Sig.* < 0.05; ** *Sig.* < 0.01.

On the other hand, after the pandemic: the self-rated physical health of a part of the population decreased, as there was a negative correlation between the gender of the participants ($p < 0.05$) and the quality of the green space of the living environment ($p < 0.01$); participants who showed a decline in self-rated of mental health were associated with three factors, namely, a negative correlation ($p < 0.05$) between whether they owned their own home and the quality of green space in their residential area, and a positive correlation between the quality of green space in parks and changes in mental health; the shift from companionship mode to solo mode was correlated with the gender of the participants ($p < 0.05$); the decrease in time spent visiting UGSs was positively correlated with the time spent living and the quality of green spaces in parks, and negatively correlated with the quality of green spaces in the participants' living environment.

4. Discussion

The COVID-19 pandemic is still present and will continue to be prevalent. The urban development, physical and mental health of residents, and social changes affected by the pandemic are likely to continue for a longer period of time. It is particularly important to study and discuss the physical and mental health of urban residents. Has the pandemic had an impact on the behavior of urban residents visiting public green spaces? For which specific groups did it have a significant impact? How have they been affected? are the causes of the impact are and the extent of the impact will be the focus of our study. As an urban park consisting of five distinctive urban green areas, East Lake Scenic Area can cover a larger sample of users. Therefore, the results obtained from the survey data we collected are credible and very meaningful. This study focuses on the changes in behavior and attitudes of urban residents using public green spaces during a pandemic to confirm the hypothesized results of the study. Figure 5 illustrates the relationship between UGSs and residents' physical and mental health, as summarized in the study. The first part deals with the use of UGSs, including frequency, purpose, and pattern, which affect the health of the residents through direct or indirect ways, with health manifested as physical, psychological, and social behaviors.

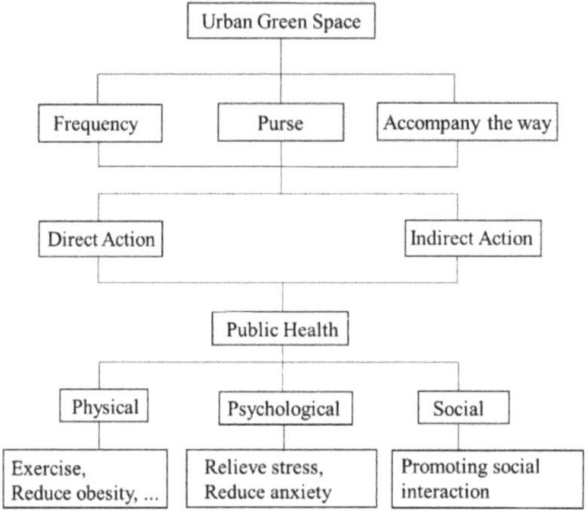

Figure 5. The relationship between the factors.

First, the pandemic proved to have an impact on the daily life of the residents. On the one hand, residents' self-ratings of their physical and mental health has changed significantly as a result of the pandemic; on the other hand, specific behaviors in using the garden are significantly different from those before the pandemic. Second, the negative impact of the pandemic on residents' daily life has been confirmed.

(1) The population visiting UGSs is relatively concentrated. There are more male visitors than female visitors, and the majority of those in the visitor group are young. Among the respondents, the participation rate of men was higher than that of women, and their ages were concentrated between 20–40 years old, indicating that young people are more health-conscious and willing to spend time visiting parks, and confirming that the physical health of some young people has improved after the outbreak. In addition, visitors' behavior was correlated with information about their status and occupation, reflecting that local residents who own their own houses, have long years

of housing, have stable jobs, and are well-educated outperformed other respondents in terms of behavior and health perceptions about visiting green spaces.

(2) The use of UGSs was specific to use behaviors and attitudes. The pre- and post-pandemic dynamics reflected that the pandemic had an impact on the use of park green spaces by urban residents. First, the improvement of participants' health levels after the pandemic was negatively correlated with age; the way of visiting green spaces was correlated with income level; and the change in the duration of visiting green spaces was significantly correlated with income. Second, after the pandemic, the decline in self-rated health status of some people was negatively correlated with gender and the quality of green space in the residential environment; and the decline in participants' self-rated mental health status was negatively correlated with home ownership and the quality of green space in the residential area, while the decline in the self-rated mental health was positively correlated with changes in mental health. Third, it is worth noting that the change in the choice of travel companion mode has a great relationship with gender.

The number of residents choosing to use park green spaces for physical exercise, stress relief, and relaxation increased, which is consistent with existing studies. The predominant focus of choice on physical exercise and relaxation [8] may be related to the negative impact of the pandemic on physical and mental health, which is well-explained by the desire to improve by visiting public green spaces. In terms of frequency of use and duration of use by residents, it is intuitive from the research data that the frequency of use, in general, shows an upward trend, and there is even the phenomenon that some people go almost every day, which may be affected by the quality of urban public spaces and residential green spaces. In terms of the duration of the visits to parks, it is evident that the overall duration of residents is increasing, and there is also a negative correlation between changes in this segment of the population and the income level of residents; there is also a decrease in duration of visits, and the study confirms that it is related to the period of residence, the quality of green spaces in parks, and the quality of the environment in residential areas. This study reveals that pandemics directly or indirectly influence the attitudes and behavioral characteristics of people using green park spaces.

The results indicate that changes in UGS visitation behavior can be inferred from participants' demand for UGSs and that visiting UGSs can be effective in improving physiological and psychological health and social interaction development, enhancing public health while promoting harmonious social development. This study highlights the critical and positive role of urban parks during pandemics [46]. However, it is undeniable that the negative effects of the lack of UGSs during quarantine are objective and important. Therefore, it is more important to consider the availability of UGSs, especially those that can be found during special periods. Urban managers can consider building small-scale green spaces in residential areas to mitigate the effects of inaccessibility and lack of access [17] and to encourage people to have more access to outdoor spaces to enhance their physical health [4].

(3) Comparing the type, distribution characteristics, quality, and other characteristics of green spaces in urban public spaces, residents evaluated the green spaces in their area and those in the East Lake Greenway. The increased frequency of visiting UGSs after the pandemic is closely related to the quality of green space distribution in urban public spaces. Additionally, the phenomenon that residents spend less time visiting UGSs shows a positive correlation with the quality of green spaces in parks and a negative correlation with the quality of green spaces in residential areas. In addition, there was a negative correlation between the decline in participants' self-rated mental health status and the quality of green spaces in residential areas. The residents' comparison of green spaces showed that they were generally less satisfied with the green spaces in their area than with the green spaces in the East Lake Greenway Park, and had a higher preference for the East Lake Green Space. The research highlighted public space issues such as uneven distribution of green spaces, differences in green

space types, differences in green space quality, and residents' potential demand for green spaces.

A more in-depth discussion of the issues related to behaviors and attitudes toward urban park use compiled in this study is necessary. Changes in visiting behavior and attitudes after the pandemic were also manifested in the duration and frequency of use of UGSs, travel patterns, and other related aspects. Increased or decreased changes in park use showed the impact of the pandemic on this. In this paper, we start from behavioral perceptions and find that the pandemic affected people's food, clothing, and housing in terms of life and work, but also changed people's habits and ways of using parks. On the one hand, the pandemic changed people's normal life and increased their willingness to visit green spaces. Especially important is that access to UGSs can be effective in improving physical and mental health. On the other hand, the attractiveness of UGSs and the need for their health make access to them inconvenient in this period. In the face of this unexpected situation, we need to: (1) raise awareness and strengthen people's understanding of pandemic and pandemic prevention and control, as well as the benefits of visiting green spaces for physical and mental health; (2) create a good environment, emphasize the positive role of visiting public green spaces, improve the urban green space environment, narrow the gap in the quality of urban public green spaces, highlight the diversity and local characteristics of UGSs, and improve the supporting infrastructure services for urban green spaces; and (3) improve infrastructure services by maintaining or increasing publicly accessible urban green spaces, reassessing our relationship with nature, and resisting future epidemics and pandemics.

5. Conclusions

COVID-19 outbreaks and epidemics pose serious challenges to people's productivity and livelihoods. The development of cities, public health, and ecological sustainability are all related to it. Overall, the sudden health crisis is not simple, independent, or minor; it needs to be given sufficient attention and studied. In this study, a quantitative and qualitative approach was used to study 302 participants in Wuhan.

It has been established that the quality of UGSs is crucial to the physical and mental health of residents' lives and the improvement of their well-being. Additionally, our study has made new findings from the association between UGS types and residents' usage patterns and attitudes. The results show that participants' visits to green spaces changed considerably after the COVID-19 outbreak. First of all, the basic information of the participants has changed, basically showing the following characteristics: (1) the gender and age of the participants are concentrated in the group of men and young people; (2) among all the participants, 78.15% are permanent residents, 57.95% own a home, and 68.54% have a bachelor's degree or higher, which allows them to be judged as well educated; (3) in addition to having a stable job, nearly half of the respondents' salaries changed after the pandemic, with 77.27% of those whose salaries changed being lower than before. Second, their perceived health ratings were more significantly related to changes in green space use behavior. Changes in visitors' physical and mental health after the outbreak were significantly correlated with several factors, including the duration and frequency of use of UGSs, travel patterns, and other related aspects.

UGS is not only an important place for regulating the physical and mental health of residents, but also an important ecological resource for the whole city and the world, and plays a regulating role in the ecological crisis of cities as well as the whole human race, and it can be said that the protection and development of UGSs is an epoch-making and historically significant initiative with im-measurable ecological significance. While we draw conclusions from the study, we should also pay attention to the more multifaceted thinking that this issue represents. Therefore, while strengthening the connection between urban residents and UGSs, it is also important to protect UGSs as ecological resources and to help urban parks play a role in sustainable urban development.

Finally, there are some limitations in the methodology, process, and conclusions of this study. This study used questionnaires and on-site research to confirm most of the conjectures, enrich the existing studies, and truly reflect the real usage of residents in Wuhan, the city with the highest concentration of epidemics, under the influence of the pandemic. However, there are still some shortcomings. (1) There are limitations in controlling the whole green space. The UGS we selected is special in that while it represents the park activity characteristics of most city residents, it does not represent the full range of activity characteristics of Wuhan residents. In order to obtain a more comprehensive picture of residents' behavior, more research on other UGSs in different areas is needed. (2) The small sample size and the low participation of the elderly in the study group are also related to the limitations of the location of the selected green spaces. (3) The high educational level of the sample size may be due to the administrative division of the Donghu green space as an educational area in The Wuchang district, with more college students living nearby. However, this study truly reflects the changes in the use of UGSs and provides an effective reference for the next in-depth study.

Author Contributions: Funding acquisition, H.W.; methodology, L.L.; project administration, Z.P.; validation, H.W. and Z.S.; writing—original draft, L.C.; writing—review and editing, Z.S. and Z.P. All authors have read and agreed to the published version of the manuscript.

Funding: This research was funded by the National Natural Science Foundation of China, grant number 52078390.

Institutional Review Board Statement: Ethical review and approval were waived for this study, due to involving no more than minimal risk.

Informed Consent Statement: Informed consent was obtained from all subjects involved in the study.

Data Availability Statement: Data sharing not applicable.

Conflicts of Interest: The authors declare no conflict of interest.

References

1.   Corley, J.; Okely, J.A.; Taylor, A.M.; Page, D.; Welstead, M.; Skarabela, B.; Redmond, P.; Cox, S.R.; Russ, T.C. Home garden use during COVID-19: Associations with physical and mental wellbeing in older adults. *J. Environ. Psychol.* **2021**, *73*, 101545. [CrossRef]
2.   Dzhambov, A.M.; Lercher, P.; Browning, M.H.E.M.; Stoyanov, D.; Petrova, N.; Novakov, S.; Dimitrova, D.D. Does greenery experienced indoors and outdoors provide an escape and support mental health during the COVID-19 quarantine? *Environ. Res.* **2021**, *196*, 110420. [CrossRef] [PubMed]
3.   Hunter, M.R.; Gillespie, B.W.; Chen, S.Y.P. Urban Nature Experiences Reduce Stress in the Context of Daily Life Based on Salivary Biomarkers. *Front. Psychol.* **2019**, *10*, 722. [CrossRef] [PubMed]
4.   Ugolini, F.; Massetti, L.; Calaza-Martinez, P.; Carinanos, P.; Dobbs, C.; Ostoic, S.K.; Marin, A.M.; Pearlmutter, D.; Saaroni, H.; Sauliene, I.; et al. Effects of the COVID-19 pandemic on the use and perceptions of urban green space: An international exploratory study. *Urban For. Urban Green.* **2020**, *56*, 126888. [CrossRef]
5.   Geary, R.S.; Wheeler, B.; Lovell, R.; Jepson, R.; Hunter, R.; Rodgers, S. A call to action: Improving urban green spaces to reduce health inequalities exacerbated by COVID-19. *Prev. Med.* **2021**, *145*, 106425. [CrossRef]
6.   Geng, D.H.; Innes, J.; Wu, W.L.; Wang, G.Y. Impacts of COVID-19 pandemic on urban park visitation: A global analysis. *J. For. Res.* **2021**, *32*, 553–567. [CrossRef]
7.   Bowler, D.E.; Buyung-Ali, L.M.; Knight, T.M.; Pullin, A.S. A systematic review of evidence for the added benefits to health of exposure to natural environments. *BMC Public Health* **2010**, *10*, 456. [CrossRef]
8.   Hartig, T.; Evans, G.W.; Jamner, L.D.; Davis, D.S.; Garling, T. Tracking restoration in natural and urban field settings. *J. Environ. Psychol.* **2003**, *23*, 109–123. [CrossRef]
9.   Bao, Y.P.; Sun, Y.K.; Meng, S.Q.; Shi, J.; Lu, L. 2019-nCoV epidemic: Address mental health care to empower society. *Lancet* **2020**, *395*, E37–E38. [CrossRef]
10.   Wolch, J.R.; Byrne, J.; Newell, J.P. Urban green space, public health, and environmental justice: The challenge of making cities 'just green enough'. *Landsc. Urban Plan.* **2014**, *125*, 234–244. [CrossRef]
11.   Nita, M.R.; Arsene, M.; Barbu, G.; Cus, A.G.; Ene, M.; Serban, R.M.; Stama, C.M.; Stoia, L.N. Using Social Media Data to Evaluate Urban Parks Use during the COVID-19 Pandemic. *Int. J. Environ. Res. Public Health* **2021**, *18*, 10860. [CrossRef] [PubMed]
12.   Chiesura, A. The role of urban parks for the sustainable city. *Landsc. Urban Plan.* **2004**, *68*, 129–138. [CrossRef]

13. De Bruin, W.B. Age Differences in COVID-19 Risk Perceptions and Mental Health: Evidence From a National US Survey Conducted in March 2020. *J. Gerontol. Ser. B-Psychol. Sci. Soc. Sci.* **2021**, *76*, E24–E29. [CrossRef] [PubMed]

14. Spano, G.; D'Este, M.; Giannico, V.; Elia, M.; Cassibba, R.; Lafortezza, R.; Sanesi, G. Association between indoor-outdoor green features and psychological health during the COVID-19 lockdown in Italy: A cross-sectional nationwide study. *Urban For. Urban Green.* **2021**, *62*, 127156. [CrossRef] [PubMed]

15. Yamazaki, T.; Iida, A.; Hino, K.; Murayama, A.; Hiroi, U.; Terada, T.; Koizumi, H.; Yokohari, M. Use of Urban Green Spaces in the Context of Lifestyle Changes during the COVID-19 Pandemic in Tokyo. *Sustainability* **2021**, *13*, 9817. [CrossRef]

16. Ren, S.Y.; Gao, R.D.; Chen, Y.L. Fear can be more harmful than the severe acute respiratory syndrome coronavirus 2 in controlling the corona virus disease 2019 epidemic. *World J. Clin. Cases* **2020**, *8*, 652–657. [CrossRef]

17. Olszewska-Guizzo, A.; Mukoyama, A.; Naganawa, S.; Dan, I.; Husain, S.F.; Ho, C.S.; Ho, R. Hemodynamic Response to Three Types of Urban Spaces before and after Lockdown during the COVID-19 Pandemic. *Int. J. Environ. Res. Public Health* **2021**, *18*, 6118. [CrossRef]

18. Pouso, S.; Borja, A.; Fleming, L.E.; Gomez-Baggethun, E.; White, M.P.; Uyarra, M.C. Contact with blue-green spaces during the COVID-19 pandemic lockdown beneficial for mental health. *Sci. Total Environ.* **2021**, *756*, 143984. [CrossRef]

19. Khalilnezhad, M.R.; Ugolini, F.; Massetti, L. Attitudes and Behaviors toward the Use of Public and Private Green Space during the COVID-19 Pandemic in Iran. *Land* **2021**, *10*, 1085. [CrossRef]

20. Kaplan, S. The Restorative Benefits Of Nature—Toward An Integrative Framework. *J. Environ. Psychol.* **1995**, *15*, 169–182. [CrossRef]

21. Zick, C.D.; Smith, K.R.; Kowaleski-Jones, L.; Uno, C.; Merrill, B.J. Harvesting More Than Vegetables: The Potential Weight Control Benefits of Community Gardening. *Am. J. Public Health* **2013**, *103*, 1110–1115. [CrossRef] [PubMed]

22. Park, S.A.; Lee, A.Y.; Park, H.G.; Son, K.C.; Kim, D.S.; Lee, W.L. Gardening Intervention as a Low- to Moderate-Intensity Physical Activity for Improving Blood Lipid Profiles, Blood Pressure, Inflammation, and Oxidative Stress in Women over the Age of 70: A Pilot Study. *Hortscience* **2017**, *52*, 200–205. [CrossRef]

23. Lachowycz, K.; Jones, A.P. Towards a better understanding of the relationship between greenspace and health: Development of a theoretical framework. *Landsc. Urban Plan.* **2013**, *118*, 62–69. [CrossRef]

24. Shin, Y.H. The effects of a walking exercise program on physical function and emotional state of elderly Korean women. *Public Health Nurs.* **1999**, *16*, 146–154. [CrossRef] [PubMed]

25. Giustino, V.; Parroco, A.M.; Gennaro, A.; Musumeci, G.; Palma, A.; Battaglia, G. Physical Activity Levels and Related Energy Expenditure during COVID-19 Quarantine among the Sicilian Active Population: A Cross-Sectional Online Survey Study. *Sustainability* **2020**, *12*, 4356. [CrossRef]

26. Lesser, I.A.; Nienhuis, C.P. The Impact of COVID-19 on Physical Activity Behavior and Well-Being of Canadians. *Int. J. Environ. Res. Public Health* **2020**, *17*, 3899. [CrossRef]

27. Jackson, S.B.; Stevenson, K.T.; Larson, L.R.; Peterson, M.N.; Seekamp, E. Outdoor Activity Participation Improves Adolescents' Mental Health and Well-Being during the COVID-19 Pandemic. *Int. J. Environ. Res. Public Health* **2021**, *18*, 2506. [CrossRef]

28. Larson, L.R.; Mullenbach, L.E.; Browning, M.; Rigolon, A.; Thomsen, J.; Metcalf, E.C.; Reigner, N.P.; Sharaievska, I.; McAnirlin, O.; D'Antonio, A.; et al. Greenspace and park use associated with less emotional distress among college students in the United States during the COVID-19 pandemic. *Environ. Res.* **2022**, *204*, 112367. [CrossRef]

29. Dodd, R.H.; Dadaczynski, K.; Okan, O.; McCaffery, K.J.; Pickles, K. Psychological Wellbeing and Academic Experience of University Students in Australia during COVID-19. *Int. J. Environ. Res. Public Health* **2021**, *18*, 866. [CrossRef]

30. Browning, M.; Larson, L.R.; Sharaievska, I.; Rigolon, A.; McAnirlin, O.; Mullenbach, L.; Cloutier, S.; Vu, T.M.; Thomsen, J.; Reigner, N.; et al. Psychological impacts from COVID-19 among university students: Risk factors across seven states in the United States. *PLoS ONE* **2021**, *16*, e0245327. [CrossRef]

31. Derks, J.; Giessen, L.; Winkel, G. COVID-19-induced visitor boom reveals the importance of forests as critical infrastructure. *For. Policy Econ.* **2020**, *118*, 102253. [CrossRef]

32. Lu, Y.; Zhao, J.T.; Wu, X.Y.; Lo, S.M. Escaping to nature during a pandemic: A natural experiment in Asian cities during the COVID-19 pandemic with big social media data. *Sci. Total Environ.* **2021**, *777*, 146092. [CrossRef]

33. Venter, Z.S.; Barton, D.N.; Gundersen, V.; Figari, H.; Nowell, M. Urban nature in a time of crisis: Recreational use of green space increases during the COVID-19 outbreak in Oslo, Norway. *Environ. Res. Lett.* **2020**, *15*, 104075. [CrossRef]

34. Harrison, E.; Monroe-Lord, L.; Carson, A.D.; Jean-Baptiste, A.M.; Phoenix, J.; Jackson, P.; Harris, B.M.; Asongwed, E.; Richardson, M.L. COVID-19 pandemic-related changes in wellness behavior among older Americans. *BMC Public Health* **2021**, *21*, 755. [CrossRef]

35. Bulfone, T.C.; Malekinejad, M.; Rutherford, G.W.; Razani, N. Outdoor Transmission of SARS-CoV-2 and Other Respiratory Viruses: A Systematic Review. *J. Infect. Dis.* **2021**, *223*, 550–561. [CrossRef]

36. Xiao, Y.J.; Becerik-Gerber, B.; Lucas, G.; Roll, S.C. Impacts of Working From Home During COVID-19 Pandemic on Physical and Mental Well-Being of Office Workstation Users. *J. Occup. Environ. Med.* **2021**, *63*, 181–190. [CrossRef]

37. Pascoe, A.; Johnson, D.; Putland, M.; Willis, K.; Smallwood, N.; Paul, E. Differential Impacts of the COVID-19 Pandemic on Mental Health Symptoms and Working Conditions for Senior and Junior Doctors in Australian Hospitals. *J. Occup. Environ. Med.* **2022**, *64*, E291–E299. [CrossRef]

38. Wood, L.; Hooper, P.; Foster, S.; Bull, F. Public green spaces and positive mental health-investigating the relationship between access, quantity and types of parks and mental wellbeing. *Health Place* **2017**, *48*, 63–71. [CrossRef]

39. Larcher, F.; Pomatto, E.; Battisti, L.; Gullino, P.; Devecchi, M. Perceptions of Urban Green Areas during the Social Distancing Period for COVID-19 Containment in Italy. *Horticulturae* **2021**, *7*, 55. [CrossRef]

40. Garrett, J.K.; White, M.P.; Huang, J.; Ng, S.; Hui, Z.; Leung, C.; Tse, L.A.; Fung, F.; Elliott, L.R.; Depledge, M.H.; et al. Urban blue space and health and wellbeing in Hong Kong: Results from a survey of older adults. *Health Place* **2019**, *55*, 100–110. [CrossRef]

41. Yu, J.; Yan, L.; Deng, J.; Li, J. Study on the influence of urban green space on the physical and mental welfare of residents. *Acta Ecol. Sin.* **2020**, *40*, 3338–3350.

42. Fongar, C.; Aamodt, G.; Randrup, T.B.; Solfjeld, I. Does Perceived Green Space Quality Matter? Linking Norwegian Adult Perspectives on Perceived Quality to Motivation and Frequency of Visits. *Int. J. Environ. Res. Public Health* **2019**, *16*, 2327. [CrossRef] [PubMed]

43. Hubei Provincial Stitistics Bureau. Wuhan Statistical Yearbook in 2021. Available online: https://tjj.hubei.gov.cn/tjsj/sjkscx/tjnj/gsztj/whs/ (accessed on 28 June 2022).

44. Orsega-Smith, E.; Mowen, A.J.; Payne, L.L.; Godbey, G. The interaction of stress and park use on psycho-physiological health in older adults. *J. Leis. Res.* **2004**, *36*, 232–256. [CrossRef]

45. Fofana, N.K.; Latif, F.; Sarfraz, S.; Bilal; Bashir, M.F.; Komal, B. Fear and agony of the pandemic leading to stress and mental illness: An emerging crisis in the novel coronavirus (COVID-19) outbreak. *Psychiatry Res.* **2020**, *291*, 113230. [CrossRef]

46. McCormack, G.R.; Rock, M.; Toohey, A.M.; Hignell, D. Characteristics of urban parks associated with park use and physical activity: A review of qualitative research. *Health Place* **2010**, *16*, 712–726. [CrossRef]

Article

Green Space for Mental Health in the COVID-19 Era: A Pathway Analysis in Residential Green Space Users

Hansen Li [1], Matthew H. E. M. Browning [2], Angel M. Dzhambov [3], Guodong Zhang [1,*] and Yang Cao [4,5]

[1]   Institute of Sports Science, College of Physical Education, Southwest University, Chongqing 400175, China; lhs510416413@swu.edu.cn
[2]   Department of Park, Recreation, and Tourism Management, Clemson University, Clemson, SC 29634, USA; mhb2@clemson.edu
[3]   Department of Hygiene, Faculty of Public Health, Medical University of Plovdiv, 4002 Plovdiv, Bulgaria; angel.dzhambov@mu-plovdiv.bg or angelleloti@gmail.com
[4]   Clinical Epidemiology and Biostatistics, School of Medical Sciences, Örebro University, 70182 Örebro, Sweden; yang.cao@oru.se
[5]   Unit of Integrative Epidemiology, Institute of Environmental Medicine, Karolinska Institutet, 17177 Stockholm, Sweden
*   Correspondence: lygd777@swu.edu.cn; Tel.: +86-15730267257

Abstract: Residential green space is among the most accessible types of urban green spaces and may help maintain mental health during the COVID-19 pandemic. However, it is insufficiently understood how residents use residential green space for exercise during the epidemic. The pathways between residential green space and mental health also merit further exploration. Therefore, we conducted an online study among Chinese residents in December 2021 to capture data on engagement with urban green space for green exercise, the frequency of green exercise, perceived pollution in green space, perceptions of residential green space, social cohesion, depression, and anxiety. Among the 1208 respondents who engaged in green exercise last month, 967 (80%) reported that green exercise primarily occurred in residential neighborhoods. The rest (20%) reported that green exercise occurred in more distant urban green spaces. The most common reasons that respondents sought green exercise in urban green spaces were better air and environmental qualities. Structural equation modeling (SEM) was then employed to explore the pathways between the perceived greenness of residential neighborhoods and mental health among respondents who used residential green space for exercise. The final model suggested that residential green space was negatively associated with anxiety ($\beta = -0.30$, $p = 0.001$) and depression ($\beta = -0.33$, $p < 0.001$), mainly through indirect pathways. Perceived pollution and social cohesion were the two mediators that contributed to most of the indirect effects. Perceived pollution was also indirectly associated with green exercise through less social cohesion ($\beta = -0.04$, $p = 0.010$). These findings suggest a potential framework to understand the mental health benefits of residential green space and its accompanying pathways during the COVID-19 era.

Keywords: neighborhood; community; physical activity; mental health; urban greening

Citation: Li, H.; Browning, M.H.E.M.; Dzhambov, A.M.; Zhang, G.; Cao, Y. Green Space for Mental Health in the COVID-19 Era: A Pathway Analysis in Residential Green Space Users. *Land* **2022**, *11*, 1128. https://doi.org/10.3390/land11081128

Academic Editors: Francesca Ugolini and David Pearlmutter

Received: 18 June 2022
Accepted: 20 July 2022
Published: 22 July 2022

Publisher's Note: MDPI stays neutral with regard to jurisdictional claims in published maps and institutional affiliations.

1. Introduction

Being exposed to nature is known to benefit mental and physical health [1–3]. However, expanding urbanization has reduced the connection between humans and nature, thus posing risks to human health [4–7]. In this context, urban green spaces have become increasingly important since they can improve the urban environment and offer residents opportunities to interact with natural elements [8]. Normally, public green spaces such as urban parks and urban woodlands are popular because they can provide cleaner environments for physical activities. For example, fresh air was a common reason mentioned by visitors to green spaces [9,10]. However, the unexpected COVID-19 pandemic has significantly changed people's visitation of, attitudes toward, and behaviors in urban green spaces [11,12]. Determining the extent to which urban residents visit green spaces for physical activity and why visitors engage in exercise, specifically in urban green space during the pandemic, is not yet well understood.

China has been making arguably the most strict policies to curb the spread of the virus [13]. Though many public facilities have become available again as the virus has been gradually controlled, communities and university campuses are still frequently sealed off during the subsequent waves of disease, which may have threatened the mental health of Chinese residents [14]. In March 2022, the emerging Omicron variant caused blockades (lockdowns) in 571 regions and cities in China [15], extending the durative impact of the pandemic on mental health.

In Western countries, nature contact in urban parks or other public green spaces has been deemed an option to maintain mental health during such periods [16–19]. However, heading to an urban park may not be easy in China because of many cities' extremely high population density and insufficient urban park cover. When a new case is confirmed, travel within a city becomes restricted or forbidden, restricting outdoor activities. In this context, residential green space may be ideal for residents since they do not have to travel far or risk the increased probability of infection. Like other green spaces, residential green spaces are associated with lower depression and anxiety [20,21]. However, the associations between exposure and mental health still need more investigation during the pandemic, and potential pathways are not fully explored.

Conceptual Framework

Green spaces may reduce mental health issues in myriad ways [22]. Aside from direct associations [23,24], we proposed the following mediators: social cohesion, perceived pollution, and green exercise.

Social cohesion has many definitions [25]. For residential neighborhoods, social cohesion can be described as the social connections, trust, and solidarity among residents [26,27] that protect mental health [28,29], especially during the pandemic [30,31]. It has been suggested that urban green space can support social cohesion by increasing the likelihood of meeting others and the feelings of comfort that connects people to places and fellow visitors [32]. Furthermore, numerous studies suggest that social cohesion may mediate the association between green space and mental health [33–36], supporting our first hypothesized pathway across urban natural environments.

Air pollutants and noise are common forms of pollution that threaten mental health [37]. Perceived air and noise pollution are usually investigated by self-reported measures, which can resemble objectively measured pollution levels [38,39]. Interestingly, perceived pollution may impact people's behaviors more strongly than objectively measured pollution [40]. This may be because perceived pollution is related to individuals' sensitivity to pollution and acts as a mediator between measured pollution and psychological responses [41,42]. Green space may remove harmful gas and inhalable particles [43], attenuate noise pollution [44], and alter individuals' susceptibility to noise [45]. Green space may reduce both measured and perceived levels of air and noise pollution [46–49], which may further contribute to mental health [50–52]. These clues support our second hypothesized pathway between green space and mental health through perceived pollution.

Green exercise is a third viable factor, which refers to the combination of physical activity and nature exposure [53]. Green exercise has been assumed to be more beneficial than physical activity in urban "grey" areas [54,55]. Green spaces may encourage physical activity by increasing the restorative quality of the environment in which someone exercises [22,50,56] and may therefore increase the mental health benefits of exercise [36,57]. The frequency of visitation is often employed to measure green space utilization [58–60] and also serves as a critical dimension of physical activity associated with green spaces [61]. Closer and greener urban spaces may be associated with more frequent green exercise among residents [62,63], which supports our third hypothesized pathway between green space and mental health through green exercise.

The three mediators above are likely to interrelate when modeled together with green space and mental health. Residents tend to perform walking and relaxation activities in urban green spaces with low noise levels [64]. In contrast, a space heavily affected by air and noise pollution may discourage participation in outdoor physical activity [65] because pollutants reduce environmental quality and are deemed harmful [50,66,67]. Pollutants may increase annoyance with urban environments and reduce social cohesion in those spaces [51,65,68]. Many studies have also shown that social cohesion may motivate physical activity [69,70], including in urban green spaces [33]. Therefore, perceived pollution may discourage green exercise by reducing social cohesion.

Our aim was to investigate the association of residential green space with depression and anxiety through three hypothesized mediators: social cohesion, perceived pollution (air and noise), and green exercise. In addition, we established a covariance link between depression and anxiety based on their strong association [71]. Given the possibility of confounding in these relationships, we control for the following:

(1) Age and gender, which may affect mental health conditions [72,73] and the chance for green exercise [74];
(2) Income, which may affect residential green space [75], physical activity [76], social cohesion [77], mental health [78], and benefits derived from green space [79];
(3) COVID-19 condition, which can impact mental health [80], social cohesion [81], and chances for green exercise [12,82].

Our conceptual framework is presented in Figure 1. We hypothesize the following:

Hypothesis 1 (H1). *Residential green space is the primary type of green space used by residents for green exercise; furthermore, high environmental quality is the stated reason to engage in green exercise in these spaces.*

Hypothesis 2 (H2). *Perceived residential green space impacts depression and anxiety directly and indirectly through perceived pollution, social cohesion, and green exercise.*

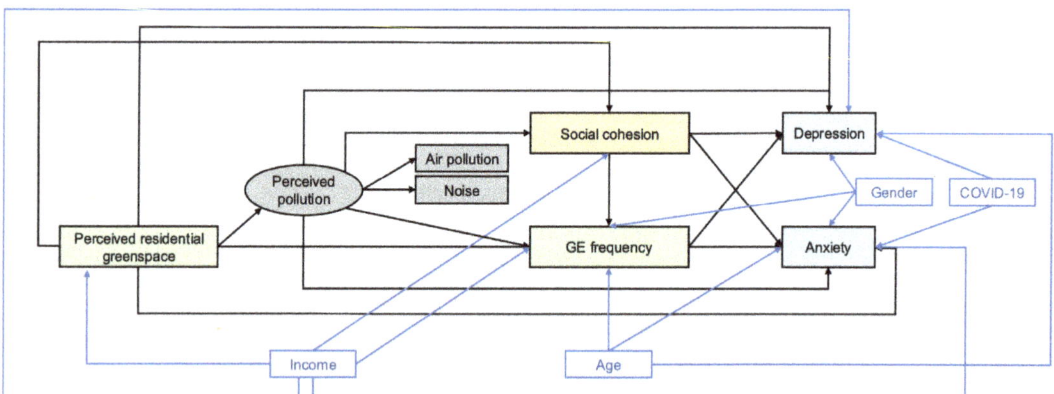

Figure 1. Conceptual framework. Note: GE, green exercise. Black lines indicate pathways between core variables (perceived residential green space, perceived pollution, social cohesion, green exercise frequency, depression, and anxiety). Blue lines indicate confounding pathways between core and control variables (income, age, gender, and COVID-19 condition).

2. Materials and Methods

2.1. Study Design and Participants

We conducted a cross-sectional survey in December 2021 when regional COVID-19 outbreaks reemerged in many areas of China. Our target population was Chinese urban residents with no restrictions on the province. The study was approved and supervised by the Ethics Review Board of University, China.

We distributed recruitment messages via online social media, including WeChat and Tencent QQ. We employed a snowball sampling technique that involved inviting middle-aged participants to spread information to people of similar ages or older. We described the study's topic as investigating the utilization of urban green space and mental health during the COVID-19 epidemic. Details of the research questions were not disclosed during participant recruitment. We offered compensation of CNY 5 (approximately USD 0.8) for completing the online. The participants used WeChat accounts linked with their personal IDs to fill out the questionnaires (the IP address, device, and account were restricted, and each participant could submit once). In total, 1329 questionnaires were collected. After removing incomplete questionnaires and those that failed human verification tests (to confirm that people carefully completed the questionnaire), a total of 1223 qualified questionnaires were included in the analysis.

2.2. Instruments and Measurements

2.2.1. Location of and Reasons for Green Exercise

The following question was used to identify the location of urban green spaces used for green exercise: "Where is the urban green space that you usually did physical activities (e.g., walking, running, biking, dancing, or ball games) in the last month?" Response options included the following: "Did not do physical activity in any urban green space in the last month"; "the green space in my residential neighborhood (in and around my community, only need a very few minutes of walk for arrival)"; or "other green places that are more distant (usually need to walk a while or even need a vehicle for arrival)."

The participants were also asked to briefly indicate their main reasons for green exercise in green space with a single sentence. An open-ended question was used to obtain their responses.

2.2.2. Frequency of Green Exercise

We measured the frequency of green exercise by asking, "How often did you do physical activities (e.g., walking, running, biking, dancing, ball games) in the mentioned green space during the last month?" Answers were provided on a 7-point Likert-type scale (1 = once per week or less to 7 = about days per week).

2.2.3. Perceived Pollution of Green Space

We measured the perception of pollution in green spaces among the respondents who claimed green exercise in residential green spaces by asking: "Please describe the general level of air pollution in the mentioned green space during the last month"; and "Please describe the general level of noise in the green space (where you usually did physical activities) during the last month." Answers were given using a 5-point Likert-type scale (1 = very low or almost imperceptible to 5 = very high or easily felt).

2.2.4. Perceived Residential Green Space

Based on the methods of Liu et al. [52] and Yang et al. [51], we asked the following to measure the availability of perceived residential green space: "How much green space (e.g., trees/plants) is there in your residential neighborhood?" Answers were given using a 5-point Likert-type scale (1 = very little or almost no greenness to 5 = very much or full of natural greenness).

2.2.5. Social Cohesion

Trust is a critical factor that affects the willingness to intervene for the common good at the neighborhood level [83]. It can be measured with the Sampson et al. [84] scale, which measures social cohesion/trust [85]. We used the highest-loading item from this scale, following another study that measured social cohesion in residential neighborhoods among Chinese populations [51]: "Do you think people in your residential neighborhood can be trusted?" on a 5-point Likert-type scale (1 = not at all, 5 = absolutely).

2.2.6. Anxiety

The Chinese version of the Generalized Anxiety Disorder 7-item (GAD-7) scale was used to measure anxiety conditions in the last month. The scale assesses how often a person is bothered by common symptoms of anxiety, such as nervousness and worries. Responses were given using a 4-point Likert-type scale (0 = not at all to 3 = nearly every day). The total score can be used for identifying anxiety symptoms (0–4 for minimal anxiety; 5–9 for mild anxiety; 10–14 for moderately severe anxiety, and 20–27 for severe anxiety). High internal reliability was observed in our sample (Cronbach $\alpha = 0.94$).

2.2.7. Depression

The Chinese version of the Patient Health Questionnaire 9-item (PHQ-9) scale was used to measure depression conditions in the last month. The scale assesses how often a person is bothered by common symptoms of depression, such as fatigue and sleep problems. Responses were given using a 4-point Likert-type scale (0 = not at all to 3 = nearly every day). The total score can be used for identifying depression symptoms (0–4 for minimal depression; 5–9 for mild depression; 10–14 for moderately severe depression, and 20–27 for severe depression). In our sample, high internal reliability was again observed ($\alpha = 0.95$).

2.2.8. Sociodemographic Characteristics

The gender of the respondents was categorized into two groups (1 = male, 2 = female). Age of respondents was categorized into eight levels (<18 year, 18–25, 26–30, 31–40, 41–50, 51–60, 61–70, >70). Monthly incomes were categorized into seven levels (no reported income, RMB $\leq$ 3000, RMB 3001–6000, RMB 6001–9000, RMB 9001–12,000, RMB 12,001–15,000, RMB >15,000). For reference, RMB 1 is approximately USD 0.8 or EUR 0.7.

2.2.9. COVID-19 Condition

The number of infections among different provinces was obtained from the China National Health Commission, provincial and municipal health commissions, provincial and municipal governments, and official channels (https://wp.m.163.com/163/page/news/virus_report/index.html, accessed on 25 December 2021). The number affects social restrictions in China and was used to control for the COVID-19 conditions for each respondent.

2.3. Analysis

2.3.1. Bivariate Correlations

Spearman's rank-order correlation (for correlations between two continuous variables) and point-biserial correlation (for correlations between a binary variable and a continuous variable) were used to detect the general pattern of associations between the variables.

2.3.2. Keywords for Green Exercise in Green Space

Natural language process (NLP) refers to the branch of artificial intelligence or AI that combines computational linguistics with statistical, machine learning, and deep learning models [86,87]. These functions enable computers to process human language in the form of text to uncover its full meaning. NPL is used for translation, sentiment analysis, text summarization, and more. Many companies have constructed AI platforms that offer NLP services.

In this study, the keywords in respondents' reasons for green exercise in green space were identified using the NLP engine empowered by Tencent Holdings Ltd. (Shenzhen, China). Since we required participants to briefly state their reasons in a single sentence, the same noun was not likely to appear twice in a single response (we manually confirmed this). After removing typos in the questionnaires, participants' responses were input together as a single file for keywords summarization without extra restriction.

2.3.3. Structural Equations Modeling

Structural equation modeling (SEM) was employed to examine the hypothesized directional paths in the conceptual framework presented in Figure 2 for the respondents who engaged in green exercise in residential green space. According to Bagozzi and Yi [88], the sample size for SEM should be twice the number of model parameters. Based on the parameters (n = 22) required to estimate in the framework, our sample size of 967 (respondents who reported green exercise in residential green space) was adequate. As performed in prior research on green space and mental health during the COVID-19 pandemic, the results of the GAD-7 and PHQ-9 were processed as continuous summary scores, while the levels of air pollution and noise were assumed to load onto one latent factor that we labeled as "perceived pollution" [51,52,71].

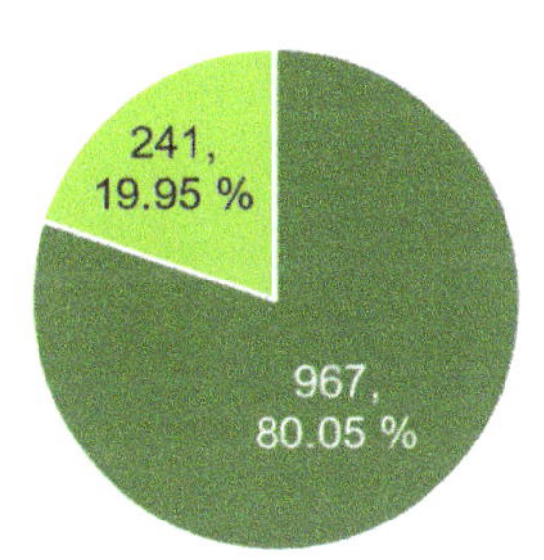

Figure 2. Proportion of respondents reporting green exercise in different urban green spaces and top five keywords for reasons to engage in green exercise (n = 1208). Note: n: times of the keywords appeared in respondents' answers.

Variance inflation factor (VIF) values smaller than 5.0 were considered evidence of the absence of multicollinearity. Based on this rule, no multicollinearity was observed among the independent variables (VIF < 3.0) [89].

Given our sample size and a high level of multivariate non-normality in the data, we used an asymptotically distribution-free (ADF)/weighted least squares (WLS) estimator for analysis [90,91]. The bootstrap method with 10,000 replications was used to generate corresponding standard errors and confidence intervals for all paths [92–94].

Based on the ADF/WLS estimator, the goodness of fit was assessed using the following indices [91,95]: standardized root mean square residual (SRMR) < 0.08; Tucker–Lewis index (TLI) > 0.95; and goodness-of-fit index (CFI) > 0.95. In addition, a root mean square error of approximation (RMSEA) < 0.05 was also considered because it is essential for maximum likelihood (ML) and generalized least squares (GLS) estimations that were later performed for sensitivity analysis [91]. We did not employ the $\chi2$ test because it is strongly affected to sample size and violation of the multivariate normality assumption [96–98].

A factor loading for latent variables of the conceptual model > 0.5 was considered acceptable. An indirect effect (i.e., a product of coefficients for the constituent links) that significantly exceeded zero was evidence of mediation [99,100].

The Akaike information criterion (AIC) and Bayesian information criterion (BIC) were used for model selection, as they are useful for selecting the model with the least overfitting [101].

2.3.4. Model Modification

The initial model M0 resulted in a low TLI value (SRMR = 0.03; CFI = 0.96; TLI = 0.89; RMSEA = 0.04 [90%CI: 0.03, 0.06]; AIC = 143.53; BIC = 377.49). Thereafter, we removed the confounding paths without at least a marginal statistical significance ($p > 0.1$). The COVID-19 condition was removed from the model due to weak impacts on the core variables ($p > 0.1$). The TLI value of the modified model (M1) was still lower than an ideal but reached an acceptable level, as indicated in past research (SRMR = 0.04; CFI = 0.96; TLI = 0.90; RMSEA = 0.04 [90%CI: 0.03, 0.06]; AIC = 122.08; BIC = 307.30) [102–104]. Therefore, the model M1 was selected as the final model, and all non-significant pathways between the core variables were retained.

2.3.5. Sensitivity Analyses

Regarding keywords scanning, we switched to another NPL engine developed by the Baidu Holdings Ltd. (Beijing, China) to examine any changes in the identified keywords as a sensitivity analysis.

Given that ML and GLS estimators may be comparative to ADF/WLS estimator in some scenarios, even where normality is violated [91,105], we also applied two estimators to re-examine the robustness of the identified pathways between the core variables. Additionally, we established competing models by re-specifying localized points of theoretical causalities based on the following two theories (Figure S1):

(1) Model 2 (M2): The frequency of green space visitation predicting social cohesion [106];
(2) Model 3 (M3): A reciprocal relationship between green exercise and social cohesion (Jennings and Bamkole, 2019).

To check if the final model could fit different gender and ages, we ran subgroup analyses that stratified the respondents by gender and median age, respectively. All the statistical analyses were conducted in SPSS 25.0 and AMOS 20.0. software (SPSS Inc., Chicago, IL, USA).

3. Results

3.1. Characteristics of Respondents

A total of 1223 respondents were eligible for the final analysis, and 50% were males. More than 50% of the respondents were between 18 and 30 years old. Around 65% of the respondents had a monthly income between CNY 3000 and 9000. Only 1% reported no green exercise during the last month, and over 50% reported green exercise two or three times per week during the last month (Table 1).

3.2. Locations and Reasons for Green Exercise

A total of 1208 respondents claimed green exercise experiences during the last month. Specifically, 967 (80%) reported that green exercises were mainly in residential green spaces, whereas 241 (20%) reported green exercise in more distant spaces (Figure 2). The top five keywords drawn from respondents' answers showed that looking for better air and environment were the main reasons for performing green exercises in both residential and distant green spaces (Figure 2).

Table 1. Respondent characteristics (*n* = 1223).

Variable	Category	Mean (SD)	Percentage (*n*)
Gender	Male	-	49.71% (608)
	Female	-	50.29% (615)
Age (year)	<18	-	2.13% (26)
	18–25	-	24.45% (299)
	26–30	-	27.64% (338)
	31–40	-	31.40% (384)
	41–50	-	10.96% (134)
	51–60	-	3.19% (39)
	61–70	-	0.16% (2)
	>70	-	0.08% (1)
Monthly income (CNY)	None	-	5.15% (63)
	$\leq$ 3000	-	12.92% (158)
	3001–6000	-	32.05% (392)
	6001–9000	-	32.71% (400)
	9001–12,000	-	11.53% (141)
	12,001–15,000	-	3.76% (46)
	>15,000	-	1.88% (23)
Residential greenness level	Little	-	2.13% (26)
	A little	-	10.79% (132)
	Moderate	-	18.72% (229)
	Much	-	40.39% (494)
	Very Much	-	27.96% (342)
Green exercise	None	-	1.23% (15)
	Once per week or less	-	14.39% (176)
	Twice per week	-	20.44% (250)
	Three times per week	-	29.35% (359)
	Four times per week	-	9.73% (119)
	Five times per week	-	8.58% (105)
	Six times per week	-	1.72% (21)
	Seven times per week	-	14.55% (178)
Anxiety	-	10.28 (3.89)	-
	Minimal anxiety		64.29% (786)
	Mild anxiety		29.60% (362)
	Moderate anxiety		4.66% (57)
	Severe anxiety		1.5% (18)
Depression	-	13.06 (5.00)	-
	Minimal depression		62.80% (768)
	Mild depression		27.80% (340)
	Moderate depression		5.81% (71)
	Moderately severe depression		1.80% (22)
	Severe depression		1.80% (22)

3.3. Correlations between Variables

Table 2 displays the correlations between the variables of interest among respondents who performed green exercise in residential green space during the last month (*n* = 967). Residential green space, physical activity in residential green space, social cohesion, and age were negatively correlated with anxiety and depression. COVID-19 conditions, perceived noise, and air pollution were positively correlated with anxiety and depression. Females showed lower levels of depression and anxiety than males.

Table 2. Bivariate Spearman or point-biserial correlations (n = 967).

Item	1	2	3	4	5	6	7	8	9	10	11
1. Gender (male = 1)	1.00										
2. Age	0.18 **	1.00									
3. Income	−0.14 **	0.22 **	1.00								
4. Infection	−0.07 *	−0.12 **	−0.02	1.00							
5. Residential greenspace	0.06	−0.05	0.11 **	−0.04	1.00						
6. Noise	−0.06	−0.06	−0.02	0.06	−0.35 **	1.00					
7. Air pollution	−0.08 *	0.03	−0.01	0.03	−0.45 **	0.49 **	1.00				
8. GE frequency	0.06	0.08 *	0.17 **	0.03	0.30 **	−0.08 *	−0.15 **	1.00			
9. Social cohesion	0.06	−0.02	0.07 *	−0.04	0.56 **	−0.35 **	−0.49 **	0.26 **	1.00		
10. Anxiety	−0.16 **	−0.12 **	−0.04	0.06 *	−0.35 **	0.33 **	0.33 **	−0.23 **	−0.41 **	1.00	
11. Depression	−0.16 **	−0.11 **	−0.03	0.08 *	−0.35 **	0.35 **	0.35 **	−0.24 **	−0.41 **	0.82 **	1.00

Note: *, $p < 0.05$; **, $p < 0.01$. GE: green exercise.

3.4. Results of the SEM Analysis

In the final SEM model (M2), we did not find a significant direct pathway between perceived residential green space and depression (β = −0.05, p = 0.325) (Figure 3). The pathway between perceived pollution and green exercise frequency was not significant either (β = −0.03, p = 0.440). All other pathways between the core variables were significant. The direct pathway between perceived residential green space and anxiety was significant but relatively weak (β = −0.09, p = 0.037).

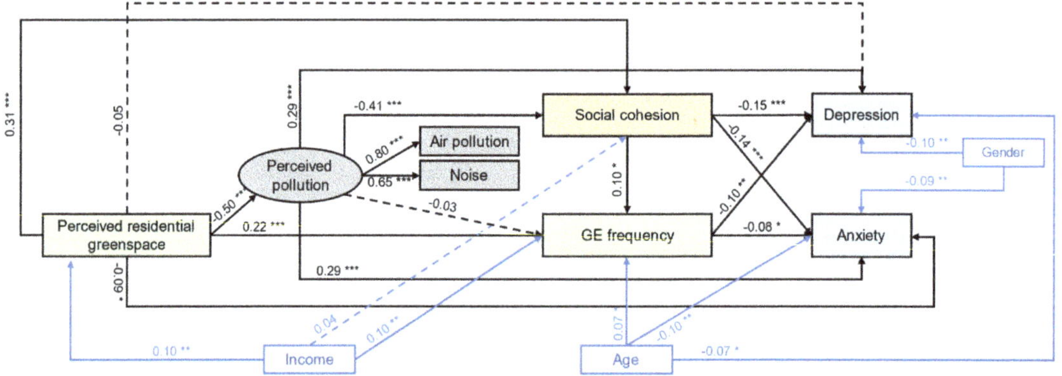

Figure 3. Final structural equation model (M2) with standardized regression weights (β) and significance levels (n = 967). **Note:** GE, green exercise. *, $p < 0.05$; **, $p < 0.01$; ***, $p < 0.001$.

Perceived residential green space had significant total effects on all variables of interest (Table 3). The total effects on social cohesion and green exercise frequency were mainly from direct effects, while those on depression and anxiety were mainly from indirect effects.

Table 3. Standardized total, direct, and indirect effects of perceived residential green space on the core variables.

Variable	Total β (95% CI)	p	Direct β (95% CI)	p	Indirect β (95% CI)	p
Perceived pollution	−0.50 (−0.57, −0.42)	<0.001	−0.50 (−0.57, −0.42)	<0.001	−	−
Social cohesion	0.50 (0.44, 0.56)	0.001	0.30 (0.21, 0.38)	<0.001	0.20 (0.15, 0.27)	<0.001
Green exercise	0.29 (0.23, 0.34)	<0.001	0.22 (0.15, 0.30)	<0.001	0.07 (0.02, 0.12)	0.007
Depression	−0.30 (−0.37, −0.22)	0.001	−0.05 (−0.14, 0.05)	0.325	−0.25 (−0.31, −0.19)	<0.001
Anxiety	−0.33 (−0.39, −0.26)	<0.001	−0.09 (−0.18, −0.01)	0.037	−0.24 (−0.30, −0.18)	<0.001

The specific indirect pathways from perceived residential green space to mental health are shown in Table 4. Pathways mediated by perceived pollution had major contributions to the total effects (Pathways 1 and 7 in Table 4). Pathways mediated by social cohesion were relatively weaker but still made considerable contributions to the total effects (Pathways 3 and 9 in Table 4).

Table 4. Standardized indirect pathways from perceived residential green space to mental health.

Pathway	β (95% CI)	*p*
1. Residential green space → Pollution → Anxiety	−0.14 (−0.21, −0.08)	<0.001
2. Residential green space → GE → Anxiety	−0.02 (−0.04, −0.003)	0.014
3. Residential green space → Cohesion → Anxiety	−0.04 (−0.08, −0.02)	0.001
4. Residential green space → Pollution → Cohesion → Anxiety	−0.03 (−0.05, −0.01)	0.001
5. Residential green space → Cohesion → GE → Anxiety	−0.003 (−0.01, 0.00)	0.021
6. Residential green space → Pollution → Cohesion → GE → Anxiety	−0.002 (−0.01, 0.00)	0.020
7. Residential green space → Pollution → Depression	−0.15 (−0.22, −0.08)	<0.001
8. Residential green space → GE → Depression	−0.02 (−0.04, −0.01)	0.002
9. Residential green space → Cohesion → Depression	−0.04 (−0.08, −0.02)	0.002
10. Residential green space → Pollution → Cohesion → Depression	−0.03 (−0.05, −0.01)	0.001
11. Residential green space → Cohesion→ GE → Depression	−0.003 (−0.01, 0.00)	0.013
12. Residential green space → Pollution → Cohesion → GE → Depression	−0.002 (−0.01, 0.00)	0.012

Note: Residential green space, perceived residential green space; Pollution, perceived pollution; GE, green exercise; Cohesion, social cohesion.

3.5. Sensitivity Analyses

Identified keywords remained unchanged in the alternative NPL engine.

Regarding structural equation modeling, Acceptable model fits were obtained through the ML and GLS estimations (Table S1). Similar results were obtained for most pathways, except for the direct pathway between residential green space and anxiety, which was not significant in the ML ($\beta = -0.07$, $p > 0.10$) and GLS estimations ($\beta = -0.07$, $p > 0.10$).

The two competing models, Model M2 (where GE predicts social cohesion) and Model M3 (where a reciprocal relationship was tested between GE and social cohesion), showed acceptable fits (Figure S1 and Table S1). However, our final model (M1) retained the lowest AIC and BIC values, indicating a better fit than these alternative models.

In models stratified by gender, we found that both the unconstrained baseline model and model with constrained structural weights showed acceptable fits to the data, and no significant difference was found between the two models ($\Delta\chi^2 = 10.53$, $p = 0.84$). Likewise, no significant differences were observed in the subgroup analysis stratified by the median age ($\Delta\chi^2 = 11.16$, $p = 0.89$), indicating the final model M2 is suitable for these subgroups.

4. Discussion

The current study aimed to explore how residential green space was used and how it could benefit mental health during the COVID-19 pandemic. The results suggest that residential green space may be the main urban green space where young and middle-aged Chinese carry out green exercise. Further, residential green space was negatively associated with depression and anxiety through direct and/or indirect pathways. These findings underline the role of residential green space in promoting urban mental health during the COVID-19 pandemic.

4.1. Findings on Green Space Utilization

We found around 80% of green space users usually performed green exercise in residential green spaces. In Italy, most urban residents switched from urban parks to gardens or other residential green spaces during the pandemic due to social distancing and other regulations or restrictions on movement [12]. So far, there is little understanding of changes in green space use in China. What is known is that the demand for public green spaces has remained large throughout the pandemic [107]. Due to active restrictions on social activities, urban residents in China may gradually have relied more on residential green spaces.

Aside from the pandemic, other reasons may explain the popularity of residential green space in our study. Before the pandemic, children in Shanghai, China, primarily carried out physical activity in residential green space [108]. Residential green spaces provide shorter travel distances than more distant public green spaces. Many studies have proved that distance is a crucial factor negatively associated with the frequency of green space visits [109,110], and the case of China appears not to be an exception.

Regarding the reasons for performing exercise in green space, better air and environmental conditions were mentioned as a rationale for both residential and other green space exercisers. This result partially supports our first hypothesis and is similar to previous studies in some Western countries, where better environmental conditions are critical reasons for visiting green space [9,10]. These findings imply that the key reasons for visiting green spaces remained unchanged during the pandemic, which underlines the role of urban green space in offering cleaner environments for residents' activities [22].

Several theories may help explain why green spaces are perceived as better/healthier environments. Green space can reduce air pollution and noise from traffic [22] by depositing air pollutants (e.g., PM10 and ozone) [111] and diffracting, absorbing, or destructing interference of sound waves [112]. These mechanisms can also apply to residential green space, making the neighborhood more suitable for physical activity. Unfortunately, residential green space is not a focus in some Chinese urban greening projects [113]. Chinese policymakers may need to prioritize residential green space construction and consider improving facilities for green exercise in neighborhoods to help fight against the mental health impacts of the COVID-19 pandemic. Moreover, since lower disturbance was the crucial reason mentioned for green exercise, using proper plant species and schemes may be a key point in residential greening. Some approaches in Chinese urban park planning can be applied to residential green space. For example, planting at least two types of small trees and shrubs in lines can enhance the aesthetic value and effectively absorb noise [114]. Recommended species include *Camellia sp.*, *Callistemon rigidus R. Br.*, and *Gordonia axillaris (Roxb. ex Ker Gawl.) D. Dietr.* These species are common shrubs or small trees in China and are usually planted for ornamental purposes. Moreover, they have been proved to have great potential in decontaminating air pollutants containing nitrogen, sulfur, and fluorine [115–117].

4.2. Findings on Pathways between Green Space and Mental Health

We did not observe a significant direct pathway between perceived residential green space and depression. When tested with different estimators, the direct association between perceived residential green space and anxiety was weak or non-significant. Nevertheless, multiple indirect pathways were identified, and perceived pollution of green space and social cohesion were the two critical mediators linking perceived residential green space and mental health issues. These results do not fully support our second hypothesis but are in line with previous analyses that perceived residential green space indirectly reduced mental health problems through enhancing social cohesion [51]. Therefore, these findings may collectively underline the mediatory role of social cohesion between residential green space and mental health.

In a previous framework by Liu et al. [52], walking behavior, social cohesion, and perceived pollution were mediators between neighborhood greenness and mental wellbeing, which is consistent with our final model. However, the previous study investigated the perceived pollution of residential neighborhoods. In contrast, we investigated the perceived pollution of green space due to our focus on green exercise. The pollution level in residential green space may also indicate the general pollution of residential neighborhoods because residential green space is part of the neighborhood and a reason for reduced pollution [22,49]. Therefore, our study may still support the framework by Liu et al. [52]

and underline a general mechanism of harvesting mental health benefits from residential green space.

Regarding relationships between the mediators, we did not observe a significant direct pathway between perceived pollution and green exercise. However, we observed a significant mediation effect between them through social cohesion. Multiple factors may explain this finding. Lower levels of urban pollutants have been associated with green space and may promote residents to be in contact with their neighbors more frequently, thus enhancing social cohesion [118]. Enhanced social cohesion may promote the adoption of healthy behaviors and utilization of community resources [70,119]. Last, a socially cohesive neighborhood may reduce conditions, such as crime and civil disorder, providing safer places for physical activity [84,120].

During the COVID-19 pandemic, various restrictions have limited people's basic human need for social interaction [121]. In this context, neighbors' interactions may have become essential because support from friends or family has been reduced for many people due to social blockades/lockdowns [122]. It is assumed that interactions among neighbors during the pandemic enhanced the sense of neighborhood cohesion among Chinese residents [123]. As discussed above, residential green space may enhance social cohesion by encouraging neighborly interactions. The pathway between residential green space and social cohesion is likely more apparent during the pandemic, making their subsequent mental health benefits more obvious. Svensson and Elntib [31] suggested that allowing community members access to green space and sensibly meeting others may help mediate the harmful effects of anxiety and stress. Based on our findings, residential green space can be an option for promoting mental health through encouraging physical activity and social interaction.

4.3. Limitations and Future Directions

This study has several limitations. The research topic was described during recruiting participants. It is possible that people who benefited from or cared about urban green spaces have responded disproportionately. This may have caused us to overestimate the connections between green spaces and mental health.

The age structure of our respondents was inconsistent with that of the general population of China. The China Bureau of Statistics has disclosed that the population over 60 years old accounted for 18.70% of the total pollution as of 2021. By contrast, only around 0.2% of our participants were in this age group. Further studies may hire professional marketing teams to recruit greater shares of older people. Offline recruitment in middle-aged and elderly communities may be helpful because many aged Chinese use less Internet than the younger generations.

We only used self-reported measures due to our limited experimental conditions, so reporting bias must have existed. Some variables, such as residential neighborhood, can be re-investigated with objective measures, such as land use/land cover (LULC), which can reflect plantation coverage and other urban properties. Using machine learning to capture data from street view maps may also be a promising method to measure exposure to green spaces [124].

Although we compared different causal relationships in sensitivity analyses, the cross-sectional feature of the data cannot verify the direction of causalities. This is an inherent limitation with using such data. To re-examine the causality in each path, longitudinal trials are warranted. For example, tracking changes in pollution, social cohesion, and green exercise in a community with constructing residential green spaces would serve this purpose.

5. Conclusions

This study aimed to investigate how the Chinese performed green exercise in urban green spaces and to propose a framework to understand associations between residential green space and mental health through green exercise, perceived pollution, and social cohesion during the COVID-19 era. To achieve this goal, we checked how many survey respondents preferred to carry out green exercise in their residential neighborhood and obtained keywords from their reasons for choosing residential green space. Meanwhile, we carried out structural equation modeling to explore potential pathways among these residential green space "exercisers" with our cross-sectional data.

Our findings suggest residential green space may be the most popular green space for green exercise during the pandemic among young and middle-aged Chinese. Moreover, residential green space was negatively associated with anxiety and depression, mainly through mediators, including social interaction, less perceived pollution, and green exercise. These findings reinforce other research showing that residential green space is a health resource for urban residents during the COVID-19 era.

Based on our findings, we call for a focus on residential greening in China to meet the demand of its residents in the COVID-19 era. Although causal relationships between studied variables could not be confirmed due to the nature of our cross-sectional data, our survey and model indicate that better environmental qualities, including less noise and air pollution, can support mental health. Future urban planning should utilize existing plant-based strategies (e.g., plant purification or phytoremediation) to protect residential environments and physical activity experiences. Recognizing the limitations of our study, we also call for future research to examine and extend our tentative theoretical conclusions in larger populations using longitudinal study designs.

Supplementary Materials: The following supporting information can be downloaded at: https://www.mdpi.com/article/10.3390/land11081128/s1, Figure S1: Competing models; Table S1: The goodness of fit of tested models.

Author Contributions: Conceptualization, H.L.; methodology, H.L.; investigation, H.L.; resources, H.L.; data curation, H.L.; writing—original draft preparation, H.L. and M.H.E.M.B.; writing—review and editing, M.H.E.M.B., A.M.D. and Y.C.; supervision, G.Z.; project administration, G.Z. and Y.C.; funding acquisition, G.Z. All authors have read and agreed to the published version of the manuscript.

Funding: This research received no external funding.

Institutional Review Board Statement: The study was conducted in accordance with the Declaration of Helsinki, and approved by the Institutional Review Board of Southwest University.

Informed Consent Statement: Informed consent was obtained from all subjects involved in the study.

Data Availability Statement: The data are available upon request from the corresponding author.

Conflicts of Interest: The authors declare no conflict of interest.

References

1. Trøstrup, C.H.; Christiansen, A.B.; Stølen, K.S.; Nielsen, P.K.; Stelter, R. The effect of nature exposure on the mental health of patients: A systematic review. *Qual. Life Res.* **2019**, *28*, 1695–1703. [CrossRef] [PubMed]
2. Stanhope, J.; Breed, M.F.; Weinstein, P. Exposure to greenspaces could reduce the high global burden of pain. *Environ. Res.* **2020**, *187*, 109641. [CrossRef] [PubMed]
3. Li, H.; Zhang, X.; Wang, H.; Yang, Z.; Liu, H.; Cao, Y.; Zhang, G. Access to Nature via Virtual Reality: A Mini-Review. *Front. Psychol.* **2021**, *12*, 725288. [CrossRef] [PubMed]
4. Grant, M.; Brown, C.; Caiaffa, W.T.; Capon, A.; Corburn, J.; Coutts, C.; Crespo, C.J.; Ellis, G.; Ferguson, G.; Fudge, C. Cities and health: An evolving global conversation. *Cities Health* **2017**, *1*, 1–9. [CrossRef]
5. Turner, W.R.; Nakamura, T.; Dinetti, M. Global urbanization and the separation of humans from nature. *Bioscience* **2004**, *54*, 585–590. [CrossRef]
6. Moore, M.; Gould, P.; Keary, B.S. Global urbanization and impact on health. *Int. J. Hyg. Environ. Health* **2003**, *206*, 269–278. [CrossRef]
7. Coutts, C.; Horner, M.; Chapin, T. Using geographical information system to model the effects of green space accessibility on mortality in Florida. *Geocarto Int.* **2010**, *25*, 471–484. [CrossRef]
8. Heidt, V.; Neef, M. Benefits of urban green space for improving urban climate. In *Ecology, Planning, and Management of Urban Forests*; Springer: Berlin/Heidelberg, Germany, 2008; pp. 84–96.
9. Schipperijn, J.; Ekholm, O.; Stigsdotter, U.K.; Toftager, M.; Bentsen, P.; Kamper-Jørgensen, F.; Randrup, T.B. Factors influencing the use of green space: Results from a Danish national representative survey. *Landsc. Urban Plan.* **2010**, *95*, 130–137. [CrossRef]

10. Fongar, C.; Aamodt, G.; Randrup, T.B.; Solfjeld, I. Does Perceived Green Space Quality Matter? Linking Norwegian Adult Perspectives on Perceived Quality to Motivation and Frequency of Visits. *Int. J. Environ. Res. Public Health* **2019**, *16*, 2327. [CrossRef]

11. Ugolini, F.; Massetti, L.; Calaza-Martínez, P.; Cariñanos, P.; Dobbs, C.; Ostoić, S.K.; Marin, A.M.; Pearlmutter, D.; Saaroni, H.; Šaulienė, I.; et al. Effects of the COVID-19 pandemic on the use and perceptions of urban green space: An international exploratory study. *Urban For. Urban Green.* **2020**, *56*, 126888. [CrossRef]

12. Ugolini, F.; Massetti, L.; Pearlmutter, D.; Sanesi, G. Usage of urban green space and related feelings of deprivation during the COVID-19 lockdown: Lessons learned from an Italian case study. *Land Use Policy* **2021**, *105*, 105437. [CrossRef]

13. Müller, O.; Lu, G.; Jahn, A.; Razum, O. COVID-19 Control: Can Germany Learn From China? *Int. J. Health Policy Manag.* **2020**, *9*, 432–435. [CrossRef]

14. Gan, Y.; Ma, J.; Wu, J.; Chen, Y.; Zhu, H.; Hall, B.J. Immediate and delayed psychological effects of province-wide lockdown and personal quarantine during the COVID-19 outbreak in China. *Psychol. Med.* **2020**, *52*, 1321–1332. [CrossRef]

15. CDC of Henan, China The List of High-Risk Areas in the National Epidemic Has Increased to 571. Available online: https://tanghe.gov.cn/thyw/61528.html (accessed on 22 March 2022).

16. Slater, S.J.; Christiana, R.W.; Gustat, J. Recommendations for Keeping Parks and Green Space Accessible for Mental and Physical Health During COVID-19 and Other Pandemics. *Prev. Chronic Dis.* **2020**, *17*, E59. [CrossRef]

17. Pouso, S.; Borja, Á.; Fleming, L.E.; Gómez-Baggethun, E.; White, M.P.; Uyarra, M.C. Contact with blue-green spaces during the COVID-19 pandemic lockdown beneficial for mental health. *Sci. Total Environ.* **2021**, *756*, 143984. [CrossRef]

18. Astell-Burt, T.; Feng, X. Time for 'Green' during COVID-19? Inequities in Green and Blue Space Access, Visitation and Felt Benefits. *Int. J. Environ. Res. Public Health* **2021**, *18*, 2757. [CrossRef]

19. Lanza-León, P.; Pascual-Sáez, M.; Cantarero-Prieto, D. Alleviating mental health disorders through doses of green spaces: An updated review in times of the COVID-19 pandemic. *Int. J. Environ. Health Res.* **2021**, 1–18. [CrossRef]

20. Dzhambov, A.M.; Hartig, T.; Tilov, B.; Atanasova, V.; Makakova, D.R.; Dimitrova, D.D. Residential greenspace is associated with mental health via intertwined capacity-building and capacity-restoring pathways. *Environ. Res.* **2019**, *178*, 108708. [CrossRef]

21. Beyer, K.M.; Kaltenbach, A.; Szabo, A.; Bogar, S.; Nieto, F.J.; Malecki, K.M. Exposure to neighborhood green space and mental health: Evidence from the survey of the health of Wisconsin. *Int. J. Environ. Res. Public Health* **2014**, *11*, 3453–3472. [CrossRef]

22. Markevych, I.; Schoierer, J.; Hartig, T.; Chudnovsky, A.; Hystad, P.; Dzhambov, A.M.; De Vries, S.; Triguero-Mas, M.; Brauer, M.; Nieuwenhuijsen, M.J. Exploring pathways linking greenspace to health: Theoretical and methodological guidance. *Environ. Res.* **2017**, *158*, 301–317. [CrossRef]

23. Nutsford, D.; Pearson, A.L.; Kingham, S. An ecological study investigating the association between access to urban green space and mental health. *Public Health* **2013**, *127*, 1005–1011. [CrossRef]

24. Cohen-Cline, H.; Turkheimer, E.; Duncan, G.E. Access to green space, physical activity and mental health: A twin study. *J. Epidemiol. Community Health* **2015**, *69*, 523. [CrossRef]

25. Chan, J.; To, H.-P.; Chan, E. Reconsidering social cohesion: Developing a definition and analytical framework for empirical research. *Soc. Indic. Res.* **2006**, *75*, 273–302. [CrossRef]

26. Comstock, N.; Dickinson, L.M.; Marshall, J.A.; Soobader, M.-J.; Turbin, M.S.; Buchenau, M.; Litt, J.S. Neighborhood attachment and its correlates: Exploring neighborhood conditions, collective efficacy, and gardening. *J. Environ. Psychol.* **2010**, *30*, 435–442. [CrossRef]

27. Jennings, V. Social Cohesion and City Green Space: Revisiting the Power of Volunteering. *Challenges* **2019**, *10*, 36. [CrossRef]

28. Kingsbury, M.; Clayborne, Z.; Colman, I.; Kirkbride, J.B. The protective effect of neighbourhood social cohesion on adolescent mental health following stressful life events. *Psychol. Med.* **2020**, *50*, 1292–1299. [CrossRef]

29. Fone, D.; White, J.; Farewell, D.; Kelly, M.; John, G.; Lloyd, K.; Williams, G.; Dunstan, F. Effect of neighbourhood deprivation and social cohesion on mental health inequality: A multilevel population-based longitudinal study. *Psychol. Med.* **2014**, *44*, 2449–2460. [CrossRef]

30. Robinette, J.W.; Bostean, G.; Glynn, L.M.; Douglas, J.A.; Jenkins, B.N.; Gruenewald, T.L.; Frederick, D.A. Perceived neighborhood cohesion buffers COVID-19 impacts on mental health in a United States sample. *Soc. Sci. Med.* **2021**, *285*, 114269. [CrossRef]

31. Svensson, S.J.; Elntib, S. Community cohesion during the first peak of the COVID-19 pandemic: A social antidote to health anxiety and stress. *J. Appl. Soc. Psychol.* **2021**, *51*, 793–808. [CrossRef]

32. Peters, K.; Elands, B.; Buijs, A. Social interactions in urban parks: Stimulating social cohesion? *Urban For. Urban Green.* **2010**, *9*, 93–100. [CrossRef]

33. Jennings, V.; Bamkole, O. The relationship between social cohesion and urban green space: An avenue for health promotion. *Int. J. Environ. Res. Public Health* **2019**, *16*, 452. [CrossRef] [PubMed]

34. Van den Berg, M.M.; van Poppel, M.; van Kamp, I.; Ruijsbroek, A.; Triguero-Mas, M.; Gidlow, C.; Nieuwenhuijsen, M.J.; Gražulevičiene, R.; van Mechelen, W.; Kruize, H. Do physical activity, social cohesion, and loneliness mediate the association between time spent visiting green space and mental health? *Environ. Behav.* **2019**, *51*, 144–166. [CrossRef] [PubMed]

35. Sugiyama, T.; Leslie, E.; Giles-Corti, B.; Owen, N. Associations of neighbourhood greenness with physical and mental health: Do walking, social coherence and local social interaction explain the relationships? *J. Epidemiol. Community Health* **2008**, *62*, e9. [CrossRef] [PubMed]

36. Zhang, R.; Zhang, C.-Q.; Rhodes, R.E. The pathways linking objectively-measured greenspace exposure and mental health: A systematic review of observational studies. *Environ. Res.* **2021**, *198*, 111233. [CrossRef]

37. Ventriglio, A.; Bellomo, A.; di Gioia, I.; Di Sabatino, D.; Favale, D.; De Berardis, D.; Cianconi, P. Environmental pollution and mental health: A narrative review of literature. *CNS Spectr.* **2021**, *26*, 51–61. [CrossRef]

38. Peng, M.; Zhang, H.; Evans, R.D.; Zhong, X.; Yang, K. Actual Air Pollution, Environmental Transparency, and the Perception of Air Pollution in China. *J. Environ. Dev.* **2019**, *28*, 78–105. [CrossRef]

39. Baranzini, A.; Schaerer, C.; Thalmann, P. Using measured instead of perceived noise in hedonic models. *Transp. Res. Part D Transp. Environ.* **2010**, *15*, 473–482. [CrossRef]

40. Xu, X.; Reed, M. Perceived pollution and inbound tourism for Shanghai: A panel VAR approach. *Curr. Issues Tour.* **2019**, *22*, 601–614. [CrossRef]

41. Kou, L.; Tao, Y.; Kwan, M.-P.; Chai, Y. Understanding the relationships among individual-based momentary measured noise, perceived noise, and psychological stress: A geographic ecological momentary assessment (GEMA) approach. *Health Place* **2020**, *64*, 102285. [CrossRef]

42. Rotko, T.; Oglesby, L.; Künzli, N.; Carrer, P.; Nieuwenhuijsen, M.J.; Jantunen, M. Determinants of perceived air pollution annoyance and association between annoyance scores and air pollution ($PM_{2.5}$, NO_2) concentrations in the European EXPOLIS study. *Atmos. Environ.* **2002**, *36*, 4593–4602. [CrossRef]

43. Selmi, W.; Weber, C.; Rivière, E.; Blond, N.; Mehdi, L.; Nowak, D. Air pollution removal by trees in public green spaces in Strasbourg city, France. *Urban For. Urban Green.* **2016**, *17*, 192–201. [CrossRef]

44. Fan, Y.; Zhiyi, B.; Zhujun, Z.; Jiani, L. The investigation of noise attenuation by plants and the corresponding noise-reducing spectrum. *J. Environ. Health* **2010**, *72*, 8–15.

45. Dzhambov, A.M.; Dimitrova, D.D. Green spaces and environmental noise perception. *Urban For. Urban Green.* **2015**, *14*, 1000–1008. [CrossRef]

46. Zapf, V. Influence of green space design on individual noise perception. In Proceedings of the INTER-NOISE and NOISE-CON Congress and Conference Proceedings, Washington, DC, USA, 1 August 2021; pp. 3130–3142.

47. Cetin, C.; Karafakı, F. The influence of green areas on city-dwellers' perceptions of air pollution: The case of Nigde city center. *J. Environ. Biol.* **2020**, *41*, 453–461. [CrossRef]

48. Margaritis, E.; Kang, J. Relationship between green space-related morphology and noise pollution. *Ecol. Indic.* **2017**, *72*, 921–933. [CrossRef]

49. Liu, H.-L.; Shen, Y.-S. The impact of green space changes on air pollution and microclimates: A case study of the Taipei metropolitan area. *Sustainability* **2014**, *6*, 8827–8855. [CrossRef]

50. Dzhambov, A.; Hartig, T.; Markevych, I.; Tilov, B.; Dimitrova, D. Urban residential greenspace and mental health in youth: Different approaches to testing multiple pathways yield different conclusions. *Environ. Res.* **2018**, *160*, 47–59. [CrossRef]

51. Yang, M.; Dijst, M.; Faber, J.; Helbich, M. Using structural equation modeling to examine pathways between perceived residential green space and mental health among internal migrants in China. *Environ. Res.* **2020**, *183*, 109121. [CrossRef]

52. Liu, Y.; Wang, R.; Grekousis, G.; Liu, Y.; Yuan, Y.; Li, Z. Neighbourhood greenness and mental wellbeing in Guangzhou, China: What are the pathways? *Landsc. Urban Plan.* **2019**, *190*, 103602. [CrossRef]

53. Rogerson, M.; Brown, D.K.; Sandercock, G.; Wooller, J.-J.; Barton, J. A comparison of four typical green exercise environments and prediction of psychological health outcomes. *Perspect. Public Health* **2016**, *136*, 171–180. [CrossRef]

54. Lee, J.-Y.; Lee, D.-C. Cardiac and pulmonary benefits of forest walking versus city walking in elderly women: A randomised, controlled, open-label trial. *Eur. J. Integr. Med.* **2014**, *6*, 5–11. [CrossRef]

55. Elsadek, M.; Liu, B.; Lian, Z.; Xie, J. The influence of urban roadside trees and their physical environment on stress relief measures: A field experiment in Shanghai. *Urban For. Urban Green.* **2019**, *42*, 51–60. [CrossRef]

56. Astell-Burt, T.; Feng, X.; Kolt, G.S. Green space is associated with walking and moderate-to-vigorous physical activity (MVPA) in middle-to-older-aged adults: Findings from 203,883 Australians in the 45 and Up Study. *Br. J. Sports Med.* **2014**, *48*, 404–406. [CrossRef]

57. McEachan, R.R.C.; Prady, S.L.; Smith, G.; Fairley, L.; Cabieses, B.; Gidlow, C.; Wright, J.; Dadvand, P.; van Gent, D.; Nieuwenhuijsen, M.J. The association between green space and depressive symptoms in pregnant women: Moderating roles of socioeconomic status and physical activity. *J. Epidemiol. Community Health* **2016**, *70*, 253. [CrossRef]

58. Holt, E.W.; Lombard, Q.K.; Best, N.; Smiley-Smith, S.; Quinn, J.E. Active and Passive Use of Green Space, Health, and Well-Being amongst University Students. *Int. J. Environ. Res. Public Health* **2019**, *16*, 424. [CrossRef]

59. Sugiyama, T.; Giles-Corti, B.; Summers, J.; du Toit, L.; Leslie, E.; Owen, N. Initiating and maintaining recreational walking: A longitudinal study on the influence of neighborhood green space. *Prev. Med.* **2013**, *57*, 178–182. [CrossRef]

60. Bloemsma, L.D.; Gehring, U.; Klompmaker, J.O.; Hoek, G.; Janssen, N.A.; Smit, H.A.; Vonk, J.M.; Brunekreef, B.; Lebret, E.; Wijga, A.H. Green space visits among adolescents: Frequency and predictors in the PIAMA birth cohort study. *Environ. Health Perspect.* **2018**, *126*, 047016. [CrossRef]

61. Hug, S.-M.; Hartig, T.; Hansmann, R.; Seeland, K.; Hornung, R. Restorative qualities of indoor and outdoor exercise settings as predictors of exercise frequency. *Health Place* **2009**, *15*, 971–980. [CrossRef]

62. Tan, C.L.Y.; Chang, C.-C.; Nghiem, L.T.P.; Zhang, Y.; Oh, R.R.Y.; Shanahan, D.F.; Lin, B.B.; Gaston, K.J.; Fuller, R.A.; Carrasco, L.R. The right mix: Residential urban green-blue space combinations are correlated with physical exercise in a tropical city-state. *Urban For. Urban Green.* **2021**, *57*, 126947. [CrossRef]

63. Akpınar, A. Green Exercise: How Are Characteristics of Urban Green Spaces Associated with Adolescents' Physical Activity and Health? *Int. J. Environ. Res. Public Health* **2019**, *16*, 4281. [CrossRef]

64. Gozalo, G.R.; Morillas, J.M.B.; González, D.M.; Moraga, P.A. Relationships among satisfaction, noise perception, and use of urban green spaces. *Sci. Total Environ.* **2018**, *624*, 438–450. [CrossRef] [PubMed]

65. Dzhambov, A.; Tilov, B.; Markevych, I.; Dimitrova, D. Residential road traffic noise and general mental health in youth: The role of noise annoyance, neighborhood restorative quality, physical activity, and social cohesion as potential mediators. *Environ. Int.* **2017**, *109*, 1–9. [CrossRef] [PubMed]

66. Lü, J.; Liang, L.; Feng, Y.; Li, R.; Liu, Y. Air Pollution Exposure and Physical Activity in China: Current Knowledge, Public Health Implications, and Future Research Needs. *Int. J. Environ. Res. Public Health* **2015**, *12*, 14887–14897. [CrossRef] [PubMed]

67. Foraster, M.; Eze, I.C.; Vienneau, D.; Brink, M.; Cajochen, C.; Caviezel, S.; Héritier, H.; Schaffner, E.; Schindler, C.; Wanner, M.; et al. Long-term transportation noise annoyance is associated with subsequent lower levels of physical activity. *Environ. Int.* **2016**, *91*, 341–349. [CrossRef]

68. Subiza-Pérez, M.; García-Baquero, G.; Babarro, I.; Anabitarte, A.; Delclòs-Alió, X.; Vich, G.; Roig-Costa, O.; Miralles-Guasch, C.; Lertxundi, N.; Ibarluzea, J. Does the perceived neighborhood environment promote mental health during pregnancy? Confirmation of a pathway through social cohesion in two Spanish samples. *Environ. Res.* **2021**, *197*, 111192. [CrossRef]

69. Kim, J.; Kim, J.; Han, A. Leisure Time Physical Activity Mediates the Relationship Between Neighborhood Social Cohesion and Mental Health Among Older Adults. *J. Appl. Gerontol.* **2019**, *39*, 292–300. [CrossRef]

70. Cradock, A.L.; Kawachi, I.; Colditz, G.A.; Gortmaker, S.L.; Buka, S.L. Neighborhood social cohesion and youth participation in physical activity in Chicago. *Soc. Sci. Med.* **2009**, *68*, 427–435. [CrossRef]

71. Dzhambov, A.M.; Lercher, P.; Browning, M.; Stoyanov, D.; Petrova, N.; Novakov, S.; Dimitrova, D.D. Does greenery experienced indoors and outdoors provide an escape and support mental health during the COVID-19 quarantine? *Environ. Res.* **2021**, *196*, 110420. [CrossRef]

72. Afifi, M. Gender differences in mental health. *Singap. Med. J.* **2007**, *48*, 385.

73. Mirowsky, J.; Ross, C.E. Age and depression. *J. Health Soc. Behav.* **1992**, *33*, 187–205. [CrossRef]

74. Bolte, G.; Nanninga, S.; Dandolo, L. Sex/Gender Differences in the Association between Residential Green Space and Self-Rated Health—A Sex/Gender-Focused Systematic Review. *Int. J. Environ. Res. Public Health* **2019**, *16*, 4818. [CrossRef]

75. Chang, C.-c.; Oh, R.R.Y.; Le Nghiem, T.P.; Zhang, Y.; Tan, C.L.; Lin, B.B.; Gaston, K.J.; Fuller, R.A.; Carrasco, L.R. Life satisfaction linked to the diversity of nature experiences and nature views from the window. *Landsc. Urban Plan.* **2020**, *202*, 103874. [CrossRef]

76. Shuval, K.; Li, Q.; Gabriel, K.P.; Tchernis, R. Income, physical activity, sedentary behavior, and the 'weekend warrior' among U.S. adults. *Prev. Med.* **2017**, *103*, 91–97. [CrossRef]

77. Delhey, J.; Dragolov, G. Happier together. Social cohesion and subjective well-being in Europe. *Int. J. Psychol.* **2016**, *51*, 163–176. [CrossRef]

78. Zimmerman, F.J.; Katon, W. Socioeconomic status, depression disparities, and financial strain: What lies behind the income-depression relationship? *Health Econ.* **2005**, *14*, 1197–1215. [CrossRef]

79. Rigolon, A.; Browning, M.H.; McAnirlin, O.; Yoon, H.V. Green space and health equity: A systematic review on the potential of green space to reduce health disparities. *Int. J. Environ. Res. Public Health* **2021**, *18*, 2563. [CrossRef]

80. Pfefferbaum, B.; North, C.S. Mental health and the COVID-19 pandemic. *N. Engl. J. Med.* **2020**, *383*, 510–512. [CrossRef]

81. Borkowska, M.; Laurence, J. Coming together or coming apart? Changes in social cohesion during the COVID-19 pandemic in England. *Eur. Soc.* **2021**, *23*, S618–S636. [CrossRef]

82. Erdönmez, C.; Atmiş, E. The impact of the COVID-19 pandemic on green space use in Turkey: Is closing green spaces for use a solution? *Urban For. Urban Green.* **2021**, *64*, 127295. [CrossRef]

83. Coleman, J. *Foundations of Social Theory*; Harvard University Press: Cambridge, UK, 1990.

84. Sampson, R.J.; Raudenbush, S.W.; Earls, F. Neighborhoods and violent crime: A multilevel study of collective efficacy. *Science* **1997**, *277*, 918–924. [CrossRef]

85. Gaither, C.J.; Aragón, A.; Madden, M.; Alford, S.; Wynn, A.; Emery, M. "Black folks do forage": Examining wild food gathering in Southeast Atlanta Communities. *Urban For. Urban Green.* **2020**, *56*, 126860. [CrossRef]

86. Raina, V.; Krishnamurthy, S. Natural language processing. In *Building an Effective Data Science Practice*; Springer: Berlin/Heidelberg, Germany, 2022; pp. 63–73.

87. Chopra, A.; Prashar, A.; Sain, C. Natural language processing. *Int. J. Technol. Enhanc. Emerg. Eng. Res.* **2013**, *1*, 131–134.

88. Bagozzi, R.P.; Yi, Y. Specification, evaluation, and interpretation of structural equation models. *J. Acad. Mark. Sci.* **2012**, *40*, 8–34. [CrossRef]

89. Rogerson, P.A. *Statistical Methods for Geography: A Student's Guide*; Sage: Newcastle, UK, 2019.

90. Byrne, B.M. *Structural Equation Modeling with Mplus: Basic Concepts, Applications, and Programming*; Routledge: London, UK, 2013.

91. Schermelleh-Engel, K.; Moosbrugger, H.; Müller, H. Evaluating the fit of structural equation models: Tests of significance and descriptive goodness-of-fit measures. *Methods Psychol. Res. Online* **2003**, *8*, 23–74.

92. Haukoos, J.S.; Lewis, R.J. Advanced statistics: Bootstrapping confidence intervals for statistics with "difficult" distributions. *Acad. Emerg. Med.* **2005**, *12*, 360–365. [CrossRef]
93. Kelley, K. The Effects of Nonnormal Distributions on Confidence Intervals Around the Standardized Mean Difference: Bootstrap and Parametric Confidence Intervals. *Educ. Psychol. Meas.* **2005**, *65*, 51–69. [CrossRef]
94. Brown, T.A. *Confirmatory Factor Analysis for Applied Research*; Guilford Publications: New York, NY, USA, 2015.
95. Hu, L.-T.; Bentler, P.M. Fit indices in covariance structure modeling: Sensitivity to underparameterized model misspecification. *Psychol. Methods* **1998**, *3*, 424. [CrossRef]
96. Curran, P.J.; West, S.G.; Finch, J.F. The robustness of test statistics to nonnormality and specification error in confirmatory factor analysis. *Psychol. Methods* **1996**, *1*, 16. [CrossRef]
97. Hu, L.-T.; Bentler, P.M.; Kano, Y. Can test statistics in covariance structure analysis be trusted? *Psychol. Bull.* **1992**, *112*, 351. [CrossRef]
98. West, S.G.; Finch, J.F.; Curran, P.J. Structural equation models with nonnormal variables: Problems and remedies. In *Structural Equation Modeling: Concepts, Issues, and Applications*; Hoyle, R.H., Ed.; Sage Publications, Inc.: New York, NY, USA, 1995; pp. 56–75.
99. Hayes, A.F. *Introduction to Mediation, Moderation, and Conditional Process Analysis: A Regression-Based Approach*; Guilford Publications: New York, NY, USA, 2017.
100. Zhao, X.; Lynch, J.G., Jr.; Chen, Q. Reconsidering Baron and Kenny: Myths and truths about mediation analysis. *J. Consum. Res.* **2010**, *37*, 197–206. [CrossRef]
101. Burnham, K.P.; Anderson, D.R. Multimodel inference: Understanding AIC and BIC in model selection. *Sociol. Methods Res.* **2004**, *33*, 261–304. [CrossRef]
102. Elkaseh, A.M.; Wong, K.W.; Fung, C.C. Perceived ease of use and perceived usefulness of social media for e-learning in Libyan higher education: A structural equation modeling analysis. *Int. J. Inf. Educ. Technol.* **2016**, *6*, 192. [CrossRef]
103. Kim, N.E.; Han, S.S.; Yoo, K.H.; Yun, E.K. The impact of user's perceived ability on online health information acceptance. *Telemed. E-Health* **2012**, *18*, 703–708. [CrossRef] [PubMed]
104. Zhou, H.; Li, L. The impact of supply chain practices and quality management on firm performance: Evidence from China's small and medium manufacturing enterprises. *Int. J. Prod. Econ.* **2020**, *230*, 107816. [CrossRef]
105. Olsson, U.H.; Foss, T.; Troye, S.V.; Howell, R.D. The Performance of ML, GLS, and WLS Estimation in Structural Equation Modeling Under Conditions of Misspecification and Nonnormality. *Struct. Equ. Modeling A Multidiscip. J.* **2000**, *7*, 557–595. [CrossRef]
106. Shanahan, D.F.; Bush, R.; Gaston, K.J.; Lin, B.B.; Dean, J.; Barber, E.; Fuller, R.A. Health benefits from nature experiences depend on dose. *Sci. Rep.* **2016**, *6*, 28551. [CrossRef] [PubMed]
107. Zhu, J.; Xu, C. Sina microblog sentiment in Beijing city parks as measure of demand for urban green space during the COVID-19. *Urban For. Urban Green.* **2021**, *58*, 126913. [CrossRef]
108. Baoxin, Z.; Wei, Z. A study on the temporal and spatial characteristics of children's outdoor activities in big cities—A case study of Shanghai. *City Plan.* **2018**, *42*, 87–96. (In Chinese)
109. Ekkel, E.D.; de Vries, S. Nearby green space and human health: Evaluating accessibility metrics. *Landsc. Urban Plan.* **2017**, *157*, 214–220. [CrossRef]
110. Coombes, E.; Jones, A.P.; Hillsdon, M. The relationship of physical activity and overweight to objectively measured green space accessibility and use. *Soc. Sci. Med.* **2010**, *70*, 816–822. [CrossRef]
111. Kroeger, T.; Escobedo, F.J.; Hernandez, J.L.; Varela, S.; Delphin, S.; Fisher, J.R.; Waldron, J. Reforestation as a novel abatement and compliance measure for ground-level ozone. *Proc. Natl. Acad. Sci. USA* **2014**, *111*, E4204–E4213. [CrossRef]
112. Van Renterghem, T.; Forssén, J.; Attenborough, K.; Jean, P.; Defrance, J.; Hornikx, M.; Kang, J. Using natural means to reduce surface transport noise during propagation outdoors. *Appl. Acoust.* **2015**, *92*, 86–101. [CrossRef]
113. Xuejin, W.; Juxin, Z. Planning and construction of livable city green space. *Resour. Environ. Yangtze River Basin* **2011**, *20*, 5. (In Chinese)
114. Li, G.; Xiong, J.; Xu, M.; Dong, L. Comprehensive study on the noise reduction ability and the visual effect on the edge of green space of Beijing urban park. *J. Beijing For. Univ.* **2017**, *39*, 12. (In Chinese) [CrossRef]
115. Miao, Y.; Chen, Z.; Chen, Y.; Du, G. Resistance to and absorbency of gaseous for 38 young landscaping plants in Zhejiang Province. *J. Zhejiang For. Coll.* **2008**, *25*, 765–771. (In Chinese) [CrossRef]
116. Pan, W.; Zhang, W.; Zhang, F.; Liu, J.; Duan, J.; Chen, J. Decontamination ability of sulfur dioxide and nitrogen dioxide for 38 young landscaping plants in Guangzhou city. *Ecol. Environ. Sci.* **2012**, *21*, 7. [CrossRef]
117. Zhang, D.; Chu, G.; Yu, Q.; Liu, S.; Lu, Y.; Hu, X.; Xue, k.; Kong, G. Decontamination Ability of Garden Plants to Absorb Sulfur Dioxide and Fluoride. *J. Trop. Subtrop. Bot.* **2003**, *11*, 336–340. (In Chinese)
118. Maas, J.; van Dillen, S.M.E.; Verheij, R.A.; Groenewegen, P.P. Social contacts as a possible mechanism behind the relation between green space and health. *Health Place* **2009**, *15*, 586–595. [CrossRef]
119. Kawachi, I.; Kennedy, B.P.; Glass, R. Social capital and self-rated health: A contextual analysis. *Am. J. Public Health* **1999**, *89*, 1187–1193. [CrossRef]
120. Molnar, B.E.; Gortmaker, S.L.; Bull, F.C.; Buka, S.L. Unsafe to play? Neighborhood disorder and lack of safety predict reduced physical activity among urban children and adolescents. *Am. J. Health Promot.* **2004**, *18*, 378–386. [CrossRef]

121. Abel, T.; McQueen, D. *The COVID-19 Pandemic Calls for Spatial Distancing and Social Closeness: Not for Social Distancing!* Springer: Berlin/Heidelberg, Germany, 2020; Volume 65, p. 231.
122. Chen, X.; Zou, Y.; Gao, H. Role of neighborhood social support in stress coping and psychological wellbeing during the COVID-19 pandemic: Evidence from Hubei, China. *Health Place* **2021**, *69*, 102532. [CrossRef]
123. Feng, Z. New urban immigrants have a "home" only when they have neighbors. *Times Mail* **2020**, *7*, 22–24. (In Chinese)
124. Helbich, M.; Yao, Y.; Liu, Y.; Zhang, J.; Liu, P.; Wang, R. Using deep learning to examine street view green and blue spaces and their associations with geriatric depression in Beijing, China. *Environ. Int.* **2019**, *126*, 107–117. [CrossRef]

Article

Visiting Peri-Urban Forestlands and Mountains during the COVID-19 Pandemic: Empirical Analysis on Effects of Land Use and Awareness of Visitors

Yuta Uchiyama [1] and Ryo Kohsaka [2,*]

[1] Graduate School of Human Development and Environment, Kobe University, Kobe 657-8501, Japan; yuta.uchiyama@landscape.kobe-u.ac.jp

[2] Graduate School of Agricultural and Life Sciences, The University of Tokyo, Tokyo 113-8654, Japan

* Correspondence: kohsaka@hotmail.com

Abstract: This research analyzed the status of visiting peri-urban forestlands and mountains during the first COVID-19 emergency period in Japan using a large-scale online questionnaire-based survey. We identified and examined the factors that correlated with visits to such areas, including respondents' social-economic attributes, environmental conditions (such as the land use patterns of their residential areas), and awareness of the functions of forestlands. The results suggest that environmental conditions are a major factor encouraging residents to visit peri-urban forestlands and mountains during the pandemic. Peri-urban areas with forestlands have such environmental conditions, and residents who visited peri-urban forestlands and mountains tended to live in peri-urban areas. Residents' expectations regarding forest functions were also strong factors influencing them to visit those places. Those who visited forests and mountains expected these areas to have mental health and educational functions. Especially, female respondents tended to be aware of forestlands as spaces for mental and physical relaxation, and respondents who have one or more children tended to be aware of the educational functions of forests. These findings imply that policy should consider the role of environmental conditions, awareness, and expectations about the function of forests and mountains, and prior interactions with nature in encouraging residents to visit such places for their health during the pandemic. These factors could also play a role in addressing the social and environmental disparities that exist between residents of different socio-economic statuses regarding access to nature. In future research, the detailed relationships between residents' environmental conditions and expectations/awareness of the functions of peri-urban forestlands and mountains need to be explored.

Keywords: peri-urban forestlands; mountains; COVID-19; land use; Japan

Citation: Uchiyama, Y.; Kohsaka, R. Visiting Peri-Urban Forestlands and Mountains during the COVID-19 Pandemic: Empirical Analysis on Effects of Land Use and Awareness of Visitors. *Land* **2022**, *11*, 1194. https://doi.org/10.3390/land11081194

Academic Editors: Francesca Ugolini and David Pearlmutter

Received: 10 June 2022
Accepted: 26 July 2022
Published: 29 July 2022

Publisher's Note: MDPI stays neutral with regard to jurisdictional claims in published maps and institutional affiliations.

1. Introduction

Recently, access to and use of green areas have largely been affected by lockdowns and other control measures established in response to the ongoing COVID-19 pandemic [1–4]. The resultant unequal access to such areas, particularly in urban contexts, is one of the emerging injustices related to land. Specifically, existing studies e.g., [5,6] have extensively discussed the physical and mental effects of the limited access to green areas due to the pandemic. Preliminary studies have identified changes in the number of green area visitors. In addition to the overall number of visitors, certain studies have also analyzed their personal attributes [7]. Still, the target green areas are limited to parks, gardens, and agricultural lands. It is necessary to identify the status of visiting other types of green areas, such as peri-urban forestlands, which can provide spaces for relaxation, education, and other activities for visitors from urban and peri-urban areas. Furthermore, peri-urban forestlands are one of the essential components of biodiversity and ecosystem management [8–10].

The values of ecosystem services and their distribution are influenced by socio-economic factors [11]. In this circumstance, ecosystem service realization is needed to enhance citizens' awareness and facilitate participatory management [12]. Such awareness can be achieved using ecosystem services visualization, which policymakers and other experts can use to share the relevant scientific information with citizens [13]. Further, policymakers need to consider the tradeoff relationships between forest ecosystem services when formulating policy [14]. Indicator-based management is also an effective method for sharing the progress of policy implementations [15]. The main challenges of policymakers and experts are involving citizens in forest ecosystem services management and identifying stakeholders and their relationships.

To understand the status of forest ecosystem use by citizens, access and use of green areas and their values should be analyzed. Since before the pandemic, evidence of merits and issues surrounding forest ecosystem services have been provided [16]. For example, Kaźmierczak [17] highlights the direct and indirect contributions of green areas to citizens, while Wolch et al. [18] and Kabisch and Haase [19] suggest that environmental justice was one of the main concerns in issues related to the supply and demand of forest ecosystem services. Thus, to achieve fair and equitable sharing of ecosystem services, the distribution and usage trends of those services need to be identified to provide a basis for policymaking.

The impact of the pandemic varies among citizens with different socio-economic attributes [20,21]. For example, those belonging to racial minority groups and those with low income have been affected disproportionately by COVID-19 [20]. Regarding mental health, a greater impact of the pandemic on income is associated with increased depression and anxiety, while higher income is associated with better mental health conditions [22]. Similarly, the pandemic's effect on income levels also influences green area access—both physical and virtual—during the pandemic period [23]. A case study shows that low-income citizens, who are the most affected by COVID-19, have the least amount of nature-based environments nearby [24]. According to Pipitone and Jović [25], urban green areas are usually used by high-income citizens, and the pandemic has widened such socio-spatial disparities. Although the existing studies have detected this effect of income on green area use, this relationship is non-linear, and knowledge about the ideal green area in terms of health-related benefits is still lacking [26]. Furthermore, the influence of income on access and use of different types of green areas and the influence of environmental conditions such as land use in residential areas are not fully examined.

Considering the limitations of existing studies, the purpose of this research is to identify the status of visits to peri-urban forestlands and mountains during the first COVID-19 emergency period in Japan and to detect factors that correlate with such visits. This paper focused on residents' socio-economic attributes, land use, and awareness of the functions of forestlands because such factors may be correlated with green area use.

2. Materials and Methods

An online questionnaire survey was conducted to identify the status of visits to peri-urban forestlands and mountains. The survey period was from 31 July to 1 August 2020. Although there are discussions regarding the merit and demerit of online surveys e.g., [27], online surveys are still a useful method, especially when face-to-face surveys are not possible, as is the case during the COVID-19 pandemic [28]. The questionnaire was distributed to respondents in Aichi prefecture, Japan, two months after the first COVID-19 emergency period (16 April–14 May 2020) in Japan. During this period, the Japanese national government discouraged the residents from leaving their homes, especially to visit areas over prefectural borders. Although not following these recommendations did not result in penalties, there were significant effects on residents' behavior, particularly as Japanese society has relatively high peer pressure for following social norms and displaying a unified behavior. For example, it was reported that 37% of residents in Aichi did not visit green areas during the emergency period in previous studies on urban green areas [7]. Some residents were discouraged from even visiting hiking trails, although others visited them (Figure 1).

Figure 1. Signage discouraging people from visiting the hiking trail in Aichi Prefecture (**left**) and visitors of the same trail (**right**).

The research site, Aichi Prefecture, is the third largest metropolitan area (having the third largest population among other Japanese metropolises) with relatively good access to mountain areas and forestlands in Japan. Its capital is Nagoya City. The total number of respondents in this study was 1244. The female and male proportions were 47.6% and 52.4%, respectively. The proportion of over-60-years-old respondents was 36.6%; that of the five-year-interval age groups ranged between 7 and 11%, except the youngest age group of 20–24 years (2.9%).

The questionnaire consisted of questions posed to identify the status of visits to peri-urban forestlands and mountains at least once in the emergency period. The questions fell under four main categories as listed below. In the previous research, we analyzed the answers to questions about the status of visits to green areas, which were frequently visited by citizens during the emergency period [7]. In that research, the sample size of respondents who visited mountains (the main green area that was frequently visited) was too small to analyze (only 22 visitors), and mountain visitors were not analyzed.

- Socio-economic attributes: Sex (Male/Female), age, annual household income (1: <200, 2: 200–400, 3: 400–600, 4: 600–800, 5: 800–1000, 6: 1000–1200, 7: 1200–1500, 8: 1500–2000, 9: ≥2000 [Unit: 10 thousand JPY]), number of children in the household (>1/0);
- Environmental conditions: Zip-code district (the sizes of the areas and the ratios of land use categories in individual zip-code districts were computed and used in the analysis);
- Status of visits to peri-urban forestlands and mountains: Whether respondents visited mountains and forestlands during the emergency period (Answer: Yes/No);
- Awareness of the functions of forestlands (Respondents were asked about their pre- and post-emergency-period awareness of the functions of forestlands).

In order to compute the ratios of land use categories in zip-code districts, data from Japan Aerospace Exploration Agency's (JAXA) High-Resolution Land Use (2014–2016) survey (https://www.eorc.jaxa.jp/ALOS/en/dataset/lulc_e.htm [accessed on 28 May 2022]) were used. Based on the ratios of the land use categories and area sizes of the zip-code districts, the respondents' environmental conditions were analyzed. The resolution of the land use data was a 30 m square grid, and 10 land use categories were included in the data. Such high-resolution data were necessary to analyze the environmental conditions of zip-code districts with relatively smaller area sizes, which are located in urban areas. In order to analyze the overall environmental conditions of the individual zip-code districts, the ratios of urban areas, agricultural lands, and forestlands were computed.

Regarding respondents' awareness of forestland function, we assume that respondents who visited peri-urban forestlands and mountains have a certain level of awareness. To analyze the characteristics of such awareness, we compared their awareness with that of respondents who did not visit these areas. We also performed a logistic regression analysis of their answers to the question related to forestland functions to identify the factors influencing their awareness.

3. Results

The results of the survey revealed that 212 (21%) respondents visited peri-urban forestlands and mountains during the emergency period. The reasons for visiting those places are shown in Figure 2. We have used the options of the reasons from the reference [29], which is research on forest recreation in the Japanese context. Considering the situation of the pandemic, we added "visiting safe place" as an option. More than 40% of the respondents who visited peri-urban forestlands and mountains visited them for relaxation. Feeling and touching nature was the second reason, and 20 to 25% of them answered with reasons including "to see a beautiful landscape," "to break from my daily routine," and "to visit a safe place." The latter reason ("to visit a safe place" [24%]) suggests that peri-urban forestlands and mountains were considered safe to a certain degree during the COVID-19 pandemic. Minor motivations for visiting included reasons directly related to physical activities, such as doing support exercises and recreational activities. The overall trend based on the given reasons for visiting mountains shows that the residents were motivated by their mental health.

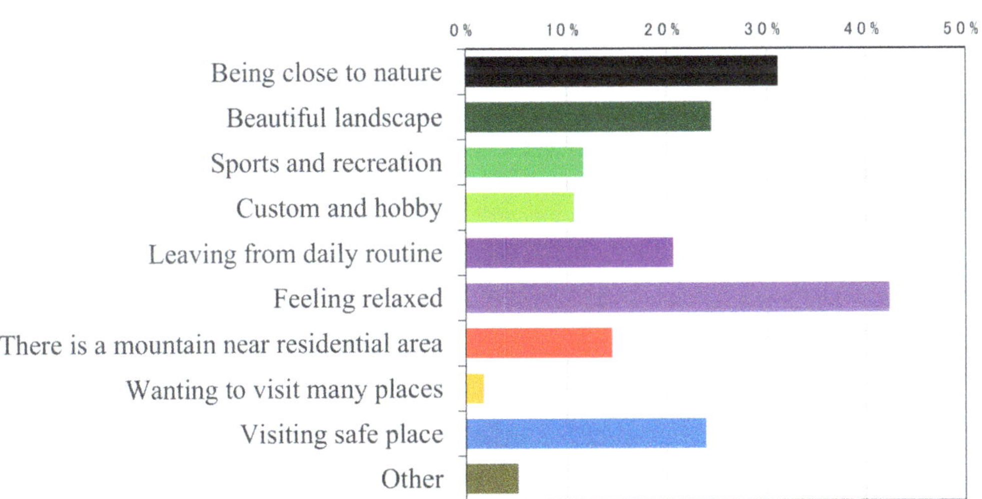

Figure 2. Reasons of visiting peri-urban forestlands and mountains.

Concerning the respondents' socio-economic attributes, there was no obvious difference between them except for gender. Figure 3 shows the male to female proportions of respondents who visited/did not visit peri-urban forestlands and mountains. The ratio of males who visited such places was high relative to that of females. Further, the over-60-years-old age group had the highest ratios of respondents in both categories (those who visited and those who did not visit), and thus it can be assumed that the older male respondents tended to visit those areas. When taken together, these results suggest that relatively older male respondents had a tendency to visit peri-urban forestlands and mountains.

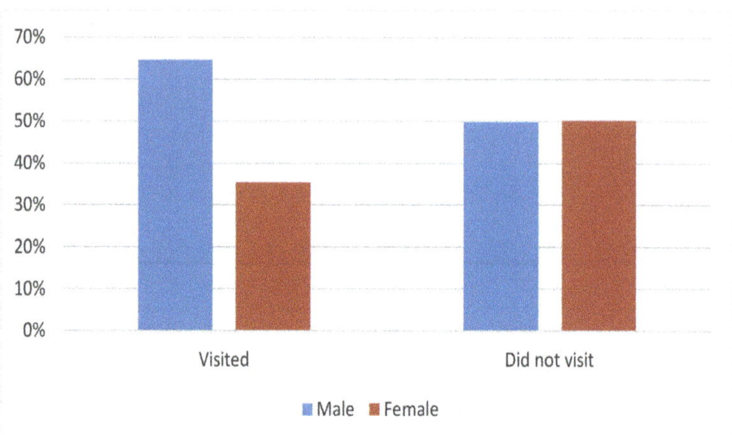

Figure 3. Ratios of respondents who visited and did not visit peri-urban forestlands and mountains by different gender groups.

Regarding residents' household income, an existing study i.e., [7] shows that residents with a higher household income had a tendency to visit green areas during the COVID-19 pandemic. However, Figure 4 indicates that in the current study, the difference in household income between the respondent groups (those who visited and those who did not visit forestlands and mountains) is unclear. That is, although relatively small differences in the ratios of respondents in individual household-income groups are evident between the two respondent groups (Figure 4), these differences are not as distinct as those shown in the existing study.

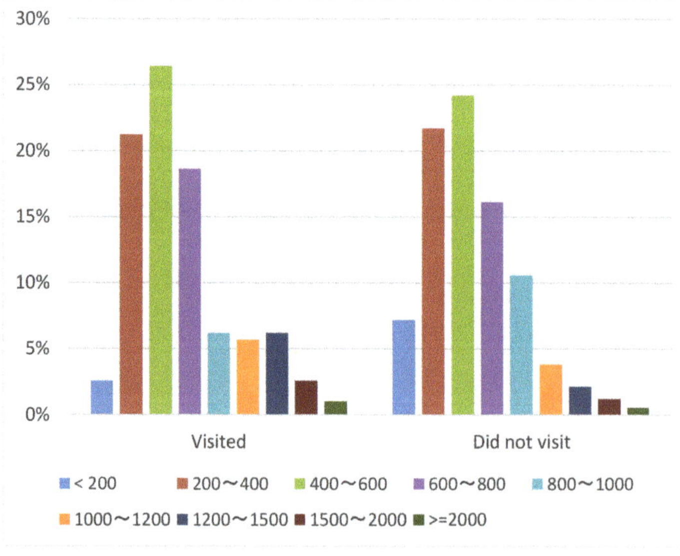

Figure 4. Ratios of respondents who visited and respondents who did not visit peri-urban forestlands and mountains by different household-income groups (Unit: million JPY).

Furthermore, the environmental conditions of respondents' residential areas were examined. It is revealed that there are statistically significant differences in the average ratio of forestland and area size of the zip-code district between the respondent groups who visited and those that did not visit peri-urban forestlands and mountains (Tables 1 and 2). Moreover, the average ratio of forestlands in the zip-code districts of respondents who visited such places is higher than that of those who did not visit (t-test, $p < 0.01$) (Table 1). This result suggests that the respondents who visited peri-urban forestlands and mountains can access forestlands in their residential areas relatively easily and might be more familiar with forest environments in their daily life. Additionally, it could be assumed that their residential places are not highly urbanized and not fully covered by built-up areas. Living in an environment that allows residents to easily access forestlands might be one of the factors encouraging residents to visit peri-urban forestlands and mountains.

Table 1. Ratios of forestlands in the zip-code districts of respondents who visited/did not visit mountains in the emergency period.

Ratio of Forestland	Visited	Did Not Visit
Average (%)	8.10	4.33
Variance	332.54	158.00
Number of respondents	212	1031
Degree of freedom	254	
t value	2.87	
p value	0.004426	

Table 2. Area sizes of the zip-code districts of respondents who visited/did not visit mountains in the emergency period.

Area Sizes of the Zip-Code Districts	Visited	Did Not Visit
Average (ha)	213.08	147.03
Variance	197,761.2	65,189.4
Number of respondents	212	1031
Degree of freedom	240	
t value	2.09	
p value	0.037421	

Regarding the area size of the zip-code district, respondents who visited forestlands and mountain areas lived in zip-code districts whose average area size was larger than that of respondents who did not visit (t-test, $p < 0.05$) (Table 2). Because zip-code area sizes of peri-urban areas are generally larger than those of urban areas, it can be assumed that the respondents who visited peri-urban forestlands and mountains live in peri-urban areas within the research site. In peri-urban areas, the residents may have relatively large residential land for their houses and gardens. In addition to access to forestlands, access to nature, such as plants in the garden, may be a condition that familiarizes the residents with nature and influences them to visit natural lands such as mountains.

In order to identify the distribution pattern of residential areas of respondents who visited peri-urban forestlands and mountains during the emergency period, the distribution of the respondents' zip-code districts was visualized; the results are shown in Figure 5. The figure also shows the zip-code districts of those who did not visit. As we mentioned previously, the smaller zip-code districts are located in urbanized areas. The central area of Nagoya City, which is the capital city of the research site, Aichi Prefecture, is indicated in Figure 5 with a red circle, and the sizes of zip-code areas are relatively small compared with those of the surrounding areas. As shown on the map, the zip-code districts of respondents who visited peri-urban forestlands and mountains (green-colored areas) are mainly located outside of the red circle, and their sizes are relatively large. This result may support the assumptions that we made based on the analysis of forestland ratios and sizes of zip-code areas, showing that residential areas of respondents who visited peri-urban forestlands and mountains are concentrated in peri-urban areas, not in urbanized areas.

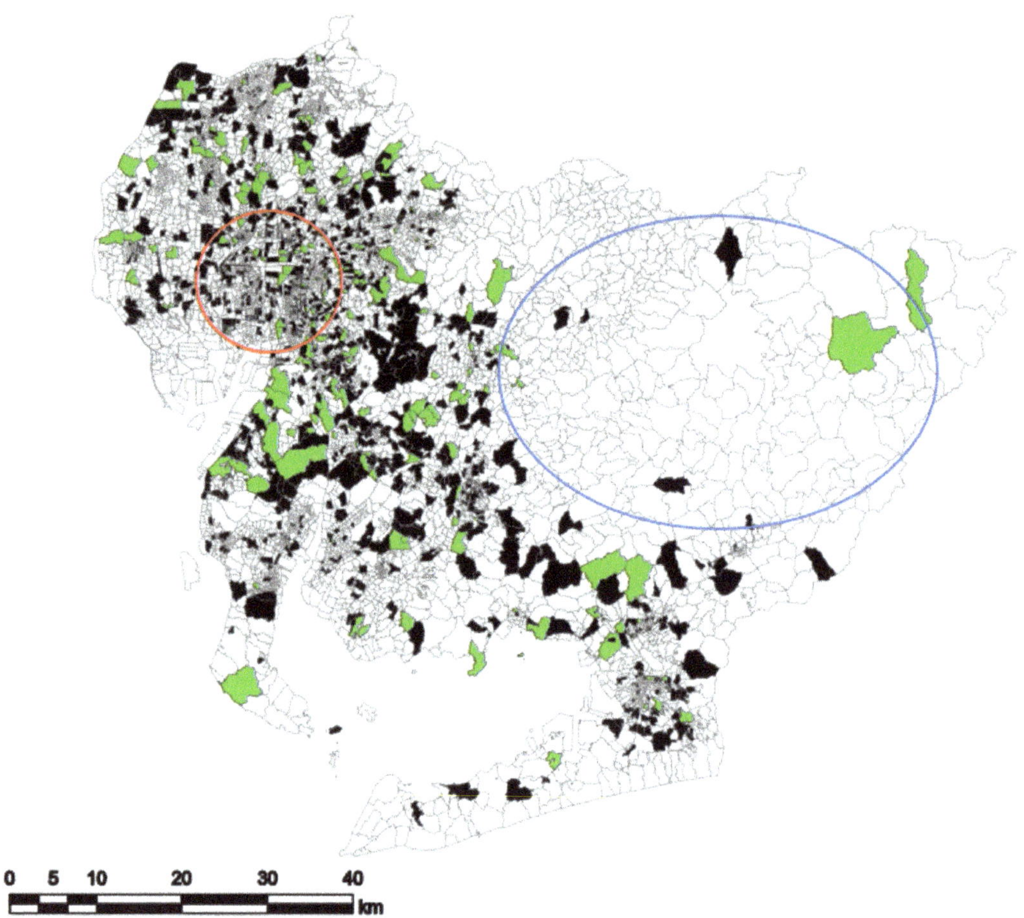

Figure 5. Distribution of zip-code districts of respondents who visited (green)/did not visit (black) forestlands and mountains during the emergency period in Aichi Prefecture. Note: Red circle shows the location of central urbanized area of the capital city (Nagoya City) of the prefecture and blue circle indicates the location of mountainous area with less population.

The analysis results of the respondents' environmental conditions imply that residents who do not have prior rich experiences of visiting forestlands and mountains might not feel encouraged to visit them during the pandemic.

Regarding awareness, we asked residents concerning their pre- and post-emergency-period awareness of forest functions. Specifically, we used the questionnaire to survey the residents' expectations concerning forest functions. We found that the ratios of respondents who expected certain functions differ between respondents who visited and those who did not visit peri-urban forestlands and mountains. As an overall trend, the difference between the residents' pre- and post-emergency-period expectations was relatively trivial. The different degrees of expectations between the two respondent groups were more evident among functions such as "providing a relaxing space," "providing an educational space," and "purification of air and reduction of noise." Before and after the emergency period, the ratios of respondents who visited forestlands and mountains and expected such functions were higher than those of respondents who did not visit. After the period, the differences became larger for functions such as "providing a relaxing space" and "providing an educational space," as shown in Figure 6.

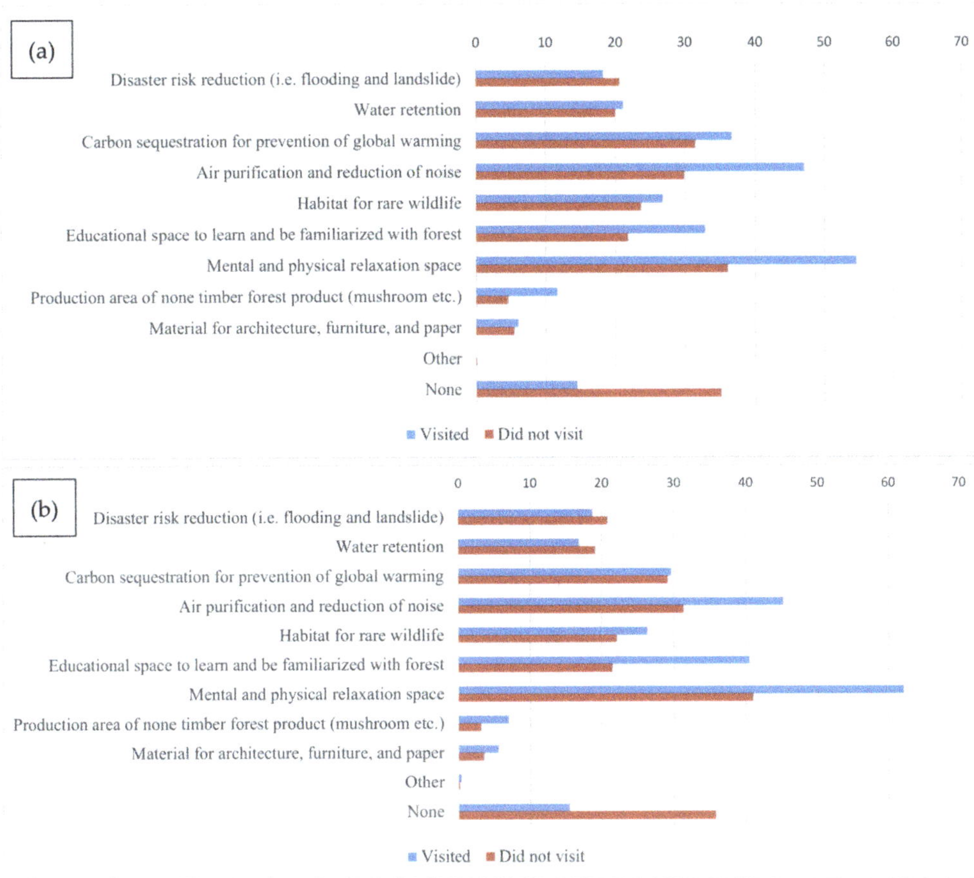

Figure 6. Awareness of the functions of forestlands and mountain areas (Respondents were asked about their awareness (**a**) before and (**b**) after the emergency period).

The results suggest that those who visited peri-urban forestlands and mountains highly expected these areas to serve mental- and education-related functions. This level of expectation may have encouraged the residents to visit those areas during the emergency period.

To identify the factors influencing awareness, we performed a binomial logistic regression analysis of the residents' responses. The results are shown in Tables 3 and 4. Statistically significant models were not detected for functions including "Disaster risk reduction [i.e., flooding and landslide]," "Habitat for rare wildlife," "Material for architecture, furniture, and paper," and "Other," meaning that awareness of these functions was not influenced by specific socio-economic attributes, environmental conditions, or whether they visited forestlands and mountains or not. The respondents' awareness of the other functions was influenced by these factors. Whether the respondents visited the forestlands and mountains or not was a statistically significant variable in most of the models for pre- and post-emergency-period awareness. This suggests that visiting such places can influence people's awareness of their functions. Several socio-economic environmental variables were not statistically significant in influencing the respondents' pre-emergency-period awareness, even if they were significant for their awareness before the period. This implies that whether the respondents visited forestlands and mountains is a relatively strong factor influencing their awareness of the functions of these places, especially after the emergency period. Regarding other variables, household income was significant only in the model of awareness related to water retention. The respondents with higher income were more aware of that function. Being female was also a major influential factor alongside whether the respondents visited forestlands and mountains or not, and its coefficient was high and positive in the model of awareness related to mental and physical relaxation space. This suggests that the female respondents tended to be aware of the mental and physical relaxation function of forestlands and mountain areas. The coefficient of age was high in the model of awareness related to water retention. The older respondents tended to be more aware of that function. The number of children was significant in the model of awareness related to "educational space where ones can learn about and familiarize with forest." The respondents with one or more children were more aware of this function. Environmental conditions were significant only in the pre-emergency-period models. The forestland ratio had a negative coefficient in the carbon sequestration model (Table 3). This implies that the respondents living in zip-code districts with fewer forestlands (such as urbanized districts) tended to be aware of carbon sequestration. Concerning the size of the zip-code area, the results suggest that the respondents living in smaller zip-code districts, which are often located near or inside urban areas, tended to be more aware of the educational function of forestlands. Moreover, the results also show that the respondents living in larger zip-code districts, which are often located in rural areas, tended to be more aware of the production of non-timber products (mushroom, etc.) function (Table 3).

Table 3. Results of logistic regression analysis on the awareness of the functions of forestlands: before the emergency period.

	Income			Sex			Age			Number of Children (>1 or 0)			Forestland Ratio			Zip Code Area size			Visited Forest or Not			Overall Model Test					
	Estimate	p	Odds Ratio	Estimate	p	Odds Ratio	Estimate	p	Odds Ratio	Estimate	p	Odds Ratio	Estimate	p	Odds Ratio	Estimate	p	Odds Ratio	Estimate	p	Odds Ratio	AIC	Deviance	R^2McF	χ^2	df	p
2	0.10380	0.033	1.109	−0.3207	0.074	0.726	0.02002	0.002	1.020													979	963	0.0243	24	7	0.001
3				0.3161	0.038	1.372	0.0131	0.018	1.013				−0.014	0.057	0.986				0.372	0.037	1.450	1204	1188	0.0141	17	7	0.017
4																			0.681	<0.001	1.975	1189	1173	0.0154	18.4	7	0.01
6				0.3159	0.061	1.371	−0.01214	0.045	0.988	0.699	<0.001	2.012				−0.001	0.014	0.999	0.691	<0.001	1.995	1018	1002	0.038	39.6	7	<0.001
7				0.4540	0.002	1.575													0.845	<0.001	2.328	1255	1239	0.0266	33.9	7	<0.001
8																0.001	0.075	1.001	0.983	0.002	2.672	396	380	0.0594	24	7	0.001
11				−0.4172	0.007	0.659													−1.135	<0.001	0.322	1178	1162	0.0331	39.8	7	<0.001

Note: Each model number shows the following functions of forestlands: 2. Water retention, 3. Carbon sequestration for prevention of global warming, 4. Air purification and reduction of noise, 6. Educational space to learn and be familiarized with forest, 7. Mental and physical relaxation space, 8. Production area of none timber forest product (mushroom etc.), 11. None. R^2McF is McFadden's R^2. The independent variable for sex is a categorical variable and its categories are male (1) and female (2). A positive estimated value of its coefficient indicates that females tend to be more aware of a function of forestlands than males. The table shows coefficients and other related values of significant variables ($p < 0.1$).

Table 4. Results of logistic regression analysis on the awareness of forestland functions: after the emergency period.

	Income			Sex			Age			Number of Children (>1 or 0)			Forestland Ratio			Zip Code Area Size			Visited Forest or Not			Overall Model Test					
	Estimate	p	Odds Ratio	Estimate	p	Odds Ratio	Estimate	p	Odds Ratio	Estimate	p	Odds Ratio	Estimate	p	Odds Ratio	Estimate	p	Odds Ratio	Estimate	p	Odds Ratio	AIC	Deviance	R^2McF	χ^2	df	p
2	0.12992	0.009	1.139				0.01982	0.003	1.020										−0.529	0.028	0.589	933	917	0.0224	21	7	0.004
4																			0.634	<0.001	1.885	1210	1194	0.0148	18	7	0.012
6				0.3735	0.026	1.453				0.579	0.002	1.784							1.092	<0.001	2.980	1032	1016	0.0478	51	7	<0.001
7				0.4957	<0.001	1.642													1.041	<0.001	2.831	1274	1258	0.0369	48.2	7	<0.001
8																			0.856	0.017	2.353	329	313	0.0377	12.3	7	0.092
11				−0.4849	0.002	0.616													−1.067	<0.001	0.344	1187	1171	0.0316	38.2	7	<0.001

Note: Each model number shows the following functions of forestlands: 2. Water retention, 4. Air purification and reduction of noise, 6. Educational space to learn and be familiarized with forest, 7. Mental and physical relaxation space, 8. Production area of none timber forest product (mushroom etc.), 11. None. R^2McF is McFadden's R^2. The independent variable for sex is a categorical variable and its categories are male (1) and female (2). A positive estimated value of its coefficient indicates that females tend to be more aware of a function of forestlands than males. The table shows coefficients and other related values of significant variables ($p < 0.1$).

4. Discussion and Conclusions

Our results suggest that residents who do not have rich experiences of visiting natural lands or access to such environments may lack the motivation to visit forestlands and mountains. This may mean that having such experiences and access to natural environments are essential factors influencing residents to visit forestlands and mountains to maintain good mental and physical health during the pandemic. Therefore, governments and local communities should establish policies and action plans to provide easy access to natural environments.

This study indicated that environmental conditions influence whether residents visit peri-urban forestlands and mountains during the pandemic. Specifically, peri-urban areas with forestlands have such environmental conditions, and the respondents who visited peri-urban forestlands and mountains tended to live in peri-urban areas.

Furthermore, the respondents' expectations concerning forest functions were also identified as factors that influenced residents to visit those places. The respondents who visited such areas had relatively high expectations that visiting forests would benefit their mental health and educational purposes. Environmental conditions may influence residents to have such expectations. The results of the logistic regression analysis show that visiting forestlands correlated with awareness of various types of forest functions, and the correlation between expectations, awareness, and whether or not the respondents visited forestlands became stronger after the emergency period. Regarding other correlations between respondents' attributes and forest function expectations, female respondents tended to be more aware of forestlands as mental and physical relaxation spaces, while respondents with one or more children tended to be aware of the educational functions of the forest. Although certain correlations between residents' environmental conditions and expectations or awareness of forestland functions were detected, in future research, these relationships need to be explored in more depth, considering that the value and meaning of forestlands differs from one country to another [30–32].

This paper contributes to the existing literature on environmental justice, specifically that related to the use of forest ecosystem services. Existing studies have examined the status of environmental justice in different socio-economic contexts [33,34], and the need to provide policy implications increased during the pandemic [35]. Additionally, extant studies e.g., [36] have also highlighted that environmental conditions are influential factors determining the behaviors of citizens in different regions. Nevertheless, studies focusing on environmental conditions are limited in this line of research. Future studies on forest visits and their environmental conditions can collaborate with the studies in different academic fields, such as public health [37], to analyze the effects of visiting forestlands on physical and mental health during the pandemic. Concerning the conservation of forest ecosystems and their biodiversity, the relationships between forest ecosystem services and biodiversity need to be identified to provide a basis for conservation policies and actions by citizens [38]. Since existing studies e.g., [39,40] have addressed the upscaling of valuations of forest ecosystem services, further research that considers visiting forestlands as use of forest ecosystem services also needs to be upscaled to provide policy implications for larger scales, such as country-level policy implications, considering ecosystem services and disservices [41].

"The extinction of experience" is a concern when discussing the essential role of having human–nature experiences in improving residents' understanding of biodiversity and ecosystem conservation. Although there are ongoing discussions about this concern in existing literature e.g., [42–44], such experience remains endangered, especially in urban areas and developed countries. An existing empirical analysis using large-scale official municipality data collected via a questionnaire-based residents' survey (cf. [45,46] for Sendai City) points to a decline in younger generations' interest in and recognition of the importance of human–nature interactions. During times of intense physical mobility limitations and mental pressures, experiences with nature can help improve residents' knowledge about the role of nature and increase their appreciation for conservation efforts.

In particular, options to address mental and physical health issues during the pandemic have been limited, especially for the vulnerable groups. Supporting, facilitating, and encouraging such groups to visit peri-urban forestlands and mountains may be a good alternative to both address their health issues and alleviate the social and environmental disparities between residents of differing socio-economic statuses regarding access to natural spaces. Further, discussions and future analyses around this issue could contribute to the debate related to the extinction of experiences and maldistribution of ecosystem services in social and environmental dimensions. Long-term analyses of similar issues in different world regions with or without continued mobility limitations are also needed.

Author Contributions: Conceptualization, Y.U. and R.K.; methodology, Y.U.; formal analysis, Y.U.; investigation, Y.U. and R.K.; writing-original draft preparation, Y.U.; writing-review and editing, Y.U. and R.K.; project administration, R.K.; funding acquisition, R.K. All authors have read and agreed to the published version of the manuscript.

Funding: This research was funded by the Japan Society for the Promotion of Science: JSPS KAKENHI Grant Numbers: JP17K02105; JP20K12398; 22H03852, JST RISTEX Grant Number JPMJRX20B3.

Data Availability Statement: The data presented in this study are available on request basis.

Acknowledgments: We would like to thank Yasushi Shoji who gave us valuable comments on the questionnaire.

Conflicts of Interest: The authors declare no conflict of interest.

References

1. Derks, J.; Giessen, L.; Winkel, G. COVID-19-induced visitor boom reveals the importance of forests as critical infrastructure. *For. Policy Econ.* **2020**, *118*, 102253. [CrossRef] [PubMed]
2. Honey-Rosés, J.; Anguelovski, I.; Chireh, V.K.; Daher, C.; Konijnendijk van den Bosch, C.; Litt, J.S.; Mawani, V.; McCall, M.K.; Orellana, A.; Nieuwenhuijsen, M.J.; et al. The impact of COVID-19 on public space: An early review of the emerging questions–design, perceptions and inequities. *Cities Health* **2021**, *5*, S263–S279. [CrossRef]
3. Ugolini, F.; Massetti, L.; Calaza-Martínez, P.; Cariñanos, P.; Dobbs, C.; Ostoić, S.K.; Marin, A.M.; Pearlmutter, D.; Saaroni, H.; Šaulienė, I.; et al. Effects of the COVID-19 pandemic on the use and perceptions of urban green space: An international exploratory study. *Urban For. Urban Green.* **2020**, *56*, 126888. [CrossRef] [PubMed]
4. Venter, Z.; Barton, D.; Gundersen, V.; Figari, H.; Nowell, M. Urban nature in a time of crisis: Recreational use of green space increases during the COVID-19 outbreak in Oslo, Norway. *Environ. Res. Lett.* **2020**, *15*, 104075. [CrossRef]
5. Freeman, S.; Eykelbosh, A. COVID-19 and outdoor safety: Considerations for use of outdoor recreational spaces. *Natl. Collab. Cent. Environ. Health* **2020**, *829*, 1–15.
6. Xie, J.; Luo, S.; Furuya, K.; Sun, D. Urban Parks as Green Buffers During the COVID-19 Pandemic. *Sustainability* **2020**, *12*, 6751. [CrossRef]
7. Uchiyama, Y.; Kohsaka, R. Access and Use of Green Areas during the COVID-19 Pandemic: Green Infrastructure Management in the "New Normal". *Sustainability* **2020**, *12*, 9842. [CrossRef]
8. Uchiyama, Y.; Hayashi, K.; Kohsaka, R. Typology of cities based on city biodiversity index: Exploring biodiversity potentials and possible collaborations among Japanese cities. *Sustainability* **2015**, *7*, 14371–14384. [CrossRef]
9. Uchiyama, Y.; Kohsaka, R. Application of the City Biodiversity Index to populated cities in Japan: Influence of the social and ecological characteristics on indicator-based management. *Ecol. Indic.* **2019**, *106*, 105420. [CrossRef]
10. Verdú-Vázquez, A.; Fernández-Pablos, E.; Lozano-Diez, R.V.; López-Zaldívar, Ó. Green space networks as natural infrastructures in PERI-URBAN areas. *Urban Ecosyst.* **2021**, *24*, 187–204. [CrossRef]
11. Ernstson, H. The social production of ecosystem services: A framework for studying environmental justice and ecological complexity in urbanized landscapes. *Landsc. Urban Plan.* **2013**, *109*, 7–17. [CrossRef]
12. Andersson, E.; Borgström, S.; Haase, D.; Langemeyer, J.; Mascarenhas, A.; McPhearson, T.; Wolff, M.; Łaszkiewicz, E.; Kronenberg, J.; Barton, D.N.; et al. A context-sensitive systems approach for understanding and enabling ecosystem service realization in cities. *Ecol. Soc.* **2021**, *26*, 35. [CrossRef]
13. García-Nieto, A.P.; García-Llorente, M.; Iniesta-Arandia, I.; Martín-López, B. Mapping forest ecosystem services: From providing units to beneficiaries. *Ecosyst. Serv.* **2013**, *4*, 126–138. [CrossRef]
14. Wang, S.; Fu, B. Trade-offs between forest ecosystem services. *For. Policy Econ.* **2013**, *26*, 145–146. [CrossRef]
15. Dobbs, C.; Escobedo, F.J.; Zipperer, W.C. A framework for developing urban forest ecosystem services and goods indicators. *Landsc. Urban Plan.* **2011**, *99*, 196–206. [CrossRef]
16. Nielsen, T.S.; Hansen, K.B. Do green areas affect health? Results from a Danish survey on the use of green areas and health indicators. *Health Place* **2007**, *13*, 839–850. [CrossRef]

17. Kaźmierczak, A. The contribution of local parks to neighbourhood social ties. *Landsc. Urban Plan.* **2013**, *109*, 31–44. [CrossRef]
18. Wolch, J.R.; Byrne, J.; Newell, J.P. Urban green space, public health, and environmental justice: The challenge of making cities 'just green enough'. *Landsc. Urban Plan.* **2014**, *125*, 234–244. [CrossRef]
19. Kabisch, N.; Haase, D. Green justice or just green? Provision of urban green spaces in Berlin, Germany. *Landsc. Urban Plan.* **2014**, *122*, 129–139. [CrossRef]
20. Green, H.; Fernandez, R.; MacPhail, C. The social determinants of health and health outcomes among adults during the COVID-19 pandemic: A systematic review. *Public Health Nurs.* **2021**, *38*, 942–952. [CrossRef]
21. Hino, K.; Asami, Y. Change in walking steps and association with built environments during the COVID-19 state of emergency: A longitudinal comparison with the first half of 2019 in Yokohama, Japan. *Health Place* **2021**, *69*, 102544. [CrossRef] [PubMed]
22. Soga, M.; Evans, M.J.; Tsuchiya, K.; Fukano, Y. A room with a green view: The importance of nearby nature for mental health during the COVID-19 pandemic. *Ecol. Appl.* **2021**, *31*, e2248. [CrossRef]
23. Nadkarni, A.; Hasler, V.; AhnAllen, C.G.; Amonoo, H.L.; Green, D.W.; Levy-Carrick, N.C.; Mittal, L. Telehealth during COVID-19—Does everyone have equal access? *Am. J. Psychiatry* **2020**, *177*, 1093–1094. [CrossRef] [PubMed]
24. Spotswood, E.N.; Benjamin, M.; Stoneburner, L.; Wheeler, M.M.; Beller, E.E.; Balk, D.; McPhearson, T.; Kuo, M.; McDonald, R.I. Nature inequity and higher COVID-19 case rates in less-green neighbourhoods in the United States. *Nat. Sustain.* **2021**, *4*, 1092–1098. [CrossRef]
25. Pipitone, J.M.; Jović, S. Urban green equity and COVID-19: Effects on park use and sense of belonging in New York City. *Urban For. Urban Green.* **2021**, *65*, 127338. [CrossRef]
26. Spencer, L.H.; Lynch, M.; Lawrence, C.L.; Edwards, R.T. A scoping review of how income affects accessing local green space to engage in outdoor physical activity to improve well-being: Implications for post-COVID-19. *Int. J. Environ. Res. Public Health* **2020**, *17*, 9313. [CrossRef] [PubMed]
27. Pforr, K.; Dannwolf, T. What do we lose with online-only surveys? Estimating the bias in selected political variables due to online mode restriction. *Stat. Polit. Policy* **2017**, *8*, 105–120. [CrossRef]
28. Islam, M.S.; Ferdous, M.Z.; Potenza, M.N. Panic and generalized anxiety during the COVID-19 pandemic among Bangladeshi people: An online pilot survey early in the outbreak. *J. Affect. Disord.* **2020**, *276*, 30–37. [CrossRef]
29. Oku, K.; Fukamachi, K. Understanding the image of forest recreation users through structural analysis of visiting motives. *Landsc. Stud.* **2003**, *66*, 693–698. (In Japanese)
30. Kohsaka, R.; Flitner, M. Exploring forest aesthetics using forestry photo contests: Case studies examining Japanese and German public preferences. *For. Policy Econ.* **2004**, *6*, 289–299. [CrossRef]
31. Kohsaka, R.; Handoh, I.C. Perceptions of "close-to-nature forestry" by German and Japanese groups: Inquiry using visual materials of "cut" and "dead" wood. *J. For. Res.* **2006**, *11*, 11–19. [CrossRef]
32. Kovács, B.; Uchiyama, Y.; Miyake, Y.; Penker, M.; Kohsaka, R. An explorative analysis of landscape value perceptions of naturally dead and cut wood: A case study of visitors to Kaisho Forest, Aichi, Japan. *J. For. Res.* **2020**, *25*, 291–298. [CrossRef]
33. Schwarz, K.; Berland, A.; Herrmann, D.L. Green, but not just? Rethinking environmental justice indicators in shrinking cities. *Sustain. Cities Soc.* **2018**, *41*, 816–821. [CrossRef]
34. Kronenberg, J.; Haase, A.; Łaszkiewicz, E.; Antal, A.; Baravikova, A.; Biernacka, M.; Dushkova, D.; Filčak, R.; Haase, D.; Ignatieva, M.; et al. Environmental justice in the context of urban green space availability, accessibility, and attractiveness in postsocialist cities. *Cities* **2020**, *106*, 102862. [CrossRef]
35. Oscilowicz, E.; Anguelovski, I.; Triguero-Mas, M.; García-Lamarca, M.; Baró, F.; Cole, H.V.S. Green justice through policy and practice: A call for further research into tools that foster healthy green cities for all. *Cities Health* **2022**, 1–16. [CrossRef]
36. Campbell, M.; Marek, L.; Wiki, J.; Hobbs, M.; Sabel, C.E.; McCarthy, J.; Kingham, S. National movement patterns during the COVID-19 pandemic in New Zealand: The unexplored role of neighbourhood deprivation. *J. Epidemiol. Community Health* **2021**, *75*, 903–905. [CrossRef]
37. Kyan, A.; Takakura, M. Socio-economic inequalities in physical activity among Japanese adults during the COVID-19 pandemic. *Public Health* **2022**, *207*, 7–13. [CrossRef]
38. Brockerhoff, E.G.; Barbaro, L.; Castagneyrol, B.; Forrester, D.I.; Gardiner, B.; González-Olabarria, J.R.; Lyver, P.O.; Meurisse, N.; Oxbrough, A.; Taki, H.; et al. Forest biodiversity, ecosystem functioning and the provision of ecosystem services. *Biodivers. Conserv.* **2017**, *26*, 3005–3035. [CrossRef]
39. Grammatikopoulou, I.; Vačkářová, D. The value of forest ecosystem services: A meta-analysis at the European scale and application to national ecosystem accounting. *Ecosyst. Serv.* **2021**, *48*, 101262. [CrossRef]
40. Orsi, F.; Ciolli, M.; Primmer, E.; Varumo, L.; Geneletti, D. Mapping hotspots and bundles of forest ecosystem services across the European Union. *Land Use Policy* **2020**, *99*, 104840. [CrossRef]
41. Dobbs, C.; Kendal, D.; Nitschke, C.R. Multiple ecosystem services and disservices of the urban forest establishing their connections with landscape structure and sociodemographics. *Ecol. Indic.* **2014**, *43*, 44–55. [CrossRef]
42. Oh, R.R.Y.; Fielding, K.S.; Carrasco, R.L.; Fuller, R.A. No evidence of an extinction of experience or emotional disconnect from nature in urban Singapore. *People Nat.* **2020**, *2*, 1196–1209. [CrossRef]
43. Soga, M.; Yamanoi, T.; Tsuchiya, K.; Koyanagi, T.F.; Kanai, T. What are the drivers of and barriers to children's direct experiences of nature? *Landsc. Urban Plan.* **2018**, *180*, 114–120. [CrossRef]
44. Miller, J.R. Biodiversity conservation and the extinction of experience. *Trends Ecol. Evol.* **2005**, *20*, 430–434. [CrossRef]

45. Imai, H.; Nakashizuka, T.; Kohsaka, R. An analysis of 15 years of trends in children's connection with nature and its relationship with residential environment. *Ecosyst. Health Sustain.* **2018**, *4*, 177–187. [CrossRef]
46. Imai, H.; Nakashizuka, T.; Kohsaka, R. A Multi-Year Investigation of the Factors Underlying Decreasing Interactions of Children and Adults with Natural Environments in Japan. *Hum. Ecol.* **2019**, *47*, 717–731. [CrossRef]

Article

New Ecological Paradigm, Leisure Motivation, and Wellbeing Satisfaction: A Comparative Analysis of Recreational Use of Urban Parks before and after the COVID-19 Outbreak

Yanju Luo [1], Jinyang Deng [2,*], Chad Pierskalla [2], Ju-hyoung Lee [3] and Jiayao Tang [4]

[1] School of Tourism, Hainan Normal University, No. 99 Longkun South Road, Haikou 571158, China; luoyanju@hainnu.edu.cn

[2] Recreation, Parks and Tourism Resources Program, School of Natural Resources, West Virginia University, Morgantown, WV 26506, USA; cpierska@wvu.edu

[3] Department of Forest Resources, Yeungnam University, 280 Daehak-ro, Gyeongsan 38541, Korea; jhlee9@yu.ac.kr

[4] School of Architecture, Academy of Art University, 16076 Blazewood Way, San Diego, CA 92127, USA; jtang27@art.edu

* Correspondence: jinyang.deng@mail.wvu.edu

Abstract: The COVID-19 pandemic has created an opportunity for us to rethink the relationship between humans and the environment. However, few studies have examined the association between environmental attitudes, motivations, wellbeing, and quality of life in the context of urban green areas before and after the outbreak of COVID-19. This paper investigated the interrelationships among these variables based on data collected in 2019 (before COVID-19) and 2021 (after COVID-19). The results show that the 2021 sample differed significantly from the 2019 sample in environmental attitudes. Respondents after the outbreak with the belief in "humans with nature" were more likely to use urban green areas for being "close to nature" than pre-pandemic respondents. In addition, stronger belief in "humans over nature" led to stronger desire for "social interactions" in 2021 than in 2019, implying a close relationship between people's perception of humankind's ability to control nature during the pandemic and their desire to interact with people in urban green areas. The study also found that there may be a pent-up satisfaction among urban dwellers after the COVID-19 outbreak.

Keywords: COVID-19; NEP; motivation; urban green areas; SEM

Citation: Luo, Y.; Deng, J.; Pierskalla, C.; Lee, J.-h.; Tang, J. New Ecological Paradigm, Leisure Motivation, and Wellbeing Satisfaction: A Comparative Analysis of Recreational Use of Urban Parks before and after the COVID-19 Outbreak. *Land* **2022**, *11*, 1224. https://doi.org/10.3390/land11081224

Academic Editors: Francesca Ugolini and David Pearlmutter

Received: 27 June 2022
Accepted: 1 August 2022
Published: 3 August 2022

Publisher's Note: MDPI stays neutral with regard to jurisdictional claims in published maps and institutional affiliations.

1. Introduction

The impacts of urban green areas on urban residents' wellbeing and health have been extensively examined in the literature. It is widely recognized that an individual's wellbeing and satisfaction are affected by factors such as accessibility, motivation, frequency of visits, duration of use, and quality and quantity of urban green areas (among others). Adding to this list is the COVID-19 pandemic, which has spread globally since it initially emerged in January 2020 in China. The pandemic has stimulated a pent-up demand not only for nature-based outdoor recreation and tourism, but also for recreational use of urban green areas. While there is a plethora of studies that have examined the association between the pandemic and residents' use and health related to urban green areas, very few studies have examined the association from the perspective of environmental attitudes in a comparative manner. Thus, this paper seeks to fill this research gap.

Humans have a biological need to be close to nature. This biophilic nature of human beings has significant implications not only for the planning and management of urban green areas, but also for the enhancement and promotion of public health and wellbeing, given that contact with nature in many communities is largely limited to local trees, parks, and green areas nearby. This is particularly true during the COVID-19 pandemic, wherein most countries/regions have imposed lockdowns and travel restrictions, which have significantly impacted people's recreational behaviors. For example, a study found that more people in the USA stayed closer to home for leisure and recreation, with 49.9% travelling within two miles as opposed to 10.8% prior to 11 March 2020 [1]. In Belgium, people tended to use urban green areas more frequently during the lockdown than prepandemic [2]. Another study involving five countries also found that people tended to use urban green areas quite often during the COVID-19 pandemic [3].

While the COVID-19 pandemic has unprecedentedly affected every aspect of our societies and people's daily lives, it has also created an opportunity for us to rethink the relationship between humans and nature and revisit models developed and used to examine human behaviors before the pandemic [4]. On the one hand, people may be more aware of the importance of urban green areas for public health and more motivated to recreate in urban green areas, which, in turn, may reinforce their attitudes toward the environment. On the other hand, reduced economic activities, less energy consumption, and less human movements and commuting following lockdowns and travel restrictions during the pandemic may "have a positive impact on the environment" [5] (p. 2) due to greenhouse gas emissions, air pollution, wastes, and noises being significantly reduced [6–9]. This "incidental" positive impact due to the pandemic may have significant implications for the long-term sustainability as it may trigger a transformative change of people's attitudes and behaviors toward nature and the environment. However, there is a lack of "detailed understanding of how large scale social-ecological upheaval impacts the values and benefits associated with human-nature relationships" [10] (p. 2). Thus, more research is needed to understand the association between these underlying values and benefits and increased nature-based activities [10].

Urban green areas can serve as an ideal platform by which the association between attitudes, motivations, benefits, and recreational use can be examined before and after the COVID-19 outbreak. This assumption follows the value-attitude-behavior model, which implies that "the influence should theoretically flow from abstract values to mid-range attitudes to specific behavior" [11] (p. 638). However, few studies, if any, have been conducted to comparatively examine the interrelationships among these variables based on samples surveyed before and after the outbreak. To this end, this study examined how environmental attitudes measured by the New Ecological Paradigm (NEP) [12] would influence leisure motivations, and how leisure motivations would influence wellbeing satisfaction measured by the Personal Wellbeing Index-Adult (PWI-A) [13], and how the latter would further influence quality of life in the context of urban green areas. Specifically, this study is guided by following questions:

(1) Did residents' recreational use of urban parks differ significantly before and after the outbreak?
(2) Did residents' environmental attitudes, leisure motivations, and PWI-A differ significantly before and after the outbreak?
(3) Did environmental attitudes influence leisure motivations, which, in turn, predicted wellbeing satisfaction measured by PWI-A?
(4) Did PWI-A further influence quality of life?
(5) Did the relationship strengths between two variables for the 2019 sample and 2021 sample differ significantly?

To answer the five questions above, Haihou, the capital city of Hainan Province, China, was chosen as the study area for this research for two reasons. First, the city has a tropical climate with four seasons being not distinct as those in many parts of China. Thus, the recreational use of urban green areas in the city is less affected by seasons. Second, as with many other cities in the country, lockdown was enforced from 24 January 2020 to 26 February 2020 in the city to control the spread of COVID-19. During this period, a total of 39 cases were confirmed. Only two cases were reported after the lockdown (one in June 2021 and one in August 2021). As of 31 December 2021, a total of 41 cases were reported with 0 deaths. Since the second survey was completed in July 2021 (where 0 cases were reported), the 2021 sample can be considered post- pandemic while the sample surveyed in 2019 can be considered pre- pandemic for the study area.

The rest of the paper is organized in a way that a review of the relevant literature is presented first (which sets the basis for the development of four hypotheses), followed by a description of the methods used and presentation of the results along with a discussion of findings. Finally, the paper is wrapped up with main conclusions along with research implications (theoretical, methodological, and managerial) as well as research limitations and future research needs. Figure 1 displays the flowchart that shows the overall structure of the remaining paper (Figure 1).

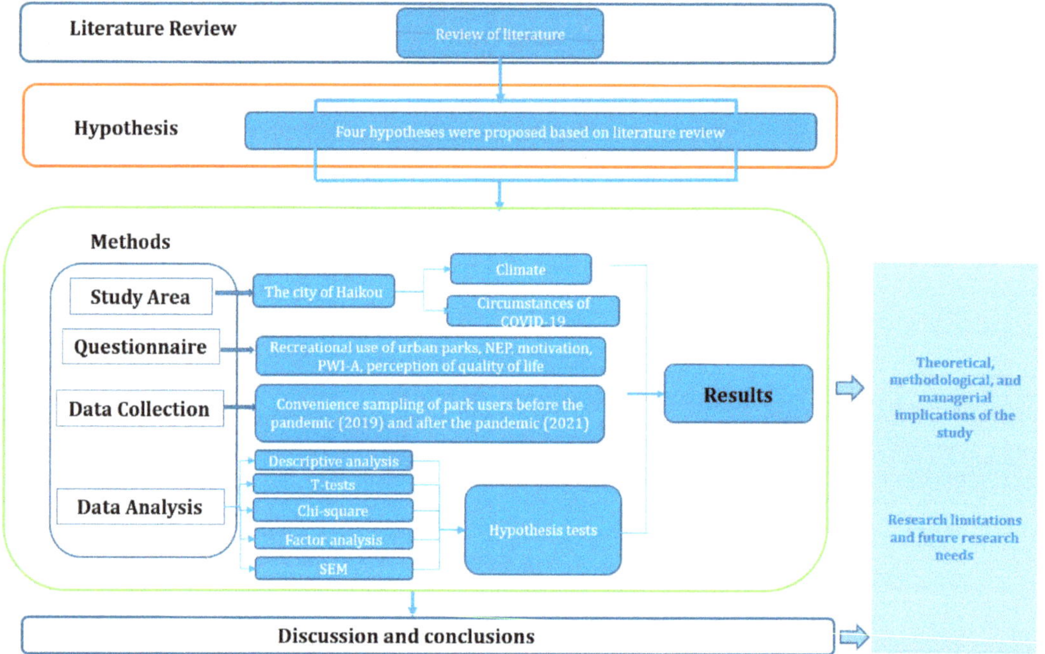

Figure 1. A schematic flowchart of the study.

2. Literature Review

2.1. COVID-19 Pandemic and Environmental Attitudes/Awareness

Several studies have examined people's environmental awareness and attitudes related to the pandemic, albeit not in the study field of urban green areas nor in the context of China. For example, one study [9] investigated public awareness of nature and the environment in 20 European countries based on online search behaviors retrieved from the Google Trends. This study found that online searches of nature-related topics (forest, birds, nature, biodiversity, gardening, and vegetable plot) increased significantly due to the COVID-19 outbreak. A second study [7] also reported a positive change of perceptions of the natural environment due to the pandemic-related lockdown in three European countries (England, Ireland, and Spain). Additionally, a third study [5] reported an increased awareness in air pollution, environmental impact, recycling, water consumption, and natural resources in Brazil and Portugal as a result of the pandemic.

There are two other studies that are specifically related to NEP. The first one [14] examined COVID-19 risk management behavior as it relates to pro-environmental attitudes measured by NEP, finding that pro-environmental behaviors are positively related to the moral obligation of protecting others from COVID-19 risks. The second study [15] found that while people's attitudes in Germany toward the environment measured by NEP tended to be more positive during the pandemic than before, there were no significant differences in subscales such as "limits of growth" and "anti-anthropocentrism". The increase in environmental concerns were mainly driven by three other subscales: "balance of nature", "anti-exemptionalism", and "eco-crisis". Thus, people's attitudes toward the environment may differ before and after the pandemic, leading to Hypothesis 1 being proposed as follows:

Hypothesis 1: *People's environmental attitudes measured by NEP would differ significantly before and after the outbreak of the pandemic.*

2.2. Environmental Attitudes and Leisure/Recreation Motivations

Attitude can be defined as "a learned predisposition to respond in a consistently favorable or unfavorable manner with respect to a given object" [16] (p. 10). Attitude is positioned in the theory of planned behavior (TPB) model as an antecedent to intention or motivation. That is, behavioral intention (motivation to act) is determined by attitude toward the behavior. Prior to the TPB model, attitude as a source of motivation was also discussed in previous studies [17–19]. It is argued that "another way to look at psychological needs is to think of them as motives rooted in systems or complexes of attitudes or values" [17] (p. 80). A study involving 372 students in the southeastern United States explored the impacts of cognitive and affective aspects of attitude on motivations and behaviors [20]. Previous studies also measured environmental attitudes in terms of three latent subscales (conservation and development, conservation priority, and leisure rights), which were found to be "significantly associated with visitors' motivations" [21] (p. 35), suggesting "nature-based visitors' environmental attitudes were related to their motivations for travelling" [21] (p. 35).

Contrary to attitude as a factor that predicts motivation as discussed above, there are studies that treated motivation as a predictor of attitude [19,22,23]. For example, one study [19], after a review of relevant literature, concluded that "attitude toward an act is determined by an individual's motivation to perform the act" (p. 285). Thus, it seems that a consensus on attitude preceding motivation or vice versa has not been reached among researchers. It seems that which one should be treated as a driving factor of the other depends on the research context and questions to be examined as the two concepts are closely related. For example, as for gambling, if one dislikes gambling (negative attitude toward gambling), they will be less likely to gamble in a casino (motivation). In this case, attitude determines motivation. That said, even basic human values which are stable and universally considered as the driving factors of attitudes can be put under the rubric of motivational continuum [24,25], thus supporting the argument of motivation determining attitude made by some researchers [19].

Nonetheless, NEP as a measure of environmental attitudes has been consistently used as a variable that predicts motivation when the relationship between attitudes, motivations and behaviors are examined. This is true in the context of environmental studies and tourism studies as well. For example, a study examined the relationships between environmental attitudes (measured by NEP), non-use values of endangered species, and underlying motivations for willingness to pay (WTP), finding that environmental attitudes significantly influenced respondents' rating of the importance of non-use motivations [26]. A significant relationship between NEP and non-use motivations was also reported in another study [27].

In the context of tourism, NEP has been examined along with motivations in several studies. NEP was found to be significantly associated with a festival motivation towards environment-related films and issues, implying that festival attendees with higher level of pro-environmental attitudes were more likely to attend the festival due to its environmental themes [28]. Another study [29] examined NEP as it relates to nature-based tourism motivations, finding that visitors who scored higher on the subscale "humans over nature" were also more motivated by seeking "novelty/self-development", suggesting that people, particularly the young with active and adventurous tourism pursuits of tourism experience, tended to emphasize the power of humans in conquering nature. In contrast, visitors who were more supportive of "limits to growth" and more concerned about "eco-crisis" were more motivated by passive and appreciative tourism pursuits with a higher desire to return to nature, to learn about nature, and to escape from routines. This suggests "a high level of environmental concern could form a high level of tourism motivation for being close to nature" [29] (p. 399). Extended from these findings, residents' environmental attitudes measured by NEP would also be closely related to their motivations to use urban parks. Thus, the second hypothesis is proposed:

Hypothesis 2: *Peoples' environmental attitudes would significantly influence their motivations to visit urban green areas.*

2.3. Motivations, Satisfaction and Quality of Life

Travel motivation is among the most researched themes in the field of tourism research [30]. It refers to factors that drive a person to participate in a recreation or tourist activity [31]. These factors can be broadly categorized into two dimensions: push and pull [32] with the former referring to internal or intrinsic motives or forces (e.g., facilitation of social interactions, desire to escape daily lives, and rest and relaxation) that cause a person to travel and the latter relating to external or extrinsic motives such as destination attributes and services [33].

Just as motivation and attitude are closely related, so are motivation and satisfaction [34]. Motivation determines people's behaviors and their expectations of products/services [35]. From the perspective of the push dimension that initiates an individual's travel desire, motivation can be defined as "state of need, a condition that exerts a push on the individual towards certain types of action that are seen as likely to bring satisfaction" [36] (p. 16). In this sense, people will feel satisfied if their needs or motives are met [37].

Motivation as an antecedent to satisfaction has been examined in the field of tourism studies [38–43]. Although a positive relationship between motivation and satisfaction was reported in most studies, there are also discrepancies. For example, no significant relationship was found between the two variables in two studies [39,42]. Another study [43] found that the pull and push motivators performed differently in a way that pull travel motivations were not significantly related to satisfaction while push motivations were. In contrast, one study [40] found that both pull and push motivations significantly influence satisfaction. A study in the context of urban parks found that experiencing nature and seeking relaxation positively affected satisfaction while other motivators on educational and cultural activities did not [27]. Thus, "findings on the relationship between motivation and satisfaction are sample specific" [44]. Given that a significant relationship exists between motivation and satisfaction in most studies, it is assumed that leisure motivations would also be significantly related to wellbeing satisfaction measured by PWI-A. In addition, previous studies conducted in both western societies and China consistently found that satisfaction significantly and positively affects quality of life [45,46]. Therefore, two additional hypotheses are proposed:

Hypothesis 3: *Leisure motivations would significantly influence PWI-A.*

Hypothesis 4: *Satisfaction of wellbeing measured by PWI-A would significantly influence quality of life.*

3. Materials and Methods

3.1. Study Area

Haikou, the capital city of China's island province Hainan (often called the "Hawaii" of China), is situated at the north coast of the Hainan island (19°31′32″ N~20°04′52″ N, 110°07′22″ E~110°42′32″ E) with a tropical maritime climate (Figure 2).

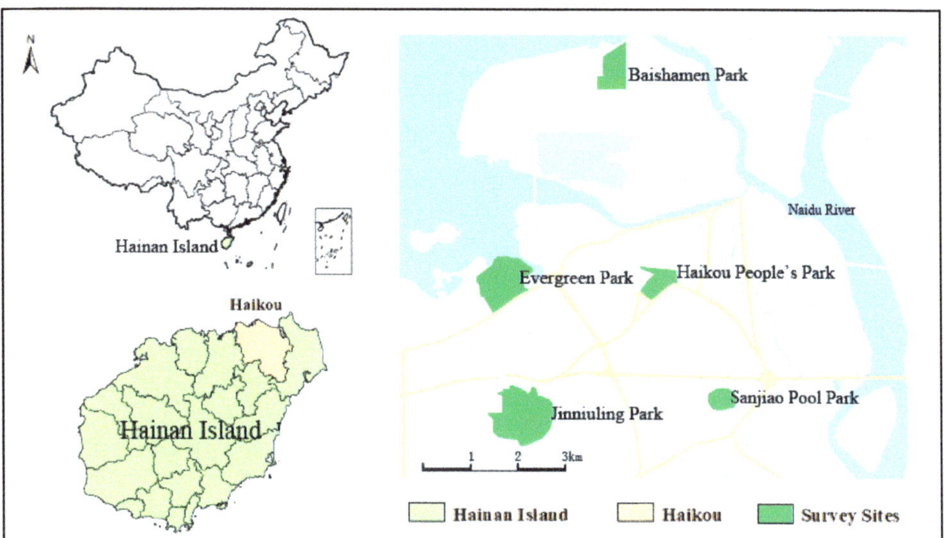

Figure 2. Location of the study area.

Haikou covers 3127 square kilometers with a population of 2.87 million. The city is ranked top in terms of its air quality among 168 major cities nationwide. It is also listed as the "national environmental protection exemplary city", "national garden city", and "international wetland city" with 38.38% of the land area being covered by tree canopy [47]. As with many other cities in the country, all of the city's urban parks and green areas are open to the public for free. Among those most popular urban parks are Evergreen Park, Haikou People's Park, Sanjiao Pool Park, Baishamen Park, and Jinniuling Park, which were selected as the survey sites for this study (Figure 2).

3.2. Questionnaire and Measures

A questionnaire was developed by drawing upon findings from the literature. This questionnaire measures participants' socio-demographics, recreational use of urban green areas, motivations for using urban green areas, attitudes toward the environment, satisfaction of wellbeing, and perceived quality of life. Motivations were measured by eight items which were adopted from a previous study [44]. Environmental attitudes were measured by the widely used and tested NEP [15,48] which consists of 15 items, whereas "agreement with the eight odd-numbered items and disagreement with the seven even-numbered items indicate pro-NEP responses" [12] (p. 433). The seven even numbered items are associated with the Dominant Social Paradigm with a focus on humans over nature, while the eight odd items reflect endorsement of the New Environmental Paradigm that emphasizes humans with nature [48,49]. Satisfaction of wellbeing was measured by seven items, five of which were adopted from the PWI-A [13]. Examples of items include "How satisfied are you with your standard of living?" and "How satisfied are you with your health?" Finally, one item "How would you rate your quality of life?" was used to measure participants' overall quality of life [50]. All of these items were measured on a 5-point Likert scale (1 = strongly disagree or extremely dissatisfied, 5 = strongly agree or extremely satisfied).

3.3. Data Collection

The survey was conducted twice (once before the COVID-19 pandemic in 2019 and once during the pandemic in 2021) using the convenience sampling method by undergraduates and graduates majoring in tourism management from Hainan Normal University under the supervision of the lead author of this paper. Both surveys were carried out during a similar time period (29 June–21 July for 2019 vs. 3 July to 20 July for 2021) and at the same sites: Evergreen Park, East Lake Park, Sanjiao Pool Park, Baishamen Park, and Jinniuling Park as aforementioned. Only local residents were surveyed while outside visitors were excluded. Specifically, prospective respondents were approached by the survey team who introduced themselves and explained the study purpose to individuals who may be willing to help out with the survey. If an individual was unwilling to participate, the survey team then approached the next individual available. If an individual showed willingness, the questionnaire was then presented for on-site completion. Once one survey was completed, the next individual was approached [51].

During the 2019 survey period, a total of 700 individuals 18 years of age or older were approached. Of this number, 635 returned their questionnaires, resulting in a return rate of 90.7%. Of the 635 returned questionnaires, 30 were removed due to systematic incomplete responses and skeptical response patterns (i.e., same rating for variables in a section of the questionnaire), resulting in 605 usable questionnaires for analysis.

For the 2021 survey period, both the hardcopy questionnaire and its digital version built in the Questionnaire Star with a QR code were offered to prospective participants. It should be noted that some residents were still concerned about the risk of being contracted with the virus even after the end of the lockdown as evidenced in other survey projects administered by the lead author of this paper. Thus, during the survey period in 2021, participants who chose to do a digital survey were asked to scan the QR code and fill the digital survey onsite. A total of 350 individuals preferred the digital survey and 321 of them submitted their filled questionnaires. In addition, a total of 350 hardcopy questionnaires were handed out to participants with 302 of them being returned. Thus, the return rate for the 2021 survey period was 89.0%. As with the 2019 survey, 28 questionnaires for the 2021 sample were removed due to the same reasons, resulting in 595 usable questionnaires for further analysis.

3.4. Data Analysis

With the removal of those questionnaires with incomplete and skeptical responses, the pattern of missing data is random, and the missing rate is quite low (between 0.2% and 0.9%). Thus, no imputation was used to replace missing data. All analyses were made based on usable questionnaires with missing data omitted using casewise deletion.

Data analyses were conducted using SPSS 28 and AMOS 28, including descriptive analysis, chi-square analysis, factor analysis, t-test analysis, and SEM. First, socio-demographics and the recreational use of urban green areas pre- and during the pandemic were described. Second, chi-square tests were conducted to see if group types, frequency of visits, and use duration are significantly different between the two samples.

Third, the principal components analysis with varimax rotation and an eigenvalue of 1.00 or more was used to derive latent variables for the 2019 sample, 2021 sample, and the aggregated data (both samples combined). A cut-off point of 0.45 was used to determine items for a factor [52], and a loading difference of 0.15 was used to separate items with cross loadings [53,54].

Fourth, t-tests were conducted to compare the similarities and differences in NEP, motivations, and PWI-A between the two samples. Fifth, a measurement model for the three datasets (the 2019 sample, 2021 sample, and the two samples combined) was tested, respectively. Three parameters such as composite reliability (CR), average variance extracted (AVE), and maximum shared variance (MSV) were used to determine internal consistency (CR > 0.70), the convergent validity (AVE > 0.50) and discriminant validity (AVE > MSV) of a construct, respectively [55]. Sixth, three individual SEMs were tested with the three datasets, respectively. The ratio of χ^2 value over the degree of freedom was used to assess the goodness of fit, with a ratio of 5 being considered acceptable and below 3 as a better fit [56].

Finally, a multiple group analysis was conducted to statistically compare the relationship strengths between two variables in the SEM for the 2019 sample and 2021 sample. The critical ratio was used to test the significant level of a regression weight, with the ratio >1.96 or <−1.96 indicating the difference between two regression weights being significant at or lower than the 0.05 level [22].

4. Results

4.1. Descriptive Analysis

4.1.1. Socio-Demographics

Participants' socio-demographic characteristics are quite comparable between the two samples. Specifically, females outnumbered males for both survey periods (48.3% males and 51.7% females in 2019 vs. 41.2% males and 58.8% females in 2021). In addition, survey participants were young (71.4% were between 21–49 years of age in 2019 vs. 75.3% in 2021) and well educated (38.2% had an undergraduate degree or above in 2019 vs. 39.3% in 2021). Finally, the majority of participants were married for both samples (51.7% in 2019 vs. 52.8% in 2021).

4.1.2. Recreational Use of Urban Parks

The majority of respondents reported visiting urban parks with family and relatives (46.1% in 2019 vs. 57.1% in 2021), followed by with friends (35.4% in 2019 vs. 23.9% in 2021), while the percentages for people who visited urban parks alone are almost the same (14.2% in 2019 vs. 14.55 in 2021). In addition, a small percentage of respondents visited parks with others (neighbors and colleagues) with 4.3% for 2019 and 4.5% for 2021, respectively. However, in terms of frequency of visits and use duration per visit, participants in 2019 were more often to use urban parks than their counterparts in 2021. For example, 11.1%, 42.7%, 28.8%, and 17.5% of respondents reported having used urban parks every day, 1–3 times per week, 1–3 times per month, and less than 11 times per year in 2019 compared to 7.2%, 35.3%, 35.8%, and 21.5% in 2021, respectively. In addition, respondents in 2019 tended to stay longer in urban parks than respondents in 2021 with 8.9%, 32.1%, 41.8%, 13.7%, and 3.5% of them reporting a use duration of ≤30 min, 30 min to 1 h, 1 to 2 h, 2 to 4 h and >4 h in 2019, respectively (vs. 14.3%, 27.5%, 39.6%, 16.9%, and 1.7%, respectively in 2021).

4.2. Chi-Square Analysis of Recreational Use of Urban Parks

Chi-square tests of group types, frequency of visits, and use duration per visit before and after the outbreak are presented in Tables 1–3, respectively. As shown, the trip characteristics were significantly different before and after the outbreak. Specifically, compared to pre-pandemic in 2019, people surveyed in 2021 were more likely to go to urban parks with family and relatives, less likely with friends (Table 1), and less often to use parks (Table 2) with a shorter stay (Table 3).

Table 1. Chi-square analysis of group types (2019 vs. 2021).

	Group Types	2019	2021	Total	χ^2	Φ
Myself	Count	86	86	172		
	% within year	14.2%	14.5%	14.3%		
	Adjusted residual	−0.1	0.1			
Family and relatives	Count	279	339	618	20.31 *	0.130
	% within year	46.1%	57.1%	51.5%		
	Adjusted residual	−3.8 **	3.8 **			
Friends	Count	214	142	356		
	% within year	35.4%	23.9%	29.7%		
	Adjusted residual	4.3 **	−4.3 **			

Table 1. *Cont.*

	Group Types	2019	2021	Total	χ^2	Φ
Others (neighbors and colleagues)	Count	26	27	53		
	% within year	4.3%	4.5%	4.4%		
	Adjusted residual	−0.2	0.2		20.31 *	0.130
Total	Count	605	594	1199		
	% within year	100.0%	100.0%	100.0%		

* $p < 0.001$, ** Absolute value of adjusted residual > 2.0.

Table 2. Chi-square analysis of frequency of visits (2019 vs. 2021).

	Frequency of Visits	2019	2021	Total	χ^2	Φ
Everyday	Count	67	33	100		
	% within year	11.1%	5.5%	8.3%		
	Adjusted residual	3.5 **	−3.5 **			
1–6 times/week	Count	258	207	465		
	% within year	42.6%	34.8%	38.8%		
	Adjusted residual	2.8 **	−2.8 **			
1–3 times/month	Count	174	219	393		25.94 *	0.147
	% within year	28.8%	36.8%	32.8%		
	Adjusted residual	−3.0	3.0			
<11 times/per year	Count	106	136	242		
	% within year	17.5%	22.9%	20.2%		
	Adjusted residual	−2.3 **	2.3 **			
Total	Count	605	595	1200		
	% within year	100.0%	100.0%	100.0%		

* $p < 0.001$, ** Absolute value of adjusted residual > 2.0.

Table 3. Chi-square analysis of length of visits (2019 vs. 2021).

	Use Duration	2019	2021	Total	χ^2	Φ
≤30 min	Count	54	85	139		
	% within year	8.9%	14.3%	11.6%		
	Adjusted residual	−2.9 **	2.9 **			
30 min–1 h	Count	194	163	357		
	% within year	32.1%	27.5%	29.8%		
	Adjusted residual	1.7	−1.7			
1–2 h	Count	253	235	488		15.63 *	0.114
	% within year	41.8%	39.6%	40.7%		
	Adjusted residual	0.8	0.8			
2–4 h	Count	83	100	183		
	% within year	13.7%	16.9%	15.3%		
	Adjusted residual	−1.5	1.5			
>4 h	Count	21	10	31		
	% within year	3.5%	1.7%	2.6%		
	Adjusted residual	1.9	−1.9			
Total	Count	605	593	1198		
	% within year	100.0%	100.0%	100.0%		

* $p < 0.01$, ** Absolute value of adjusted residual > 2.0.

4.3. Factor Analysis

As previously mentioned, leisure motivations, NEP, and PWI-A were factor analyzed for each sample and the two samples combined. Interestingly, each dataset shares exactly the same subscale patterns for each measure. To save the paper length, only the factor analysis results for the aggregated data (Tables 4 and 5) are presented while results for each single sample are included as an Appendix (see Appendix A). Two factors—"close to nature" (pull motivations) and "social interactions" (push motivations)—were obtained from the eight items measuring leisure motivations. These two latent variables explained 59.57% of the total variance, with a Cronbach's alpha value for "close to nature" being 0.87 and "social interactions" being 0.55, respectively (Table 4). While the Cronbach's alpha value of 0.70 [57] has been typically used as the threshold to determine a factor's reliability, the alpha value less than the threshold was also considered acceptable for a factor with fewer items [58,59] since the value is sensitive to the number of items. For example, a reliability of 0.454 was reported in one study examining visitors' environmental attitudes measured by NEP [60].

Table 4. Summary results of exploratory factor analysis for motivation (aggregated data).

Code	Factor (Proportion): Scale Name and Items	*M*	*SD*	Factor 1	Factor 2
Factor 1	**Close to nature**	**4.18**	**0.66**		
L1	Experience nature	4.17	0.81	**0.785**	0.165
L3	Relaxation	4.30	0.78	**0.855**	0.084
L4	Enjoy the natural tranquility	4.25	0.84	**0.854**	0.156
L5	Enjoy the fresh air of open space	4.19	0.86	**0.829**	0.117
Factor 2	**Social interactions**	**3.25**	**0.90**		
L6	With friends	3.44	1.10	0.103	**0.746**
L7	With kids	3.35	1.37	0.144	**0.643**
L8	Picnics	2.96	1.23	0.050	**0.728**
	Eigenvalues			3.41	1.35
	% of variance			42.71	16.86
	Cumulative %			-	59.57
	Standardized Cronbach's *a*			0.87	0.55

KMO = 0.82, *p* < 0.001. Note. Item L2 "fitness and jogging" was removed from further analysis due to it being cross loaded on two factors.

Table 5. Summary results of exploratory factor analysis for New Ecological Paradigm (aggregated data).

Code	Factor (Proportion): Scale Name and Items	*M*	*SD*	Factor 1	Factor 2
Factor 1	**Humans with nature**	**3.82**	**0.64**		
NEP1	We are approaching the limit of the number of people the earth can support	3.72	0.95	−0.074	**0.520**
NEP3	When humans interfere with nature it often produces disastrous consequences	3.85	0.97	−0.017	**0.651**
NEP5	Humans are severely abusing the environment	3.62	1.02	−0.033	**0.634**
NEP7	Plants and animals have as much right as humans to exist	4.10	0.91	0.152	**0.655**
NEP9	Despite our special abilities humans are still subject to the laws of nature	3.81	0.90	−0.018	**0.633**
NEP11	The earth is like a spaceship with very limited room and Resources	3.97	0.92	0.109	**0.704**
NEP13	The balance of nature is very delicate and easily upset	3.82	0.99	−0.016	**0.682**
NEP15	If things continue on their present course, we will soon experience a major ecological catastrophe	3.64	0.94	−0.108	**0.667**

Table 5. *Cont.*

Code	Factor (Proportion): Scale Name and Items	*M*	*SD*	Factor	
				1	2
Factor 2	**Humans over nature**	**2.84**	**0.85**		
NEP2	Humans have the right to modify the natural environment to suit their needs	3.18	1.17	**0.605**	-0.077
NEP4	Human ingenuity will ensure that we do NOT make the earth unlivable	3.21	1.08	**0.652**	−0.155
NEP6	The earth has plenty of natural resources if we just learn how to develop them	2.88	1.22	**0.729**	−0.053
NEP8	The balance of nature is strong enough cope with the impacts of modern industrial nations	2.77	1.23	**0.779**	0.042
NEP10	The so-called "ecological crisis" facing humankind has been greatly exaggerated	2.86	1.11	**0.711**	0.028
NEP12	Humans were meant to rule over the rest of nature	2.48	1.23	**0.764**	0.118
NEP14	Humans will eventually learn enough about how nature works to be able to control it	2.54	1.22	**0.788**	0.076
	Eigenvalues			3.70	3.39
	% of variance			24.63	22.57
	Cumulative %			-	47.20
	Standardized Cronbach's *a*			0.80	0.84

KMO = 0.87, *p* < 0.001.

Two factors were extracted from the 15 NEP items with all eight odd-numbered items loaded on one factor–"humans with nature" (with a Cronbach's *a* of 0.87) and all seven even-numbered items on another factor—"humans over nature" (with a Cronbach's *a* of 0.84). A total of 47.20% of variance was explained by the two factors (Table 5). Factor analysis results for PWI-A are not tabulated as all seven items were loaded on one single factor.

4.4. T-Tests

t-tests were conducted to examine the similarities and differences in leisure motivations, NEP, and PWI-A between the two samples. Results are presented in Table 6. As shown, both groups were not significantly different in their leisure motivations (*p* > 0.05). However, they differed significantly in NEP and PWI-A with the 2021 sample being more positive toward "humans with nature", while also more supportive of "humans over nature", and more satisfied with their wellbeing than the 2019 sample counterparts. Thus, hypothesis 1 (people's environmental attitudes measured by NEP would differ significantly before and after the outbreak of the pandemic) is fully supported.

Table 6. T-tests of subscales of the three measures (2019 vs. 2021).

Subscales		Mean		Mean Difference	*t*	*p*	95% Confidence Interval of the Difference	
		2019	2021					
Leisure motivations	Close to nature	4.29	4.22	0.07	1.73	0.085	−0.01	0.15
	Social interactions	3.27	3.23	0.04	0.72	0.470	−0.06	0.14
NEP	Humans with nature	3.76	3.87	−0.11	−2.97 *	0.003	−0.17	−0.04
	Humans over nature	2.74	2.96	0.23	−4.58 **	0.000	−0.32	−0.13
PWI-A	Satisfaction of wellbeing	3.35	3.52	−0.17	−4.84 **	0.000	−0.24	−0.10

* *p* < 0.01, ** *p* < 0.001.

4.5. Structural Equation Modeling

4.5.1. Measurement Model

Data skewness and kurtosis for observed variables for the three datasets were examined prior to the test of the measurement model. A sample is considered not to deviate too much from the normal distribution if absolute values of univariate skewness and univariate kurtosis are less than 2 and 3, respectively [61]. The normality assessment indicated that the absolute values of all observed variables met the criteria (Appendix B), suggesting the appropriateness of the datasets for SEM analyses.

Table 7 presents CR, AVE, and MSV for the three datasets. CR is consistently above 0.70 for the two NEP subscales, PWI-A and one leisure motivation subscale "close to nature" for the three datasets. While CR is less than 0.70 for another leisure motivation subscale "social interactions", it is close or equal to 0.60. In terms of AVE, the value for "close to nature" is above 0.50 for all three datasets, while between 0.30 and 0.50 for the rest. An AVE close to 0.50 is still adequate if CR is greater than 0.70 [55]. It is worth noting that all AVE values are close to or higher than MSV, except the pair "humans over nature" and "social interactions" which has a MSV of 0.37. AVE values of between 0.30 and 0.50 and CR around 0.6 were also reported in other studies [62]. Thus, the three measures for each analysis group have a moderate to good composite reliability, convergent validity, and discriminant validity.

Table 7. Composite reliability, average variance extracted, and maximum shared variance.

	2019	2021	Aggregated
Composite reliability (CR)			
Leisure motivations			
Close to nature	0.84	0.87	0.87
Social interactions	0.60	0.56	0.56
NEP			
Harmony with nature	0.76	0.83	0.80
Humans over nature	0.81	0.87	0.85
PWI-A	0.80	0.86	0.84
Average variance extracted (AVE)			
Leisure motivations			
Close to nature	0.58	0.67	0.62
Social interactions	0.33	0.31	0.30
NEP			
Harmony with nature	0.30	0.38	0.34
Humans over nature	0.38	0.50	0.44
PWI-A	0.37	0.49	0.43
Maximum Shared Variance (MSV)			
Humans with nature↔Humans over nature	0.122	0.028	0.001
Humans with nature↔Close to nature	0.037	0.310	0.147
Humans with nature↔Social interactions	0.015	0.081	0.051
Satisfaction of wellbeing↔Humans with nature	0.057	0.047	0.055
Humans over nature↔Close to nature	0.002	0.003	0.000
Humans over nature↔Social interactions	0.044	0.372	0.192
Satisfaction of wellbeing↔Humans over nature	0.011	0.108	0.063
Close to nature↔Humans over nature	0.177	0.144	0.183
Satisfaction of wellbeing↔Close to nature	0.019	0.046	0.031
Satisfaction of wellbeing↔Social interactions	0.039	0.239	0.130

While CR and AVE can be improved by deleting items with low loadings, this study did not choose to do so for three reasons. First, the purpose of this study is not to develop and test a measure, but to obtain latent variables for the sake of simplicity of analysis. Second, keeping all items allows for comparison analysis being consistent with the same configurations between the two samples. Third, the measurement model for each dataset has a χ^2/df less than 3, RMSEA less than 0.05 and IFI and CFI greater than 0.90, indicating a good model fit of data (Appendix C).

4.5.2. Structural Model

Overall Structural Model

For the structural model, two variables—duration of recreational use of urban parks and frequency of visits were added. Figures 3–5 present the structural models for the 2019 sample, 2021 sample, and two samples combined, respectively. The model fit parameters are presented in Table 8. The ratio χ^2/df less than or slightly over 3, RMSEA close to or slightly over 0.05, IFI and CFI close to or equal to 0.90 indicate each dataset fits the model well.

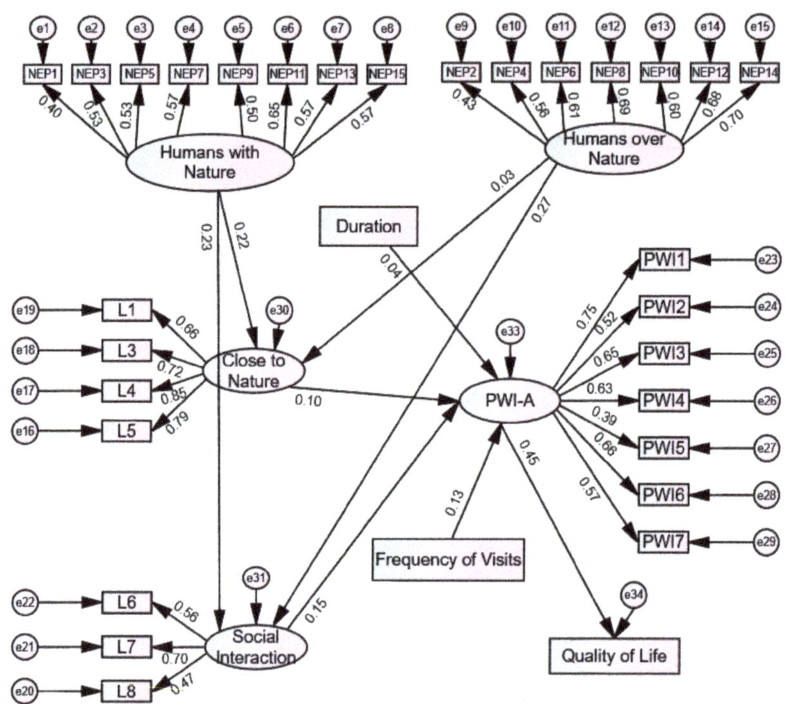

Figure 3. Structural equation model for the 2019 sample.

The results show that the relationship patterns among the five latent variables, frequency of visits, duration, and quality of life are amazingly consistent across all three models. Specifically, "human with nature" is significantly and positively related to the two leisure motivation subscales: "close to nature" ($p < 0.001$) and "social interactions" ($p < 0.001$), which, in turn, are significantly and positively related to wellbeing satisfaction ($p < .05$ for the former and $p < 0.01$ for the latter), PWI-A ($p < 0.05$ for the former and $p < 0.01$ for the latter), which further significantly influences quality of life ($p < 0.001$). While frequency of visits significantly and positively predicts PWI-A ($p < 0.05$) which further significantly contributes to quality of life ($p < 0.001$), duration does not. Finally, "humans over nature" significantly and positively predicts social interactions ($p < 0.001$). However, its relationship with "close to nature" is not significant ($p > 0.05$). Thus, Hypothesis 2 (residents' environmental attitudes would significantly influence their motivations to visit urban parks) and hypothesis 4 (satisfaction of wellbeing measured by PWI-A would significantly influence quality of life) are fully supported, while hypothesis 3 (leisure motivations would significantly influence PWI-A) is partially supported.

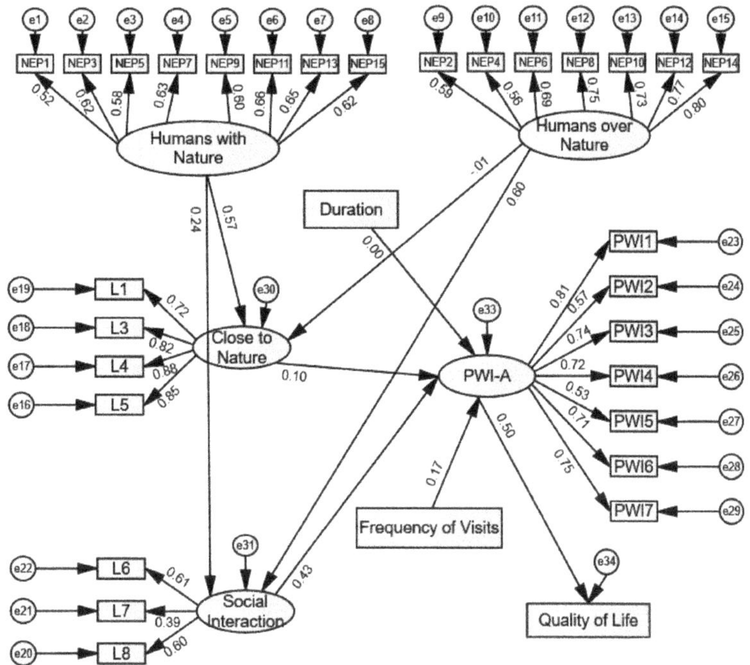

Figure 4. Structural equation model for the 2021 sample.

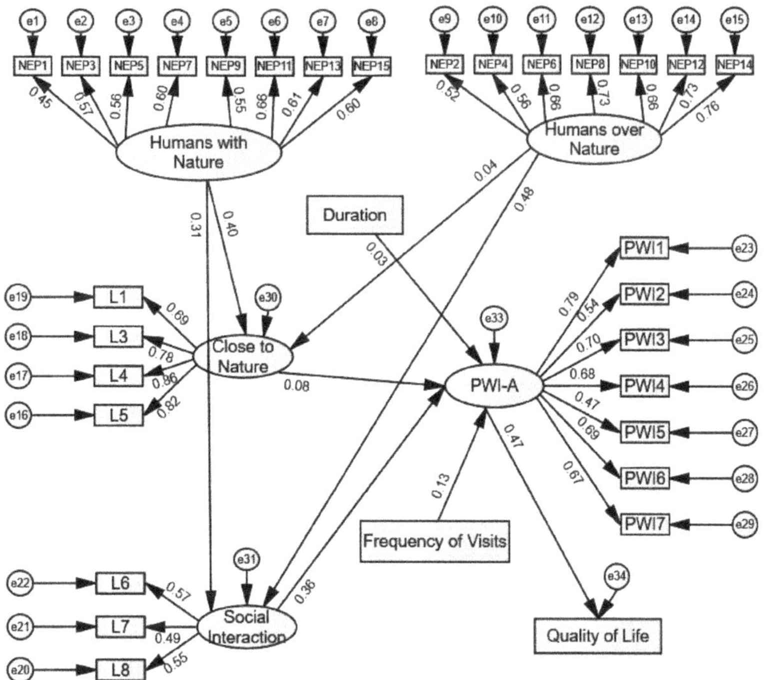

Figure 5. Structural equation model for the aggregated data (two samples combined).

Table 8. Assessment indices.

Model Fit Indices	χ^2	df	χ^2/df	RMSEA	IFI	CFI
2019 sample	1077.59	458	2.35	0.047	0.87	0.87
2021 sample	1239.72	458	2.70	0.054	0.89	0.89
Aggregated (two samples combined)	1616.04	458	3.53	0.046	0.90	0.90

Multiple Group Comparison Analysis

Measurement invariance across groups needs to be supported prior to the comparative analysis being carried out between groups [63]. There are four types of measurement invariance commonly reported in the literature, including configural invariance, metric invariance, scalar invariance, and residual invariance [64]. It is argued that "since full measurement invariance in all four steps is often not supported, it is becoming common practice to accept some violations of measurement invariance" [64] (p. 79).

Table 9 presents results of measurement invariance test for the two groups: 2019 sample and 2021 sample. The measurement model test reported earlier in Appendix C exhibits the same measurement pattern with good model fit between the two samples, supporting configural invariance. However, the metric invariance, if judged by the change of χ^2 would not be supported (Table 9) ($p < 0.001$). Since χ^2 is sensitive to sample size, a large sample size may lead to over-rejection of measurement invariance if the change of χ^2 is used as the only criterion to judge model fit [63,64]. Alternatively, change in alternative fit indices (AFIs) which is less sensitive to sample size has been used to evaluate fit [65]. AFIs criteria include a 0.01 change in CFI, 0.015 in RMSEA and 0.030 in SRMR (for metric invariance) or 0.015 (for scalar or residual invariance) [66]. The ΔCFI of 0.007, ΔRMSEA of 0.00 and ΔSRMR of 0.008 (Table 9) support metric, scalar, or residual invariance, suggesting the appropriateness of the data for cross group comparison analysis.

Table 9. Measurement invariance test between the 2019 sample and 2021 sample.

	χ^2	df	χ^2/df	RMSEA	SRMR	CFI
Unconstrained measurement model	1618.76	734	2.21	0.032	0.0449	0.924
Constrained measurement model	1728.24	763	2.27	0.032	0.0529	0.917
Difference	109.48 *	29	-	0.00	0.008	0.007

* $p < 0.001$.

The multiple group comparison analysis indicates that the three pairs of regression weights are significantly different between the two samples, including paths linking one NEP subscale "humans with nature" and one leisure motivation subscale "close to nature" (0.57 in 2021 vs. 0.22 in 2019, z = 2.889, $p < 0.001$), another NEP subscale "humans over nature" and another leisure motivation subscale "social interactions" (0.60 in 2021 vs. 0.27 in 2019, z = 3.468, $p < 0.001$), and "social interactions" and PWI-A (0.43 in 2021 vs. 0.15 in 2019, z = 2.283, $p < 0.01$) with the relationships being consistently stronger for the 2021 sample than for the 2019 sample. No significant differences were found for all other path coefficients between the two samples.

5. Discussion

Human beings are increasingly facing uncertainties that may have severe impacts on our health. The COVID-19 outbreak that has inflicted China and many other countries is such an example. The scope of restriction measures enforced by a country or region has largely affected the level of recreational use of urban green areas during the pandemic. More recreational uses of urban green areas were reported in countries/regions where people were still allowed to use urban green areas during the pandemic. For example, residents in the city of Freiburg visited the city's urban forests more often during the lockdown (4.2 visits per week) than before the pandemic (2.7 visits per week) [67]. A global study on the impacts of COVID-19 pandemic on urban park visitation using data recorded from the Google Community Mobility Report found that as of 26 May 2020, the number of park visits increased in all 48 countries selected with some countries experiencing more visits than others [68].

It seems that more visits experienced in other cities outside China as a result of the pandemic did not happen in Haikou as the city's residents visited urban parks less often and with shorter duration during the pandemic than before the pandemic. Less use of urban parks during the pandemic in the city seems to endorse another study in Hong Kong [69] where lockdown was not practiced during the pandemic and where people tended to participate outdoor activities less often than before the pandemic. Thus, people in China or at least in Haikou and Hong Kong maybe more cautious and preventive than people in the western societies in dealing with the pandemic. This can also be reflected by the finding that respondents surveyed in 2021 were more likely to visit urban parks with family/relatives and less likely with friends than their counterparts surveyed in 2019. Decreased visits to urban forests and parks were also reported in Cambodia, Indonesia, and Myanmar during the pandemic [70].

Interestingly, it is the frequency of visits and not their duration that contributes to wellbeing satisfaction. This is true for each individual sample and the two samples combined. This finding endorses a study which found that frequency of visits, not amount of time spent in urban green areas, significantly and positively predicts life satisfaction for residents in Daejeon City, South Korea [71]. Thus, frequent visits to urban green areas mean more than duration in increasing positive emotions, "leading to a feeling of happiness in daily life" [71] (p. 2). Previous studies involving tourism and destination satisfaction also found that frequency of visits is positively and significantly related to satisfaction [72–76].

In terms of the two NEP factors "humans with nature" and "humans over nature", the 2021 sample was more likely than the 2019 sample to emphasize the importance of "humans with nature" as well as "humans over nature", which seems to be paradoxical as items in "humans with nature" represent pro-environmental attitudes while the opposite is true for the items measuring "humans over nature". People may think it is more important than before for humans to maintain a harmonious relationship with nature due to the pandemic, while in the meantime, they may also believe that humans can eventually learn how to control nature with the advance of science and technology and human wisdoms. The effective control of the spread of COVID-19 in China in 2021 as a whole and in the city specifically may have reinforced this line of thoughts. Thus, two mindsets (i.e., "humans with nature" and "humans over nature") may coexist simultaneously, though the former still weighed more than the latter as indicated by the average factor score being higher for the former than for the latter (M = 3.77 for "humans with nature" vs. M = 2.74 for "humans over nature" in 2019 and M = 3.87 for "humans with nature" vs. M = 2.96 for "humans over nature" in 2021). A further t-test analysis of the seven Dominant Social Paradigm items shows that the most significant differences came from three items: "the earth has plenty of natural resources if we just learn how to develop them" (NEP4), "the balance of nature is strong enough cope with the impacts of modern industrial nations" (NEP6), and "humans will eventually learn enough about how nature works to be able to control it" (NEP14).

It should be noted that a consensus about the dimensionality of the NEP construct has not been achieved among researchers. Although the 15 items of the NEP were initially used to represent five aspects of the environmental attitudes (balance of nature, eco-crisis, limits to growth, anti-exemptionalism, anti-anthropocentrism, each with three items) [12], the developers of the scale also argue that NEP can be treated as a one-dimension measure. However, they further emphasized that "future research will be needed to address the issue of the revised NEP Scale's dimensionality, and on some samples a clearer pattern of multidimensionality will no doubt emerge and warrant creation of two or more subscales measuring distinct dimensions of the NEP" [12] (p. 439). Indeed, many following studies conducted in varying socio-cultural contexts have reported two or more subscales out of either the earlier 12-item measure or the revised measure with 15 items [48,77]. However, few, if any, have obtained five subscales through exploratory factor analysis that match exactly the five aspects of the 15 items. This led some researchers to speculate that the NEP subscales are sample specific [29]. Nevertheless, the two subscales from this study are, to a large extent, resemble a previous study on a national park in China [29], whereas all even-numbered items were loaded on one factor-"humans over nature", and all odd-numbered items were loaded on two other subscales termed "limits to growth" and "eco-crisis", which, if combined, correspond to "humans with nature" in this study.

The fact that NEP subscales vary with samples makes it difficult to compare findings of this study with those from previous studies involving NEP. For example, a study [15] found that not all NEP subscales are significantly more positive during the pandemic than pre-pandemic in the context of Germany, with respondents being more concerned about three of them—"balance of nature", "anti-exemptionalism", and "eco-crisis" while less concerned about two other subscales—"limits of growth" and "anti-anthropocentrism". Given that the four items in the subscale "humans over nature" (i.e., NEP1, NEP2, NEP11, NEP12) and other four items in the subscale "humans with nature" (i.e., NEP4, NEP8, NEP10, and NEP14) of this study are the same as reported in the study involving Germany [15], findings of this study partially endorse the study in Germany [15]. That is, people may hold views of the Dominant Social Paradigm that emphasizes "humans over nature" and views of the New Environmental Paradigm that endorses "humans with nature" during the pandemic in both countries.

While the two NEP factors differed significantly pre- and during-pandemic, the two leisure motivation factors were not significantly different from each other. However, when the relationships between NEP and leisure motivations were examined using SEM, some interesting patterns emerged. That is, the impact patterns of NEP on leisure motivations are consistent across all three datasets as shown in Figures 3–5, with the relationships between one NEP subscale "humans with nature" and the two motivation subscales "close to nature" and "social interactions" being significant, so is the relationship between another NEP subscale "humans over nature" and one motivation subscale "social interactions". The significant relationship between "humans with nature" represented by all seven odd-numbered items and "close to nature" endorses a previous study [29] which reported that all those odd-numbered items significantly influence nature-based tourism motivations to return to nature, to learn about nature, and to escape from routines. Interestingly, similar to "humans over nature" being closely related to active and adventurous tourism pursuits of nature-based tourism reported in that study [29], respondents in this study who scored higher on "humans over nature" expressed a higher motivation on social interactions. This is true for the 2019 sample, 2021 sample, and the two samples combined.

Both motivation subscales positively and significantly predicted satisfaction of wellbeing, which is consistent with previous studies on tourism motivations and destination satisfaction [38,41]. It should be noted that the push aspect of the motivation measure—"social interactions" had a larger effect on wellbeing satisfaction than the pull motivation—"close to nature" which corroborates, to some degree, a study [43] which also found that it is the push motivators, not the pull motivators, that contributed more to satisfaction. It is worth noting that the relationship between "social interactions" and wellbeing satisfaction for the 2021 sample was significantly stronger than that for the 2019 sample, implying that people were more likely to emphasize the importance of being united with family/relatives for mental and physical wellbeing during the pandemic than pre-pandemic. In other words, they felt more satisfied to socialize with family and relatives when recreating in the urban parks than before the pandemic. This is consistent with the t-test results that respondents surveyed during the pandemic scored significantly higher on PWI-A than their counterparts surveyed pre-pandemic. Thus, there may be a pent-up satisfaction among urban dwellers after the COVID-19 outbreak. This finding supports the posttraumatic growth theory that explains positive psychological change as a result of experiencing highly stressful life circumstances (i.e., disasters, crises, or traumas) [78,79].

The stronger relationship between "humans with nature" and "close to nature" for the 2021 sample suggests that respondents during the pandemic with the belief in "humans with nature" were more likely to utilize urban green areas for being "close to nature" than respondents surveyed before the pandemic in 2019, albeit both samples were not significantly different from each other in their leisure motivations. The multigroup comparison analysis also shows that stronger belief in "humans over nature" led to stronger desire for "social interactions" in 2021 than in 2019, implying a close relationship between people's perception of humankind's ability to control nature during the pandemic and their desire to interact with people in urban green areas.

6. Conclusions

This paper not only addressed the use of urban green areas as affected by the COVID-19 pandemic before and after the outbreak of the COVID-19 pandemic, but also the perception of other variables as they relate to COVID-19 (attitudes, motivation, satisfaction, etc.). The examination of the use and perception of urban green areas based on two samples (one before the pandemic and one after the outbreak of the pandemic) using the same instrument and questionnaire making this study unique among existing studies on COVID-19 and urban green areas, as it allows for a meaningful comparison of use and perception of urban green areas as affected by COVID-19. Moreover, the use and perception were examined not separately, but simultaneously using the structural equation modeling (SEM).

6.1. Research Implications

This study is of significant theoretical, methodological, and managerial implications. Theoretically, this study, for the first time, empirically examined people's environmental attitudes measured by the NEP before and after the outbreak of COVID-19, with respondents after the outbreak being more supportive of "humans with nature" than respondents before the outbreak. This stronger belief in "humans with nature" during-pandemic proves that NEP is useful "in tracking possible increases in endorsement of an ecological worldview, as well as in examining the effect of specific experiences and types of information in generating changes in this worldview" [12] (p. 439). While recognizing the increased belief in "humans with nature" as a result of the pandemic, it is "humans over nature" that contributed more than "humans with nature" to "social interactions", which further led to a higher level of wellbeing satisfaction, resulting in a pent-up satisfaction during the pandemic. This pent-up satisfaction of wellbeing endorses the posttraumatic growth theory and thus deserves more research on the relationships between environmental attitudes, motivations, satisfaction, and quality of life as they relate to the pandemic in the context of urban green areas.

Methodologically, the duality of pro- and anti-environmental attitudes challenges the practice of treating the NEP as one composite measure of environmental attitudes. Doing so would cover the nuances of people's environmental attitudes pre- and post-pandemic. The change of environmental attitudes due to the pandemic also addresses the concern raised by other scholars who state that "it remains an open, empirically unaddressed question whether the pandemic has actually shifted our environmental concerns" [15] (p. 4), thus filling the research hiatus. Additionally, the use of multiple statistical methods (i.e., t-tests, chi-square tests, SEM, and SEM multi-group comparison analysis) allows for a better understanding of findings with cross validations [80]. For instance, the chi-square analysis shows people surveyed in 2021 were more likely to visit urban parks with family and relatives (social interactions) than their counterparts surveyed in 2019 while the t-test shows the 2021 respondents were more supportive of "humans over nature" and more satisfied than the 2019 respondents. Furthermore, "humans over nature" and "social interactions" were significantly related with each other with the relationship strength for the 2021 sample being significantly stronger than that for the 2019 sample. Thus, a clear picture appears when these findings are linked together, that is, the successful control of the COVID 19 pandemic in 2021 in the city of Haikou not only led people to believe in "humans over nature", but also provided much-needed opportunities for social interactions with family and relatives in urban green areas, resulting in a higher level of satisfaction.

This intricacy among these interrelated variables cannot be revealed if a single method was used.

Managerially, one previous study [43] recommended that destination managers pay attention to family togetherness given the positive contribution from push motivators to destination satisfaction, we would like urban park managers to do the same by providing more opportunities for social interactions in the city's urban parks given the significant contribution of social interactions in urban parks to the wellbeing satisfaction of the city dwellers. In addition, urban parks as important and convenient venues for being close to nature should engage locals in the restoration and conservation of urban ecosystems through citizen science projects and activities, which may enhance their attitudes toward nature and the environment, and which, in turn, may facilitate more recreational uses of urban parks with higher levels of satisfaction and quality of life. Consequently, a virtuous circle or feedback loop for the long-term sustainability of urban green areas can be achieved. Finally, given the role of urban green areas in promoting and improving residents' wellbeing and quality of life, it is important to increase the accessibility of urban green areas to meet residents' increasing desire of going to urban green space for leisure after the pandemic. However, the planning of urban green areas lagged behind the social development in the city, in that the quantity of green spaces in neighborhoods is insufficient. Thus, more public green spaces, especially neighborhood green spaces, should be rationally planned and constructed so that residents can access and use green spaces as frequently as they wish. It is worth noting that frequency of visits was closely related to public wellbeing as shown in this study.

6.2. Research Limitations and Future Research Needs

Three research limitations need to be addressed. First, while the two samples, one in 2019 and one in 2021, were used to analyze the similarities and differences before and after the COVID-19 outbreak, these two samples were not from the same group in an experimental design with the use of control group and randomized participants. Thus, the differences between participants surveyed may not 100% result from the impact of COVID-19. That said, the use of a pure experiment design is rare in the field of social science studies in general and human dimensions of urban green areas in particular while a "survey is the most common method of data collection" in "parks, recreation, tourism, sport, and leisure studies" [81] (p.171). Second, the survey period was not spread around each season and may not represent the whole picture of recreational uses of urban green areas in the city. However, this bias may not be a big issue given that seasonality has a little impact on outdoor recreation participation in the city for the reason explained earlier in this paper. Finally, although the convenience sampling method has been commonly used in the literature, survey results using the method in this study may be biased due to participants being not sampled in a random manner. A household survey may need to be conducted in the future whereas a random survey with a larger sample size can be implemented.

While recognizing the change of environmental attitudes measured by NEP before and after the outbreak of the pandemic, no one knows if this change is transformative and permanent without follow-up surveys. Thus, future research needs to focus on the long-term monitoring of residents' use of urban green areas along with their leisure motivations and attitudes toward the environment. Future research should also utilize big data from social media or mobile phone call detail record (CDR) or other big data platforms (e.g., Google, Baidu, Sina Weibo, etc.) to analyze the recreational use of urban green areas spatially and temporally using GIS [82,83] to supplement the traditional way of using pencil-paper questionnaires. Compared with the traditional approach of data collection using questionnaires, field observations, or interviews, geotagged big data provide geographical and contextual information about people's spatial movement and thus are more useful and effective to understand the spatial movements/behaviors of individuals as well as their comments, perceptions, and momentary experience associated with urban green areas during different stages of the COVID-19 pandemic. Information obtained from big data or social media on where people visit, why they visit, how long they stay, what they experience, and how satisfied they are with urban green areas visited is tremendously useful for better planning and management of urban green areas. In addition, it is recommended that more comparative studies be conducted in other countries/regions where the pandemic still prevails so that the links between environmental attitudes, leisure motivations, wellbeing, and quality of life can be examined and compared internationally and cross-culturally.

Author Contributions: Conceptualization, J.D., J.-h.L. and J.T.; funding acquisition, Y.L.; investigation, Y.L.; methodology, J.D.; supervision, J.D.; visualization, J.D., Y.L. and J.T.; writing—original draft, J.D.; writing—review and editing, J.D., Y.L., C.P., J.-h.L. and J.T. All authors have read and agreed to the published version of the manuscript.

Funding: This study was funded by Hainan Provincial Social Science Foundation, NO.HNSK (YB)22-20; Hainan Provincial Natural Science Foundation, NO.720RC611; and Construction Fund of Key Disciplines of Hainan Provincial Education Department (Business Administration).

Institutional Review Board Statement: This research was reviewed and approved by the Institutional Review Board of School of Tourism, Hainan Normal University (No. 2022051901).

Informed Consent Statement: Not Applicable.

Data Availability Statement: Data will be made available on request.

Acknowledgments: Thanks go to those postgraduates and undergraduates who helped with the survey.

Conflicts of Interest: The authors declare no conflict of interest.

Appendix A. Factor Analysis for the 2019 Sample and 2021 Sample

Table A1. Summary results of exploratory factor analysis for leisure motivations ($n = 605$, 2019 sample).

Code	Factor (Proportion): Scale Name and Items	*M*	*SD*	Factor 1	Factor 2
Factor 1	**Close to nature**	**4.28**	**0.66**		
L1	Experience nature	4.25	0.75	**0.750**	0.214
L3	Relaxation	4.35	0.73	**0.825**	0.053
L4	Enjoy the natural tranquility	4.25	0.85	**0.837**	0.167
L5	Enjoy the fresh air of open space	4.30	0.88	**0.817**	0.087
Factor 2	**Social interactions**	**3.27**	**0.85**		
L6	With friends	3.49	1.07	0.121	**0.718**
L7	With kids	3.41	1.22	0.078	**0.790**
L8	Picnics	2.90	1.14	0.113	**0.649**
	Eigenvalues			3.27	1.36
	% of variance			40.88	17.03
	Cumulative %			-	57.91
	Standardized Cronbach's *a*			0.84	0.59

KMO = 0.81, $p < 0.001$. Note. Item L2 "fitness and jogging" was excluded from further analysis due to it being cross loaded on two factors.

Table A2. Summary results of exploratory factor analysis for New Ecological Paradigm ($n = 605$, 2019 sample).

Code	Factor (Proportion): Scale Name and Items	*M*	*SD*	Factor 1	Factor 2
Factor 1	**Humans with nature**	**3.77**	**0.58**		
NEP1	We are approaching the limit of the number of people the earth can support	3.81	0.91	−0.022	**0.504**
NEP3	When humans interfere with nature it often produces disastrous consequences	3.81	0.94	0.038	**0.626**

Table A2. *Cont.*

Code	Factor (Proportion): Scale Name and Items	*M*	*SD*	Factor 1	Factor 2
NEP5	Humans are severely abusing the environment	3.59	1.00	−0.013	**0.635**
NEP7	Plants and animals have as much right as humans to exist	4.06	0.96	0.280	**0.586**
NEP9	Despite our special abilities humans are still subject to the laws of nature	3.72	0.89	0.013	**0.585**
NEP11	The earth is like a spaceship with very limited room and Resources	3.91	0.93	0.209	**0.664**
NEP13	The balance of nature is very delicate and easily upset	3.74	0.99	0.104	**0.636**
NEP15	If things continue on their present course, we will soon experience a major ecological catastrophe	3.52	0.94	0.019	**0.648**
Factor 2	**Humans over nature**	**2.74**	**0.77**		
NEP2	Humans have the right to modify the natural environment to suit their needs	3.08	1.15	**0.537**	−0.027
NEP4	Human ingenuity will ensure that we do NOT make the earth unlivable	2.94	1.05	**0.664**	−0.038
NEP6	The earth has plenty of natural resources if we just learn how to develop them	2.72	1.19	**0.685**	0.055
NEP8	The balance of nature is strong enough cope with the impacts of modern industrial nations	2.68	1.13	**0.738**	0.143
NEP10	The so-called "ecological crisis" facing humankind has been greatly exaggerated	2.87	1.02	**0.666**	0.030
NEP12	Humans were meant to rule over the rest of nature	2.40	1.18	**0.693**	0.219
NEP14	Humans will eventually learn enough about how nature works to be able to control it	2.43	1.15	**0.729**	0.144
	Eigenvalues			4.00	2.43
	% of variance			26.64	16.22
	Cumulative %			-	42.86
	Standardized Cronbach's *a*			0.77	0.81

KMO = 0.85, *p* < 0.001.

Table A3. Summary results of exploratory factor analysis for leisure motivations (*n* = 595, 2021 sample).

Code	Factor (Proportion): Scale Name and Items	*M*	*SD*	Factor 1	Factor 2
Factor 1	**Close to nature**	**4.22**	**0.73**		
L1	Experience nature	4.09	0.87	**0.811**	0.132
L3	Relaxation	4.25	0.82	**0.877**	0.115
L4	Enjoy the natural tranquility	4.26	0.83	**0.870**	0.167
L5	Enjoy the fresh air of open space	4.29	0.84	**0.850**	0.150
Factor 2	**Social interactions**	**3.22**	**0.95**		
L6	With friends	3.38	1.13	0.060	**0.804**
L7	With kids	3.30	1.51	0.218	**0.464**
L8	Picnics	3.02	1.31	0.110	**0.783**
	Eigenvalues			3.57	1.39
	% of variance			44.63	17.40
	Cumulative %			-	62.03
	Standardized Cronbach's *a*			0.89	0.52

KMO = 0.80, *p* < 0.001. Note. Item L2 "fitness and jogging" was excluded from further analysis due to it being cross loaded on two factors.

Table A4. Summary results of exploratory factor analysis for New Ecological Paradigm (*n* = 595, 2021 sample).

Code	Factor (Proportion): Scale name and items	*M*	*SD*	Factor 1	Factor 2
Factor 1	**Humans with nature**	**3.87**	**0.63**		
NEP1	We are approaching the limit of the number of people the earth can support	3.63	0.98	−0.150	**0.567**
NEP3	When humans interfere with nature it often produces disastrous consequences	3.90	0.99	−0.050	**0.676**
NEP5	Humans are severely abusing the environment	3.65	1.04	−0.054	**0.645**
NEP7	Plants and animals have as much right as humans to exist	4.15	0.85	0.043	**0.697**
NEP9	Despite our special abilities humans are still subject to the laws of nature	3.91	0.91	−0.014	**0.674**
NEP11	The earth is like a spaceship with very limited room and Resources	4.03	0.90	0.036	**0.722**
NEP13	The balance of nature is very delicate and easily upset	3.90	0.98	−0.104	**0.705**
NEP15	If things continue on their present course, we will soon experience a major ecological catastrophe	3.76	0.93	−0.201	**0.666**
Factor 2	**Humans over nature**	**2.96**	**0.92**		
NEP2	Humans have the right to modify the natural environment to suit their needs	3.28	1.19	**0.653**	−0.122
NEP4	Human ingenuity will insure that we do NOT make the earth unlivable	3.48	1.04	**0.630**	−0.266
NEP6	The earth has plenty of natural resources if we just learn how to develop them	3.04	1.23	**0.749**	−0.142
NEP8	The balance of nature is strong enough cope with the impacts of modern industrial nations	2.86	1.31	**0.800**	−0.037
NEP10	The so-called "ecological crisis" facing humankind has been greatly exaggerated	2.85	1.18	**0.767**	0.014
NEP12	Humans were meant to rule over the rest of nature	2.57	1.27	**0.807**	0.034
NEP14	Humans will eventually learn enough about how nature works to be able to control it	2.65	1.28	**0.826**	0.021
	Eigenvalues			4.53	3.20
	% of variance			30.20	21.36
	Cumulative %			-	51.55
	Standardized Cronbach's *a*			0.83	0.87

KMO = 0.88, *p* < 0.001.

Appendix B. Descriptive Analysis for the 2019 Sample, 2021 Sample, and the Two Samples Combined

Table A5. Descriptive analysis of leisure motivations (*n* = 605, 2019 sample).

Item	*Mean*	*SD*	Skewness	Kurtosis
1. Experience nature	4.25	0.75	−1.07	2.04
2. Fitness and jogging	3.96	0.92	−0.61	−0.13
3. Relaxation	4.35	0.73	−1.19	2.22
4. Enjoy the natural tranquility	4.25	0.85	−1.27	1.92
5. Enjoy the fresh air of open space	4.30	0.88	−1.38	1.92
6. With friends	3.49	1.07	−0.26	−0.56
7. With kids	3.41	1.22	−0.37	−0.76
8. Picnics	2.90	1.14	0.15	−0.68

Table A6. Descriptive analysis of the New Ecological Paradigm (*n* = 605, 2019 sample).

Item	*Mean*	*SD*	Skewness	Kurtosis
1. We are approaching the limit of the number of people the earth can support	3.81	0.91	−0.66	0.31
2. Humans have the right to modify the natural environment to suit their needs	3.08	1.15	0.12	−0.89
3. When humans interfere with nature it often produces disastrous consequences	3.81	0.94	−0.74	0.32
4. Human ingenuity will insure that we do NOT make the earth unlivable	2.94	1.05	0.02	−0.54
5. Humans are severely abusing the environment	3.59	1.00	−0.52	−0.34
6. The earth has plenty of natural resources if we just learn how to develop them	2.72	1.19	−0.28	−0.84
7. Plants and animals have as much right as humans to exist	4.06	0.96	−1.21	1.37
8. The balance of nature is strong enough cope with the impacts of modern industrial nations	2.68	1.13	−0.27	−0.78
9. Despite our special abilities humans are still subject to the laws of nature	3.72	0.89	−0.63	0.29
10. The so-called "ecological crisis" facing humankind has been greatly exaggerated	2.87	1.02	−0.16	−0.48
11. The earth is like a spaceship with very limited room and Resources	3.91	0.93	−1.90	0.56
12. Humans were meant to rule over the rest of nature	2.40	1.18	−0.57	−0.60
13. The balance of nature is very delicate and easily upset	3.74	0.99	−0.85	0.39
14. Humans will eventually learn enough about how nature works to be able to control it	2.43	1.15	-0.50	-0.57
15. If things continue on their present course, we will soon experience a major ecological catastrophe	3.52	0.94	-0.41	0.17

Table A7. Descriptive analysis of wellbeing satisfaction (*n* = 605, 2019 sample).

Item	*Mean*	*SD*	Skewness	Kurtosis
1. How satisfied are you with your standard of living?	3.39	0.84	−0.28	−0.01
2. How satisfied are you with your health?	3.42	0.89	−0.39	0.12
3. How satisfied are you with what you are achieving in life?	3.27	0.84	−0.17	0.08
4. How satisfied are you with your personal relationships?	3.48	0.81	−0.36	0.05
5. How satisfied are you with how safe you feel?	3.14	0.86	−0.14	0.01
6. How satisfied are you with where you live in your community?	3.32	0.84	−0.27	−0.13
7. How satisfied are you with your leisure pursuits?	3.42	0.87	−0.23	0.02

Table A8. Descriptive analysis of leisure motivations (*n* = 595, 2021 sample).

Item	*Mean*	*SD*	Skewness	Kurtosis
1. Experience nature	4.09	0.87	−1.23	2.22
2. Fitness and jogging	3.78	1.02	−0.63	0.02
3. Relaxation	4.25	0.82	−1.40	2.95
4. Enjoy the natural tranquility	4.26	0.83	−1.42	2.91
5. Enjoy the fresh air of open space	4.29	0.84	−1.45	2.76
6. With friends	3.38	1.13	−0.23	−0.58
7. With kids	3.30	1.51	−0.44	−1.25
8. Picnics	3.02	1.31	−0.07	−1.02

Table A9. Descriptive analysis of the New Ecological Paradigm (*n* = 595, 2021 sample).

Item	*Mean*	*SD*	Skewness	Kurtosis
1. We are approaching the limit of the number of people the earth can support	3.63	0.98	−0.54	0.13
2. Humans have the right to modify the natural environment to suit their needs	3.28	1.19	0.41	−0.76
3. When humans interfere with nature it often produces disastrous consequences	3.90	0.99	−0.91	0.62
4. Human ingenuity will insure that we do NOT make the earth unlivable	3.48	1.04	0.32	−0.36
5. Humans are severely abusing the environment	3.65	1.04	−0.72	0.18
6. The earth has plenty of natural resources if we just learn how to develop them	3.04	1.23	0.09	−0.95
7. Plants and animals have as much right as humans to exist	4.15	0.85	−1.11	1.53

Table A9. *Cont.*

Item	*Mean*	*SD*	Skewness	Kurtosis
8. The balance of nature is strong enough cope with the impacts of modern industrial nations	2.86	1.31	−0.09	−1.13
9. Despite our special abilities humans are still subject to the laws of nature	3.91	0.91	−0.78	0.67
10. The so-called "ecological crisis" facing humankind has been greatly exaggerated	2.85	1.18	−0.16	−0.90
11. The earth is like a spaceship with very limited room and Resources	4.03	0.90	−1.16	1.82
12. Humans were meant to rule over the rest of nature	2.57	1.27	−0.42	−0.91
13. The balance of nature is very delicate and easily upset	3.90	0.98	−0.90	0.49
14. Humans will eventually learn enough about how nature works to be able to control it	2.65	1.28	−0.34	-0.96
15. If things continue on their present course, we will soon experience a major ecological catastrophe	3.76	0.93	−0.54	0.30

Table A10. Descriptive analysis of wellbeing satisfaction (n = 595, 2021 sample).

Item	*Mean*	*SD*	Skewness	Kurtosis
1. How satisfied are you with your standard of living?	3.49	0.90	−0.31	0.05
2. How satisfied are you with your health?	3.37	0.90	−0.37	0.00
3. How satisfied are you with what you are achieving in life?	3.50	0.88	−0.12	−0.11
4. How satisfied are you with your personal relationships?	3.54	0.86	−0.14	−0.14
5. How satisfied are you with how safe you feel?	3.66	0.89	−0.24	−0.12
6. How satisfied are you with where you live in your community?	3.53	0.91	−0.31	−0.07
7. How satisfied are you with your leisure pursuits?	3.57	0.89	−0.27	−0.08

Table A11. Descriptive analysis of leisure motivations (n = 1200, two samples combined).

Item	*Mean*	*SD*	Skewness	Kurtosis
1. Experience nature	4.17	0.81	−1.20	2.35
2. Fitness and jogging	3.87	0.97	−0.65	0.03
3. Relaxation	4.30	0.78	−1.33	2.80
4. Enjoy the natural tranquility	4.25	0.84	−1.34	2.38
5. Enjoy the fresh air of open space	4.29	0.86	−1.341	2.29
6. With friends	3.44	1.10	−0.25	−0.56
7. With kids	3.36	1.37	−0.45	−0.98
8. Picnics	2.96	1.23	0.04	−0.87

Table A12. Descriptive analysis of the New Ecological Paradigm (n = 1200, two samples combined).

Item	*Mean*	*SD*	Skewness	Kurtosis
1. We are approaching the limit of the number of people the earth can support	3.72	0.95	−0.60	0.22
2. Humans have the right to modify the natural environment to suit their needs	3.18	1.17	−0.25	−0.86
3. When humans interfere with nature it often produces disastrous consequences	3.85	0.97	−0.82	0.46
4. Human ingenuity will insure that we do NOT make the earth unlivable	3.21	1.08	−0.15	−0.54
5. Humans are severely abusing the environment	3.62	1.02	−0.62	−0.08
6. The earth has plenty of natural resources if we just learn how to develop them	2.88	1.22	0.10	-0.96
7. Plants and animals have as much right as humans to exist	4.10	0.91	−1.19	1.54
8. The balance of nature is strong enough cope with the impacts of modern industrial nations	2.77	1.23	0.20	−0.97
9. Despite our special abilities humans are still subject to the laws of nature	3.81	0.90	-0.69	0.43
10. The so-called "ecological crisis" facing humankind has been greatly exaggerated	2.86	1.11	0.16	−0.70
11. The earth is like a spaceship with very limited room and Resources	3.97	0.92	−1.02	1.10
12. Humans were meant to rule over the rest of nature	2.48	1.23	0.50	−0.76

Table A12. *Cont.*

Item	Mean	SD	Skewness	Kurtosis
13. The balance of nature is very delicate and easily upset	3.82	0.99	−0.87	0.42
14. Humans will eventually learn enough about how nature works to be able to control it	2.54	1.22	0.43	−0.77
15. If things continue on their present course, we will soon experience a major ecological catastrophe	3.64	0.94	−0.47	0.19

Table A13. Descriptive analysis of wellbeing satisfaction (n = 1200, two samples combined).

Item	Mean	SD	Skewness	Kurtosis
1. How satisfied are you with your standard of living?	3.44	0.87	−0.28	0.02
2. How satisfied are you with your health?	3.39	0.90	−0.38	0.05
3. How satisfied are you with what you are achieving in life?	3.39	0.87	−0.12	−0.02
4. How satisfied are you with your personal relationships?	3.51	0.83	−0.24	−0.04
5. How satisfied are you with how safe you feel?	3.40	0.91	−0.15	−0.12
6. How satisfied are you with where you live in your community?	3.42	0.88	−0.25	−0.10
7. How satisfied are you with your leisure pursuits?	3.50	0.88	−0.24	−0.05

Appendix C. Measurement Model Assessment Indices

Table A14. Measurement model assessment indices.

Model Fit Indices	χ^2	df	χ^2/df	RMSEA	IFI	CFI
2019 sample	718.01	367	1.96	0.040	0.92	0.92
2021 sample	881.41	367	2.40	0.049	0.93	0.93
Aggregated (two samples combined)	1067.42	367	2.91	0.040	0.94	0.94

References

1. Rice, W.L.; Lawhon, B.; Taff, B.D.; Matter, T.; Reigner, N.; Newman, P. The COVID-19 Pandemic is Changing the Way People Recreate Outdoors. 2020. Available online: https://lnt.org/wp-content/uploads/2020/04/GeneralPublic-CoronaSurvey-FINAL.pdf (accessed on 2 January 2022).
2. Da Schio, N.; Phillips, A.; Fransen, K.; Wolf, M.; Haase, D.; Ostoic, S.K.; Zivojinovic, I.; Valetic, D.; Derks, J.; Davies, C.; et al. The impact of the COVID-19 pandemic on the use of and attitudes towards urban forests and green spaces: Exploring the instigators of change in Belgium. *Urban For. Urban Green.* **2021**, *65*, 127305. [CrossRef] [PubMed]
3. Ugolini, F.; Massetti, L.; Calaza-Martínez, P.; Carinanos, P.; Dobbs, C.; Ostoi, S.K.; Marin, A.M.; Pearlmutterf, D.; Saaroni, I.; Simoneti, M.; et al. Effects of the COVID-19 pandemic on the use and perceptions of urban green space: An international exploratory study. *Urban For. Urban Green.* **2020**, *56*, 126888. [CrossRef]
4. Ramkissoon, H. COVID-19 place confinement, pro-social, pro-environmental behaviors, and residents' wellbeing: A new conceptual framework. *Front. Psychol.* **2020**, *11*, 2248. [CrossRef] [PubMed]
5. Severo, E.A.; De Guimaraes, J.C.F.; Dellarmelin, M.L. Impact of the COVID-19 pandemic on environmental awareness, sustainable consumption and social responsibility: Evidence from generations in Brazil and Portugal. *J. Clean. Prod.* **2021**, *286*, 12497. [CrossRef] [PubMed]
6. Buck, J.C.; Weinstein, S.B. The ecological consequences of a pandemic. *Biol. Lett.* **2020**, *16*, 20200641. [CrossRef]
7. Garrido-Cumbrera, M.; Foley, R.; Brace, O.; Correa-Fernandez, J.; Lopez-Lara, E.; Guzman, V.; Marin, A.G.; Hewlett, D. Perceptions of change in the natural environment produced by the first wave of the COVID-19 pandemic across three European countries. Results from the Green COVID study. *Urban For. Urban Green.* **2021**, *64*, 127260. [CrossRef]
8. Kanniah, K.D.; Kamarul Zaman, N.A.F.; Kaskaoutis, D.G.; Latif, M.T. COVID-19's impact on the atmospheric environment in the Southeast Asia region. *Sci. Total Environ.* **2020**, *736*, 139658. [CrossRef]
9. Rousseau, S.; Deschacht, N. Public awareness of nature and the environment during the COVID-19 crisis. *Environ. Resour. Econ.* **2020**, *76*, 1149–1159. [CrossRef]
10. Morse, J.W.; Gladkikh, T.M.; Hackenburg, D.M.; Gould, R.K. COVID-19 and human-nature relationships: Vermonters' activities in nature and associated nonmaterial values during the pandemic. *PLoS ONE* **2020**, *15*, e0243697. [CrossRef]
11. Homer, P.M.; Kahle, L.R. A structural equation test of the value-attitude behaviour hierarchy. *J. Pers. Soc. Psychol.* **1988**, *54*, 638–646. [CrossRef]
12. Dunlap, R.E.; Van Liere, K.D.; Mertig, A.G.; Jones, R.E. Measuring endorsement of the new ecological paradigm: A revised NEP scale. *J. Soc. Issues* **2000**, *56*, 425–442. [CrossRef]

13. The International Wellbeing Group. *Personal Wellbeing Index*, 5th ed.; Australian Centre on Quality of Life, Deakin University: Melbourne, Australia, 2013; Available online: https://www.acqol.com.au/uploads/pwi-a/pwi-a-english.pdf (accessed on 21 March 2019).

14. Shrum, T.R.; Nowak, S.A.; Merrill, S.C.; Hanley, J.P.; Liu, T.; Clark, E.M.; Zia, A.; Koliba, C. Pro-environmental attitudes, altruism, and COVID-19 risk management behaviour. *SocArXiv* **2021**. [CrossRef]

15. Schiller, B.; Tonsing, D.; Kleinert, T.; Bohm, R.; Heinrichs, M. Effects of the COVID-19 pandemic nationwide lockdown on mental health, environmental concern, and prejudice against other social groups. *Environ. Behav.* **2022**, *54*, 516–537. [CrossRef]

16. Fishbein, M.; Ajzen, I. *Belief, Attitude, Intention, and Behavior: An Introduction to Theory and Research*; Addison-Wesley: Boston, MA, USA, 1975.

17. Lindgren, H.C. *An Introduction to Social Psychology*; John Wiley: New York, NY, USA, 1969.

18. Hollander, P.E. *Principles and Methods of Social Psychology*; Oxford University: New York, NY, USA, 1971.

19. Hsu, C.H.C.; Cai, L.A.; Li, M. Expectation, motivation, and attitude: A tourist behavioral model. *J. Travel Res.* **2010**, *49*, 282–296. [CrossRef]

20. Ragheb, M.; Tate, R.L. A behavioral model of leisure participation, based on leisure attitude, motivation and satisfaction. *Leis. Stud.* **1993**, *12*, 61–70. [CrossRef]

21. Cheung, L.T.O.; Fok, L. The motivations and environmental attitudes of nature-based visitors to protected areas in Hong Kong. *Int. J. Sustain. Dev. World Ecol.* **2014**, *21*, 28–38. [CrossRef]

22. Li, J.; Pierskalla, C. Impact of attendees' motivation and past experience on their attitudes toward the National Cherry Blossom Festival in Washington, D.C. *Urban For. Urban Green.* **2018**, *36*, 57–67. [CrossRef]

23. Pereira, V.; Gupta, J.J.; Hussain, S. Impact of travel motivation on tourist's attitude toward destination: Evidence of mediating effect of destination image. *J. Hosp. Tour. Res.* **2022**, *46*, 946–971. [CrossRef]

24. Ponizovskiy, V.; Grigoryan, L.; Kuhnen, U.; Boehnke, K. Social construction of the value-behavior relation. *Front. Psychol.* **2019**, *10*, 00934. [CrossRef] [PubMed]

25. Schwartz, S.H.; Cieciuch, J.; Vecchione, M.; Davidov, E. Refining the theory of basic individual values. *J. Pers. Soc. Psychol.* **2012**, *103*, 663–688. [CrossRef]

26. Kotchen, M.J.; Reiling, S.D. Environmental attitudes, motivations, and contingent valuation of nonuse values: A case study involving endangered species. *Ecol. Econ.* **2000**, *32*, 93–107. [CrossRef]

27. Halkos, G.; Leonti, A.; Sardianou, E. Activities, motivations and satisfaction of urban parks visitors: A structural equation modeling analysis. *Econ. Anal. Policy* **2021**, *70*, 502–513. [CrossRef]

28. Kim, H.; Borges, M.C.; Chon, J. Impacts of environmental values on tourism motivation: The case of FICA, Brazil. *Tour. Manag.* **2006**, *27*, 957–967. [CrossRef]

29. Luo, Y.; Deng, J. The new environmental paradigm and nature-based tourism motivation. *J. Trav. Res.* **2008**, *46*, 392–402. [CrossRef]

30. Correia, A.; Moital, M. The antecedents and consequences of prestige motivation in tourism: An expectancy-value motivation. In *Handbook of Tourism Behaviour: Theory & Practice*; Kozak, M., Decrop, A., Eds.; Routledge: New York, NY, USA, 2008; pp. 16–34.

31. Pizam, A.; Neumann, Y.; Reichel, A. Tourist satisfaction. *Ann. Tour. Res.* **1979**, *6*, 195–197. [CrossRef]

32. Dann, G. Anomie, ego-enhancement and tourism. *Ann. Tour. Res.* **1977**, *4*, 184–194. [CrossRef]

33. Gnoth, J. Tourism motivation and expectation formation. *Ann. Tour. Res.* **1997**, *4*, 283–304. [CrossRef]

34. Ross, E.L.D.; Iso-Ahola, S.E. Sightseeing tourists' motivation and satisfaction. *Ann. Tour. Res.* **1991**, *18*, 226–237. [CrossRef]

35. Mill, R.; Morrison, A. *The Tourism System: An Introductory Text*, 3rd ed.; Kendall Hunt Pub Co: Dubuque, IA, USA, 1997.

36. Moutinho, L. Consumer behavior in tourism. *Eur. J. Mark.* **1987**, *21*, 3–44. [CrossRef]

37. Cole, S.; Crompton, J. A conceptualization of the relationships between service quality and visitor satisfaction, and their links to destination selection. *Leis. Stud.* **2003**, *22*, 65–80. [CrossRef]

38. Albayrak, T.; Caber, M. Examining the relationship between tourist motivation and satisfaction by two competing methods. *Tour. Manag.* **2018**, *69*, 201–213. [CrossRef]

39. Barbeitos, I.M.; Oom do Valle, P.; Guerreiro, M.; Mendes, J. Visitors' motivations, satisfaction and loyalty towards Castro Marim Medieval Fair. *J. Spat. Organ. Dyn.* **2014**, *2*, 89–104.

40. Battour, M.M.; Battor, M.M.; Ismail, M. The mediating role of tourist satisfaction: A study of Muslim tourists in Malaysia. *J. Travel Tour. Mark.* **2012**, *29*, 279–297. [CrossRef]

41. Hosany, S.; Buzova, D.; Sanz-Blas, S. The influence of place attachment, ad-evoked positive affect, and motivation on intention to visit: Imagination proclivity as a moderator. *J. Travel Res.* **2020**, *59*, 477–495. [CrossRef]

42. Langgat, J.; Marzuki, K.M.; Fabeil, N.F.; Dahnil, I. Visitor motivation, expectation and satisfaction of local cultural event in Sabah: A case study of Tamu Besar Kota Belud. *Int. J. Cult. Tour. Res.* **2012**, *5*, 48–59.

43. Yoon, Y.; Usyal, M. An examination of the effects of motivation and satisfaction on destination loyalty: A structural model. *Tour. Manag.* **2005**, *26*, 45–56. [CrossRef]

44. Yuan, J.; Li, J.; Deng, J.; Arbogast, D. Past experience and willingness to pay: A comparative examination of destination loyalty in two national parks, China. *Sustainability* **2021**, *13*, 8774. [CrossRef]

45. Liao, X.; So, A.; Lam, D. Residents' perceptions of the role of leisure satisfaction and quality of life in overall tourism development: Case of a fast growing tourism destination-Macao. *Asia Pac. J. Tour. Res.* **2016**, *21*, 1100–1113. [CrossRef]

46. Zhou, B.; Zhang, Y.; Dong, E.; Ryan, C.; Li, P. Leisure satisfaction and quality of life of residents in Ningbo, China. *J. Leis. Res.* **2021**, *52*, 469–486. [CrossRef]
47. Haikou Forestry Administration. Available online: http://lyj.haikou.gov.cn/ywdt/xwdt/202006/t20200617_1516404.html (accessed on 16 December 2021).
48. Anderson, M. New Ecological Paradigm (NEP) Scale. Available online: https://umaine.edu/soe/wp-content/uploads/sites/19 9/2013/01/NewEcologicalParadigmNEPScale1.pdf (accessed on 21 March 2019).
49. Putrawan, I.M. Measuring New Environmental Paradigm based on students' knowledge about ecosystem and locus of control. *Eurasia J. Math. Sci. Technol. Educ.* **2015**, *11*, 325–333. [CrossRef]
50. Celestine, N. 5 Quality of Life Questionnaires and Assessment. Available online: https://positivepsychology.com/quality-of-life-questionnaires-assessments/ (accessed on 29 July 2022).
51. Parsons, E.C.M.; Warburton, C.A.; Woods-Ballard, A.; Hughes, A.; Johnston, P.; Bates, H.; Luck, M. Whale-watching tourists in west Scotland. *J. Ecotour.* **2003**, *2*, 93–113. [CrossRef]
52. Comrey, A.; Lee, H. *A First Course in Factor Analysis*, 2nd ed.; Lawrence Erlbaum: Hillsdale, NJ, USA, 1992.
53. Deng, J.; Walker, G.J.; Swinnerton, G. Comparison of environmental values and attitudes between Chinese in Canada and Anglo-Canadians. *Environ. Behav.* **2006**, *38*, 22–47. [CrossRef]
54. Hair, J.F.; Black, W.C.; Babin, B.J.; Anderson, R.E. *Multivariate Data Analysis*, 7th ed.; Prentice Hall: Upper Saddle River, NJ, USA, 2010.
55. Fornell, C.; Larcker, D.F. Evaluating structural equation models with unobserved variables and measurement error. *J. Mark. Res.* **1981**, *18*, 39–50. [CrossRef]
56. Wheaton, B.; Muthén, B.; Alwin, D.F.; Summers, G.F. Assessing reliability and stability in panel models. In *Sociological Methodology*; Heise, D.R., Ed.; JosseyBass: San Francisco, CA, USA, 1977; pp. 84–136.
57. Nunnally, J.C. *Psychometric Theory*, 2nd ed.; McGraw-Hill: New York, NY, USA, 1978.
58. Carmines, E.; Zeller, R. Quantitative Applications in the Social Sciences. In *Reliability and Validity Assessment*; Sage: Newbury Park, CA, USA, 1979; Volume 17.
59. Hinton, P.R.; Brownlow, C.; McMurray, I.; Cozens, B. *SPSS Explained*, 2nd ed.; Routledge: London, UK, 2014.
60. Uysal, M.; Jurowski, C.; Noe, F.P.; McDonald, C.D. Environmental attitude by trip and visitor characteristics. *Tour. Manag.* **1994**, *15*, 284–294. [CrossRef]
61. Kline, R.B. *Principles and Practice of Structural Equation Modeling*, 2nd ed.; Guilford Press: New York, NY, USA, 2005.
62. Lam, L.W. Impact of competitiveness on salespeople's commitment and performance. *J. Bus. Res.* **2012**, *65*, 1328–1334. [CrossRef]
63. Xu, H.; Tracey, T.J.G. Use of multi-group confirmatory factor analysis in examining measurement invariance in counselling psychology research. *Eur. J. Couns. Psychol.* **2017**, *6*, 75–82. [CrossRef]
64. Putnick, D.; Bornstein, M.H. Measurement invariance conventions and reporting: The state of the art and future directions for psychological research. *Dev. Rev.* **2016**, *41*, 71–90. [CrossRef]
65. Cheung, G.W.; Rensvold, R.B. Testing factorial invariance across groups: A reconceptualization and proposed new method. *J. Manag.* **1999**, *25*, 1–27. [CrossRef]
66. Chen, F.F. Sensitivity of goodness of fit indexes to lack of measurement invariance. *Struct. Equ. Model.* **2007**, *14*, 464–504. [CrossRef]
67. Weinbrenner, H.; Breithut, J.; Hebermehl, W.; Kaufmann, A.; Klinger, T.; Palm, T.; Wirth, K. "The forest has become our new living room"—The critical importance of urban forests during the COVID-19 pandemic. *Front. For. Glob. Chang.* **2021**, *4*, 672909. [CrossRef]
68. Geng, D.; Innes, J.; Wu, W.; Wang, G. Impacts of COVID-19 pandemic on urban park visitation: A global analysis. *J. For. Res.* **2021**, *32*, 553–567. [CrossRef] [PubMed]
69. Yang, Y.; Lu, Y.; Yang, L.; Gou, Z.; Liu, Y. Urban greenery cushions the decrease in leisure-time physical activity during the COVID-19 pandemic: A natural experimental study. *Urban For. Urban Green.* **2021**, *62*, 127136. [CrossRef] [PubMed]
70. Lee, J.; Cheng, M.; Syamsi, M.N.; Lee, K.H.; Aung, T.R.; Burns, R.C. Accelerating the nature deficit or enhancing the nature-based human health during the pandemic era: An international study in Cambodia, Indonesia, Japan, South Korea, and Myanmar, following the Start of the COVID-19 Pandemic. *Forests* **2022**, *13*, 57. [CrossRef]
71. Hong, S.; Lee, S.; Jo, H.; Yoo, M. Impact of frequency of visits and time spent in urban green space on subjective well-being. *Sustainability* **2019**, *11*, 4189. [CrossRef]
72. Chi, C.G. An examination of destination loyalty: Differences between first-time and repeat visitors. *J. Hosp. Tour. Res.* **2012**, *36*, 3–24. [CrossRef]
73. Coskun, G. Investigating the relationship between values, satisfaction and intention to return: The case of Clemson International Food Festival. *J. Tour. Manag. Res.* **2018**, *3*, 187–199. [CrossRef]
74. Huang, S.; Hsu, C.H.C. Effects of travel motivation, past experience, perceived constraints, and attitude on revisit intention. *J. Travel Res.* **2009**, *48*, 29–44. [CrossRef]
75. Petrick, J.F.; Sirakaya, E. Segmenting cruisers by loyalty. *Ann. Tour. Res.* **2004**, *31*, 472–475. [CrossRef]
76. Yuksel, A. Managing customer satisfaction and retention: A case of tourist destinations Turkey. *J. Vacat. Mark.* **2001**, *7*, 153–168. [CrossRef]
77. Luck, M. The 'New Environmental Paradigm': Is the scale of Dunlap and Van Liere applicable in a tourism context? *Tour. Geogr.* **2003**, *5*, 228–240. [CrossRef]

78. Tedeschi, R.G.; Calhoun, L.W. The posttraumatic growth inventory: Measuring the positive legacy of trauma. *J. Trauma Stress* **1996**, *9*, 455–471. [CrossRef] [PubMed]
79. Tedeschi, R.G.; Calhoun, L.W. Posttraumatic growth: Conceptual foundations and empirical evidence. *Psychol. Inq.* **2004**, *15*, 1–18. [CrossRef]
80. Deng, J.; Andrada, R., II; Pierskalla, C. Visitors' and residents' perceptions of urban forests for leisure in Washington, D.C. *Urban For. Urban Green.* **2017**, *28*, 1–11. [CrossRef]
81. Henderson, K.A.; Bialeschki, M.D.; Browne, L.P. *Evaluating Recreation Services: Making Enlightened Decisions*, 4th ed.; Sagamore Venture: Urbana, IL, USA, 2017.
82. Liu, S.; Su, C.; Yang, R.; Zhao, J.; Liu, K.; Ham, K.; Takeda, S.; Zhang, J. Using crowdsourced big data to unravel urban green space utilization during COVID-19 in Guangzhou, China. *Land* **2022**, *11*, 990. [CrossRef]
83. García-Ayllón, S.; Kyriakidis, P. Spatial analysis of environmental impacts linked to changes in urban mobility patterns during COVID-19: Lessons learned from the Cartagena case study. *Land* **2022**, *11*, 81. [CrossRef]

 *land*

Article

Inequitable Changes to Time Spent in Urban Nature during COVID-19: A Case Study of Seattle, WA with Asian, Black, Latino, and White Residents

Audryana Nay [1], Peter H. Kahn, Jr. [1,2,*], Joshua J. Lawler [1] and Gregory N. Bratman [1,2,3]

[1] School of Environmental and Forest Sciences, University of Washington, Seattle, WA 98195, USA
[2] Department of Psychology, University of Washington, Seattle, WA 98195, USA
[3] Department of Environmental and Occupational Health Sciences, Seattle, WA 98195, USA
* Correspondence: pkahn@uw.edu

Abstract: The COVID-19 pandemic has impacted everyone in urban areas. Some of these impacts in the United States have negatively affected People of Color more than their White counterparts. Using Seattle, Washington as a case study, we investigated whether inequitable effects appear in residents' interactions with urban nature (such as urban green space). Using a 48-question instrument, 300 residents were surveyed, equally divided across four racial/ethnic groups: Asian, Black and African American, Latino/a/x, and White. Results showed that during the span of about 6 months after the onset of the pandemic, Black and Latino residents experienced a significant loss of time in urban nature, while Asian and White residents did not. The implications of these findings, including inequities in the potential buffering effects of urban nature against COVID-19 and the future of urban nature conservation, are discussed. Multiple variables were tested for association with the changes to time spent in urban nature, including themes of exclusion from urban nature spaces found throughout the existing literature. Findings show that decreases in time spent in urban nature among Black and Latino residents may be associated with their feeling as though they did not belong in urban nature. We provide recommendations based on these findings for how government agencies can promote more equitable access to urban nature during the pandemic and beyond. The results of this study have implications that extend beyond the US and are relevant to the international scholarly literature of inequities and urban nature interaction during the COVID-19 pandemic.

Keywords: urban nature; green space; equity; sense of belonging; COVID-19; BIPOC

Citation: Nay, A.; Kahn, P.H., Jr.; Lawler, J.J.; Bratman, G.N. Inequitable Changes to Time Spent in Urban Nature during COVID-19: A Case Study of Seattle, WA with Asian, Black, and White Residents. *Land* **2022**, *11*, 1277. https://doi.org/10.3390/land11081277

Academic Editors: Francesca Ugolini, David Pearlmutter and Yang Xiao

Received: 6 July 2022
Accepted: 4 August 2022
Published: 9 August 2022

Publisher's Note: MDPI stays neutral with regard to jurisdictional claims in published maps and institutional affiliations.

1. Introduction

This research lies at the intersection of three large conditions that are restructuring human lives and social systems. The first is recent: the COVID-19 pandemic. The second, and here we speak about the United States specifically, is the longstanding structural racism within society that continues to harm People of Color[1]. The third is the increasing diminishment of nature on this planet, and in the lives of people. Because interacting with nature can help people physically and psychologically, it seems plausible that being in nature can buffer some of the pandemic's negative effects. Yet, if so, and given existing structural racism, it is also plausible that People of Color have not equally benefited from this potential.

Examining this disparity is the motivation for this study. During COVID-19, People of Color in US cities have experienced more negative outcomes (compared with the White population), with a higher likelihood of COVID-19 infection, poorer COVID-19 outcomes, higher stress and anxiety levels, and larger unemployment rates [1–6]. Thus, in this study we investigated whether the amount of time residents of Seattle, Washington spent in urban nature changed after 6 months of the pandemic, and if so, how those changes varied across four racial/ethnic groups: Asian, Black and African American, Hispanic and Latino/a/x, and White[2][3][4].

1.1. The Importance of Assessing Race/Ethnicity When Investigating Urban Nature Visitation

One of the reasons why there may be inequities by race/ethnicity in changes in nature access during the pandemic is because of the history of racism in the US, which has led to historical differences in perceptions and use of urban nature. Roberts et al. [11] described how current environmental racism reflects racist policies in the past and how many historic policies were tied to the natural landscape. For example, after enslaved people in the US were declared free, former slave-owners "employed" freed Black Americans and sold them farm equipment, livestock, and other necessities in advance. Under sharecropping contracts, the Black workers were to pay the farm owners a share of their crop yield to pay for those assets making the accumulation of wealth extremely challenging [11]. This is just one way in which the relationship between Black Americans and the natural landscape has been intertwined with racism throughout US history.

Today, Black Americans' relationship with nature continues to be shaped by historic racism. Finney [12] showed, for example, how urban parks were historically often places for acts of racism, and how these places can convey racist sentiments for some Black individuals. Natural landscapes can be associated with lynchings, slavery, segregation policies, and events of conflict and violence [12–14]. By contrast, for many White individuals more "wild" or "untouched" nature harkens to a "simpler time" before industrialization and represents a nostalgic longing for the past [12,14][5].

The deep history of racism intertwining with the natural world is reflected in the racism in natural spaces that continues today [16]. A recent example of racism in urban nature spaces that made national headlines occurred in Central Park, New York. In May 2020, a White woman called the police on a Black man who was birdwatching in the park because the man had asked the woman to leash her dog in accordance with the law [17]. As Newsome [18] writes: "For far too long, Black people in the United States have been shown that outdoor exploration activities are not for us, whether it be because the way the media chooses to present who is the 'outdoorsy type' or the racism experienced by Black people when we do explore the outdoors, as we saw recently in Central Park."

As demonstrated in multiple studies, racial/ethnic inequities in access to urban nature appear in a multiplicity of ways: For one, White neighborhoods typically contain a higher density of urban nature areas compared with neighborhoods consisting primarily of People of Color due to redlining and other discriminatory housing practices [19–23]. Second, the quality of urban nature is generally lower in Communities of Color [24]. Third, the upkeep of urban nature is generally lower in Communities of Color [25,26]. Fourth, social barriers to urban nature accessibility for people in Communities of Color can exist at the personal, institutional, or systemic level. These barriers include lack of multilingual signage, safety concerns, lack of free time, transportation limitations, cultural expectations and norms, and historically segregated park design [19,27,28].

1.2. Effects of COVID-19 on Urban Nature Access Disaggregated by Race/Ethnicity

If measured as a homogenous group, some literature suggests that urban residents have increased their urban nature use during the pandemic [29]. For example, in a survey of land managers of urban parks across 12 US cities, 83% reported an increase in visitation to the spaces they manage [30].

However, when disaggregating residents by race/ethnicity, the emerging studies show conflicting results. Larson et al. [31], for example, found that Black and Hispanic (compared with White) residents of cities across North Carolina, USA, experienced a greater decrease in urban nature visitation 6 months after the start of the COVID-19 pandemic. Similarly, a study of New York City residents found that Black and Native American participants were more likely to experience a decrease in urban nature visitation during COVID-19 compared with Asian and White participants [32]. By contrast, other studies have found that People of Color living in cities actually increased their time spent in urban nature during the COVID-19 pandemic. Pipitone and Jović [33], for example, found that non-White New York City residents increased their frequency of urban nature visitation during the first lockdown in New York City and again about 4 months after the pandemic started. Thus, to date, the research is not clear on the effects of the pandemic on nature interaction when disaggregating by race.

We sought to understand the nuances between racial/ethnic groups in terms of the changes to urban nature during the pandemic. To perform this, we disaggregated results by race/ethnicity rather than collapsed participants into White and non-White groups. This study sought to help clarify the existing literature by means of an opening that we saw when reviewing the (above) literature on race/ethnicity and urban nature visitation. Namely, it appeared to us that one overarching construct that might help explain differences in nature access by racial/ethnic group is what we call *sense of belonging in urban nature*.

1.3. Sense of Belonging in Urban Nature

We mentioned earlier the incident of a White woman calling the police on a Black man who was birdwatching in Central Park. As Roberts writes [28], this example carries forward a long history of racism that existed at the time when Central Park was built in the 1850's, when the park became an urban oasis for White people with privilege, and largely excluded People of Color. Roberts [28] writes that it is not just Central Park where this exclusion continues to occur, but in many urban parks and green spaces nationwide. The empirical literature supports this proposition (e.g., Hoover and Lim [16], Joassart-Marcelli [34], and Wolch et al. [35]). For example, Byrne [36] conducted focus groups with Latina women living in Los Angeles, California, near an urban national park. Most participants in this study expressed feeling 'out of place' and/or 'unwelcome' there. One Latina woman expressed worry that a resident would call the sheriff if they saw a Latino in a part of the park that was too close to the White neighborhoods.

Thus, it may be the case that inequities in sense of belonging may play a large role in differences in urban nature visitation across racial/ethnic groups. To date, however, most assessments of sense of belonging have been on people's perceptions of their place within a broader community or social group. Hagerty et al. [37], for example, described belongingness as perceiving oneself as a part of, and integral to, the collective whole. Hagerty and Patusky [38] went on to develop the Sense of Belonging Instrument (SOBI), which includes items with imagery evoking social alienation. One example is an item that reads: "I feel like a square peg trying to fit into a round hole".

Other lines of investigation that touch on sense of belonging are studies that focus on sense of place and place attachment. Sense of place is an overarching construct that describes one's feelings towards a place [39]. Place attachment, a subset of sense of place, more specifically refers to the positive connection between an individual and a specific place [40]. Peters et al. [41] used the idea of place attachment to better understand whether urban parks encourage social cohesion within a neighborhood. The study, which took place in the Netherlands, found that establishing an attachment to urban nature was associated with increased social cohesion amongst non-Western Dutch immigrants.

The literature on People of Color's feelings of exclusion from urban nature spaces shows a relationship between three dimensions: The self, the social, and nature. Sense of belonging, as it is currently characterized in the literature, largely centers around one's place within society, capturing the self and the social dimensions, and lacks the relationship to nature dimension. Sense of place and place attachment both focus on the relationship between the individual and nature, but do not include a larger social dimension. A sense of belonging in urban nature measurement has the potential to bridge the self, the social, and nature to understand the intricate relationships between these dimensions. For our purposes, the limitation with the existing bodies of literature related to belongingness is that they do not focus directly enough on experiences of exclusion, especially those due to a historical legacy of racism, in the context of urban nature.

Jennings et al. [42] described how addressing one instantiation of environmental injustice is not sufficient to tackle systemic environmental racism. In the case of urban nature, seeking to address inequitable feelings of belonging alone cannot deal with the deep history of racism in the US that has fed into today's exclusion of People of Color from urban nature spaces. Nonetheless, we view sense of belonging in urban nature as a significant theme seen throughout the environmental justice literature worthy of investigation.

One study that made a substantial contribution to this concept is from Pipitone and Jović [33]. They measured participants' sense of belonging in urban green space before and during the COVID-19 pandemic through a single 4-point ordinal scale question (with response options of "very strong", "somewhat strong", "somewhat weak", "very weak", and "I don't know") adapted from Rugel et al. [43] which reads" "How would you describe your sense of belonging to local parks or urban green space?" This study found no significant difference in sense of belonging between White and non-White participants before COVID-19. Four months into the pandemic, White participants' sense of belonging was marginally significantly higher than that of non-White participants [33].

To the best of our knowledge, there is no existing scale or multi-item measurement that directly assesses sense of belonging in urban nature. Thus, this study sought to initiate the creation of such a measurement, and then to use it in our present investigation.

1.4. The Present Investigation

In this study, we investigated whether there were inequitable effects during the early period of COVID-19 in terms of the experience of urban nature across four racial/ethnic groups residing in Seattle, Washington: Asian, Black, Latino, and White. More specifically we sought:

1. To assess the frequencies of urban nature interactions before and during COVID-19 across the four racial/ethnic groups;
2. To test for differences in average change in frequency of urban nature interaction before and during COVID-19 across the four racial/ethnic groups;
3. To test whether there were inequitable effects of COVID-19 on the frequency of urban nature interaction before and during COVID-19, and to test for a significant effect of three variables: perceived coronavirus threat, perceived quality of nearby urban nature, and sense of belonging in urban nature, controlling for age, gender, income, and pre-pandemic frequency of urban nature interaction;
4. To gather themes of exclusion from urban nature among People of Color in the US to develop an exploratory new measure for sense of belonging in urban nature, and to employ the measure across the four racial/ethnic groups;
5. To characterize the types of urban nature interaction that residents engaged in before and during the pandemic to add depth to discussions of urban nature interaction.

2. Materials and Methods

2.1. Study Site

Seattle is located in the state of Washington in the Pacific Northwest of the United States. Seattle has a population of 737,015 [44]. Of Seattle's population, 67.3% identifies as White, 15.4% Asian, 7.3% Black or African American, 6.7% Hispanic or Latino (of any race), 0.5% American Indian and Alaska Native, and 0.3% Native Hawaiian and other Pacific Islander. Those who identify as some other race constitute 0.3% of the Seattle population and those who identify as two or more races constitute 6.9%. The median household income of Seattle is USD 92,263 [44].

Williams et al. [45] found Seattle to have less inequity in urban nature access compared with other major cities in the US including Atlanta, GA; Baltimore, MD; Detroit, MI; and Los Angeles, CA. Nonetheless, inequities in urban nature are prevalent. In Seattle, the amount of urban canopy cover in a given census tract is inversely correlated with the proportion of People of Color living in the census tract [46].

Seattle has a long history of racial segregation that has shaped the city. From 1910 to 1960, many Seattle housing property deeds contained clauses that explicitly prohibited People of Color or other discriminated communities from renting or buying the property. By the 1920's, certain areas of what were called the Central District and Chinatown were the only "open neighborhoods" available to People of Color [47]. Today, Seattle's Central District and International District (formerly Chinatown) are composed of 35.5% and 66.8% non-White residents, respectively. Both the Central District and International District have significant gaps in urban nature accessibility [48].

2.2. Participant Recruitment

Participants were primarily recruited online via the social media platforms Facebook and Instagram (both owned by the company Meta). Pay-per-click Facebook and Instagram ads were run by study researchers. These ads provided a short description of the study, advertised participant compensation, and provided a link to the study's eligibility questionnaire. These ads resulted in about 108,939 'impressions' (the number of times the ad is seen by a user. Users may have seen the ad more than one time). Ads were shown to Facebook and Instagram users aged 18 and over. Ads were dispersed equally to users residing within a 10-mile radius of downtown Seattle, encompassing the entire city of Seattle.

A Facebook post with identical information to the ad was shared in various community Facebook Groups. The posts were shared in general Seattle groups (such as 'Mt. Baker Neighborhood, Seattle' and 'Beacon Hill Social Club'), as well as in Facebook groups for Seattle residents of a specific race/ethnicity (such as 'Seattle Latinx Pride' and 'Families of Color Seattle'). The study description and eligibility questionnaire link were also distributed by study researchers to personal and professional connections via email to reach more potential participants. Although we do not know the exact reach of the free Facebook ads or emails, we may reasonably assume the reach of the paid Facebook ads far surpassed the reach achieved through free Facebook posts and email sharing.

Due to the recruitment method, participants were primarily drawn from Facebook and Instagram users, introducing a bias and excluding those who do not use those social media sites. Despite this drawback, Facebook advertising has become a common research recruitment technique and has been shown to result in fairly representative samples [49]. Reagan et al. [50] conducted a review of 18 studies that implemented Facebook advertising and free Facebook posts as recruitment techniques. Many of the included studies focused on sampling vulnerable populations. Reagan et al. concluded that Facebook and Instagram ads were effective recruitment methods stating, " . . . paid ads may increase the likelihood of reaching the target population and maximizing sample accrual."

Individuals interested in participating in the study were directed to an eligibility questionnaire hosted on Qualtrics where they provided responses used to determine eligibility. Prior to beginning the eligibility questionnaire, participants viewed a consent form and provided acknowledgement of consent. To be eligible for participation, individuals must reside within Seattle city limits, have lived in the same residence since at least fall 2019, be at least 18 years of age, be able to read and write English, and identify as Asian, Black or African American, Hispanic or Latino, or White. Individuals who identified as any other race/ethnicity or more than one race/ethnicity were not eligible. Eligible participants received a link to take the main survey after completing the eligibility questionnaire. Participants who completed the main survey received a USD 10 Amazon gift card via email. This study was approved by the University's Institutional Review Board (IRB ID: STUDY00011290).

2.3. Data Collection

Data collection began in January of 2021 and concluded March 2021. A quota sampling technique was used to achieve an equal number of participants in each of the four included racial/ethnic groups (Asian, Black, Latino, and White). After receiving 75 responses from White participants, White participant recruitment ceased and the survey was modified so that only Asian, Black, and Latino individuals were eligible. Similarly, after 75 Asian responses were collected, Asian participant recruitment closed. A total of 78 responses from Latino individuals and 80 responses from Black individuals were received before the survey fully closed. A total of 75 participant responses were randomly sampled from each of these two groups to achieve an equal sample size between the four racial/ethnic groups. It was important for this study to strive for equal representation of the included four racial/ethnic groups. Equal representation, something that is not common among urban nature studies, allows for comparisons between racial/ethnic groups to be made with more confidence, enabling the results to speak to any urban nature inequities that are found.

2.4. Participant Characteristics

The sample consisted of 300 participants with 75 participants in each racial/ethnic group (Asian, Black, Latino, and White). It is worth noting that participants declared their race as part of this survey but were not asked to share any information about their cultural background, which may impact one's urban nature interaction habits. Participants indicated their total household income via income categories. The median annual household income category for the sample was USD 75,000–USD 99,000, encompassing the Seattle median of USD 92,263 [44]. There were slightly more females (56%) than males (42%) in the sample. Participants provided their age according to age categories, with the median age category being 25–34 years of age. This is just under the median Seattle age of 34.7 years of age [44].

2.5. Survey Instrument

A 25-minute survey consisting of 48 multiple-choice, Likert scale, and open-ended questions was administered to participants. This online survey was hosted on Qualtrics. The survey sought to capture changes in urban nature visitation 6 months into the pandemic as well as evaluate participants' urban nature perceptions and values. The following definition of urban nature was provided to participants before and throughout the survey: "Urban nature refers to parks, green areas, open spaces, and places with water, vegetation, and/or animals within the city of Seattle. Urban nature does not include things you may pass by briefly, such as trees along a sidewalk. Nature elements which one may pass by briefly, such as urban street trees, were not included as urban nature for this study to place more emphasis on urban nature spaces one may intentionally seek to spend time in."

To understand how participants' urban nature use changed during the COVID-19 pandemic, a set of questions was provided twice within the survey. The first time, participants were asked to reflect to their experiences in fall 2019 (before the COVID-19 pandemic). Participants were then provided the same set of questions and asked to respond according to their recent experiences in fall 2020 (about 6 months into the COVID-19 pandemic). Fall was chosen as the reference period for both before the pandemic and during the pandemic. One reason for this was to reduce variability that may be due to different levels of outdoor activity throughout the year. The second time point was positioned 6 months into the COVID-19 pandemic in an attempt to accurately represent how the pandemic may affect urban residents long term. The immediate changes to urban nature interaction during the early months of the COVID-19 pandemic may be more extreme or different than those changes seen further into the pandemic. With future disruptive events possibly affecting urban life for extended periods of time, a moderately long time span of 6 months may be most appropriate when assessing the impacts of the event.

2.6. Measures

The key measurements and scales included in this survey are below:

Types of Urban Nature Interaction: To attain some specificity in Seattle residents' urban nature interactions, this survey asked about the types of urban nature activities participants engaged in. Participants were provided a list of 20 common activities in urban nature such as walking a dog and having a picnic. They were asked to indicate all urban nature activities which they had enacted in fall 2019 and fall 2020. These data draw upon Interaction Pattern Theory, a way of characterizing the meaningful and instantiated ways in which people interact with nature [51–54]. These data were collected to add more depth to our understanding of how urban nature interaction changed during the pandemic, if at all. Although we consider these data exploratory, they allow for greater nuance when considering how resiliency might be increased in urban communities.

Frequency of Urban Nature Interaction: Within this group of pre- and during COVID-19 questions, participants were asked about how frequently they spent time in urban nature in fall 2019 and fall 2020. This question read: "Over the course of fall [2019 or 2020], how frequently did you spend time in or around urban nature?" Participants responded to this multiple-choice question with how many days per week, on average, they spent time in or around urban nature in fall 2019 (before the pandemic) and fall 2020 (about 6 months into the pandemic). Multiple-choice response options included: "Less than once per month", "1–3 times per month", "Once per week", "2–3 days per week", "4–5 days per week", "6 days per week", and "Daily". These responses were converted to days per month. The average of each response option was used (e.g., "Less than once per month" was replaced with 0 days per month, "2–3 days per week" was replaced with 10 days per month, and "Daily" was replaced with 28 days per month).

Urban Nature Conservation Values: A single Likert scale question was used to measure participants' perceived level of importance of urban nature conservation to test for association with sense of belonging in urban nature. One's urban nature conservation values were assessed through the question: "How important to you is the protection of urban nature?" The 5-point Likert question response options ranged from not at all important (1) to very important (5).

Perceived Coronavirus Threat Questionnaire (short): Three measures were explored as possible explanatory variables for the differences in the effects of COVID-19 on urban nature interaction frequency. The first was the short version of the Perceived Coronavirus Threat Questionnaire, developed and validated by Conway et al. [55]. This scale is used to assess the level of which participants were fearful of the COVID-19 virus. This measure was included as spending time in urban nature may mean being in close proximity to other people. During the COVID-19 pandemic, this may motivate city residents to spend less time in urban nature. This shortened scale was modified from a 7-point Likert scale to a 5-point Likert scale ranging from not at all true of me (1) to very true of me (5). The scale includes three items which read: "Thinking about the coronavirus (COVID-19) makes me feel threatened."; "I am afraid of the Coronavirus (COVID-19)."; "I am stressed around other people because I worry I'll catch the coronavirus (COVID-19)." Cronbach's alpha for the Perceived Coronavirus Threat Questionnaire in this study was 0.74.

Perceived Green Space Quality Scale: The second measure explored as a possible explanatory variable for the differences in the effects of COVID-19 on urban nature interaction frequency was the Perceived Greenspace Quality Scale [56]. As previously noted, urban nature quality is a key dimension of accessibility, and can provide insight into why an urban nature spot might not be visited. This scale was adapted by replacing the term "greenspace" with "urban nature". The 10-point Likert scale was converted to a 5-point Likert scale ranging from completely disagree (1) to completely agree (5). Examples of items in this scale include: "My neighborhood has safe urban nature spots."; "My neighborhood has well-maintained urban nature spots."; "My neighborhood has beautiful urban nature spots." Cronbach's alpha for the Perceived Greenspace Quality Scale in this study was 0.84.

Sense of Belonging in Urban Nature Questionnaire: The third measure we sought to explore as a possible independent variable for changes in frequency of urban nature interaction was sense of belonging in urban nature. No existing measurement fit the requirements for this measure, so an exploratory questionnaire was developed for this study. This measure, which we call the Sense of Belonging in Urban Nature Questionnaire, was intended to better understand experiences of inequity in urban nature, specifically as it relates to one's sense of belonging. Each of the six items in this questionnaire correspond to a larger overall theme of inequity distilled from the existing literature. (See Table 1 for all items and corresponding literature). Participants responded to each item on a 5-point Likert scale ranging from completely disagree (5) to completely agree (1). Internal reliability of this questionnaire was high (Cronbach's alpha = 0.84); however, this questionnaire remains unvalidated. The Sense of Belonging in Urban Nature Questionnaire is composed of six themes that characterize six racial/ethnic inequities in the pursuit of characterizing one facet of environmental racism. White people (of any nationality) form the majority in the US and do not, for the most part, face racial discrimination or racial inequities in the US, and by extension, in urban nature spaces. It is for this reason that the sources for the Sense of Belonging in Urban Nature Questionnaire are all grounded in the experiences of People of Color in the US.

Table 1. The Sense of Belonging in Urban Nature Questionnaire.

Theme	Item
Ease of Access [22,36,57–62]	"It is not easy for me to get to a park or other urban nature spot near my home."
Safety [13,25,36,60,61,63,64]	"When in an urban nature spot near my residence, I fear for my own safety or the safety of others around me."
Feeling Out of Place [25,36,64,65]	"I feel out of place in the urban nature spots I visit."
Unwelcomeness [13,36,61]	"I feel unwelcome by others when in urban nature."
Institutional Acceptance [13,26,28,36,65]	"I feel uncomfortable when I see a park management employee when in urban nature."
Different Ways of Interacting with Nature Acceptance [13,36,66–68]	"I feel that the way I use urban nature is unwelcome or unaccepted by other visitors."

The six themes of racial/ethnic inequities that form the Sense of Belonging in Urban Nature Questionnaire were identified through a literature review of social barriers to urban nature use among People of Color in the US and the resulting feelings of exclusion. The themes included in the sense of belonging measurement are by no means the only ways in which Communities of Color feel excluded from urban nature. Nor are they likely uniform for experiences across all Communities of Color. The themes of inequity included in the exploratory Sense of Belonging in Urban Nature Questionnaire are intended to characterize broad ways in which exclusion presents. The sense of belonging inequity themes are Ease of Access, Safety, Feeling Out of Place, Unwelcomeness, Institutional Acceptance, and Different Ways of Interacting with Nature Acceptance. See Table 1 for source literature for each theme. Descriptions of these inequity themes are below:

- **Ease of Access:** this theme characterizes difficulties in spending time in urban nature due to socioeconomic inequities including proximity to nearby urban nature, poor quality of nearby urban nature, lack of free time, and transportation limitations;
- **Safety:** one is less likely to feel a sense of belonging in urban nature if spending time in urban nature poses a risk to personal safety or the safety of others;
- **Feeling Out of Place:** This theme seeks to capture feelings of not belonging or fitting in within the landscape. There are several factors that may lead to one feeling out of place in urban nature spaces. Some include having very limited representation of People of Color in nature spaces, cultural expectations and norms, and being the only Person of Color in an urban nature space;
- **Unwelcomeness:** Feelings of not belonging in urban nature can arise from external exclusion from those in the White majority. Overt and covert messages from White individuals in urban nature spaces can send a clear message of unwelcomeness to People of Color in the space;
- **Institutional Acceptance:** If People of Color are not accepted in urban nature on an institutional level, urban nature spaces and management practices reflect that. People of Color may feel that urban nature areas were not created for them, with the design catering to typically Eurocentric ways of interacting with urban nature. People of

Color also experience conflicts with those who manage urban nature spaces due to their presence in these spaces;

- **Different Ways of Interacting with Nature Acceptance:** People of Color may feel that the way they use urban nature is not deemed acceptable or welcome by others.

2.7. Analysis

Participants' frequencies of urban nature interaction in fall 2019 (before COVID-19) and fall 2020 (6 months into COVID-19) were first compared between racial/ethnic groups. The Kruskal–Wallis non-parametric equivalent to ANOVA was conducted to test whether any pair(s) of racial/ethnic groups had significantly different frequencies of urban nature interaction in fall 2019. The Kruskal–Wallis test uses ranked data points to test for differences in the mean rank of each group in the independent variable. Dunn's test for stochastic dominance, a common post-hoc test following the Kruskal–Wallis test, was then used to identify which pair(s) of racial/ethnic groups significantly differed in frequency of urban nature interaction in fall 2019. The "dunnTest()" function in R was used with the specification that the comparisons were one-sided. One-sided post-hoc tests allow for the results to speak of directionality. The Bonferroni method was used to adjust the p-values of this post-hoc test to reduce the familywise error rate associated with multiple testing. The same process was then conducted to compare groups' 2020 frequencies.

We used two linear regression analyses to test whether sense of belonging was significantly associated with 2019 frequency of urban nature interaction and/or 2020 urban nature interaction. Control variables were included in each regression model for race/ethnicity, age, gender, and income.

To test whether a given racial/ethnic group experienced a significant change in frequency of urban nature interaction, the average 2019 and 2020 frequencies were first calculated for each group. One-tailed paired sample t-tests were then conducted within each racial/ethnic group to compare their 2019 and 2020 average frequencies. Although the distributions for 2019 and 2020 reported frequencies of urban nature interaction are mildly non-normal, the sample size (300 total, 75 participants in each racial/ethnic group) is large enough to justify the use of Student's t-test. The tests were directional because each group's 2020 average frequency was observed, descriptively, to be either greater or less than their 2019 frequency. Control variables such as age, gender, and income were not included in these tests since the "before" and "during" data were for the same set of participants. If these t-tests were significant, it meant the racial/ethnic group experienced a significant increase or decrease (depending on the directionality of the test) in frequency of urban nature interaction from fall 2019 to fall 2020.

We then tested whether the COVID-19 pandemic impacted the frequency of urban nature interaction differently across racial/ethnic groups. The Kruskal–Wallis test was conducted to test whether the observed changes in frequency were different across racial/ethnic groups. Given that this test was significant, one or more pairs of racial/ethnic groups experienced significantly different effects of COVID-19 on their frequency of urban nature interaction. Dunn's test for stochastic dominance was then used to identify which pair(s) of racial/ethnic groups significantly differed in observed change to frequency of urban nature interaction. A one-sided Dunn's test was used for post-hoc comparisons in order to speak about directionality of significant differences. The Bonferroni method was used to adjust the p-values of this post-hoc test.

A stepwise regression analysis was conducted to test whether perceived coronavirus threat, perceived urban nature quality, or sense of belonging in urban nature can partially explain differences in the effects of COVID-19 on frequency of urban nature interaction. The automated stepwise regression analysis was chosen because we were interested in testing which variable(s), of the several that were of interest, significantly contributed to the inequities in change in frequency of urban nature interaction. Control variables (age, gender, income, and pre-pandemic frequency) were introduced to better isolate the effects of racial/ethnic inequities. Age data were converted from categorical responses to integers by taking the average of the multiple response options for age (e.g., 18–24 years of age was replaced with 21). Average annual income categories were similarly replaced with the average for that response category and rounded to the nearest whole dollar (e.g., USD 50,000 to USD 74,999 was replaced with 62500). Less than USD 25,000 was replaced with 24999 and USD 200,000 or more was replaced with 200000. Pre-pandemic frequency of urban nature interaction was included as a control variable as those with a high 2019 (pre-pandemic) frequency have the potential for a larger decrease in average days per month than those with a lower 2019 frequency (and vice versa for those who start with a low 2019 frequency). Perceived coronavirus threat, perceived quality of urban nature, and sense of belonging variables were added to a regression formula with the control variables. The dependent variable of this regression formula was change in frequency of urban nature interaction. A forward and backward variable selection process was automated using the "step()" command in R to select a formula-based linear regression model based on the Akaike Information Criterion (See Table 2 for the automated variable selection steps). This stepwise regression analysis removes any independent variables which do not significantly contribute to partially predicting the outcome variable for the specified sample. Both control and explanatory variables were permitted to be removed in this process. The "step()" function returns the regression formula that includes the independent variables which produce the lowest Akaike Information Criterion (AIC). The AIC value was used to compare regression models with different independent variables and indicate which set of variables best predict the outcome for that specific data set. This means that if another sample were tested in the same way, a different combination of independent variables may be returned by the stepwise analysis.

Table 2. The automated stepwise variable selection process.

Step	Variable Removed	Df	Deviance	Resid. Df	Resid. Dev	AIC
0	NA	NA	NA	287	12,408.43	1142.71
1	Gender	3	58.69	290	12,467.13	1138.12
2	Income	1	0.05	291	12,467.17	1136.12
3	Perceived coronavirus threat	1	3.17	292	12,470.34	1134.20
4	Perceived urban nature quality	1	13.56	293	12,483.90	1132.52

We used Kruskal–Wallis tests to observe whether sense of belonging in urban nature, perceived quality of nearby urban nature, level of importance of nearby urban nature conservation, and perceived COVID-19 threat level significantly varied across racial/ethnic groups. If the Kruskal–Wallis test was significant, we then used Dunn's test for stochastic dominance to identify which pair(s) of racial/ethnic groups significantly differed. A one-sided Dunn's test was used for post-hoc comparisons in order to speak about directionality of significant differences. The Bonferroni method was used to adjust the p-values of these post-hoc tests.

The association between the level of importance one assigns to urban nature conservation and their sense of belonging in urban nature was explored using a linear regression model. Importance of urban nature conservation was regressed onto several control variables (race/ethnicity, age, gender, and income) and responses to the Sense of Belonging in Urban Nature Questionnaire. The covariate p values were used to assess whether sense of belonging significantly predicted urban nature conservation values.

To analyze the types of urban nature interactions participants engaged in, frequencies of occurrence for each activity in fall 2019 and fall 2020 were descriptively compared. Comparisons were made for the entire sample as well as within each racial/ethnic group. Since this data are exploratory and was collected via two 'check all that apply' questions, the authors found descriptive analysis to be sufficient in this case.

We analyzed all data in RStudio version 1.4.1103. Statistical significance was $\alpha = 0.05$ for all inferential analyses.

3. Results

3.1. Frequencies of Urban Nature Interaction before and during COVID-19

In fall 2019 (before the COVID-19 pandemic), White participants spent time in or around urban nature most frequently with an average of 11.20 days per month (SD = 8.43) (see Table 3). This was followed by Black (M = 8.32, SD = 6.50) and Latino participants (M = 7.65, SD = 6.40). White participants did not spend time in urban nature significantly more frequently than Black (adjusted $p = 0.584$) or Latino participants (adjusted $p = 0.064$) during this time. Average frequency of urban nature interaction for Asian participants (M= 7.81) was significantly lower than that of White participants (M = 7.81, adjusted $p = 0.004$).

Table 3. Average measurement values across racial/ethnic groups.

	Measurement Average (SD)			
	Asian	**Black**	**Latino**	**White**
2019 Frequency [1]	7.81 (8.21)	8.32 (6.50)	7.65 (6.40)	11.20 (8.43)
2020 Frequency	7.09 (8.47)	4.56 (6.24)	5.55 (7.12)	12.35 (8.99)
Change in Frequency (δ) [2]	−0.72, −9.22% (8.08)	**−3.76, −45.19% (6.42)**	**−2.12, −27.71% (6.75)**	1.15, +10.27% (8.56)
Perceived Coronavirus Threat (Low: 3, High: 15)	12.15 (2.08)	12.88 (1.70)	12.25 (1.99)	11.87 (2.61)
Perceived Urban Nature Quality (Low: 6, High: 30)	24.07 (4.09)	21.76 (4.78)	22.31 (4.15)	24.61 (4.99)
Sense of Belonging in Urban Nature (Low: 6, High: 30)	23.24 (4.62)	18.56 (4.75)	20.12 (5.21)	25.31 (4.41)
Level of Importance of Urban Nature Conservation (Low: 1, High: 5)	4.80 (0.40)	4.44 (0.78)	4.64 (0.63)	4.97 (0.16)

[1] All frequency values are provided in average number of days per month. [2] The bolded Change in Frequency values indicate that the racial/ethnic group experienced a significant change in frequency from fall 2019 to fall 2020.

About 6 months after the start of the pandemic (in fall 2020), White participants still had the most frequent urban nature interaction with an average frequency of 12.35 days per month (SD = 8.99). Asian participants had the next most frequent urban nature use with an average of 7.09 days per month (SD = 8.47). This was followed by Latino participants (M = 5.55, SD = 7.12) and Black participants (M = 4.56, SD = 6.24). About 6 months after the pandemic began, the average frequency of urban nature interaction among White participants was significantly higher than that of Asian (adjusted $p < 0.01$), Latino (adjusted $p < 0.001$), and Black participants (adjusted $p < 0.001$). The Asian, Latino, and Black frequencies did not significantly differ from each other.

3.2. Average Change in Frequency of Urban Nature Interaction

This study investigated whether each racial/ethnic group experienced a change in frequency of urban nature interaction from before the pandemic to during the pandemic. There was no significant difference in the frequency of urban nature interaction for White and Asian participants from fall 2019 to fall 2020. Latino and Black participants, however, experienced a significant decrease in frequency of urban nature interaction 6 months into the COVID-19 pandemic (see Figure 1).

White participants experienced **no significant change** in average days per month spent in urban nature from fall 2019 to fall 2020 ($\delta = 1.15$; $p = 0.125$; H_0: $\delta \not> 0$; 95%CI$_{low}$: −0.50).

Asian participants experienced **no significant change** in average days per month spent in urban nature from fall 2019 to fall 2020 ($\delta = -0.72$; $p = 0.222$; H_0: $\delta \not< 0$; 95%CI$_{high}$: 0.84).

Latino participants experienced a **significant decrease** in average days per month spent in urban nature from fall 2019 to fall 2020 ($\delta = -2.12$; $p < 0.004$; H_a: $\delta \leq 0$; 95%CI$_{high}$: −0.81).

Black participants experienced a **significant decrease** in average days per month spent in urban nature from fall 2019 to fall 2020 ($\delta = -3.76$; $p < 0.001$; H_a: $\delta \leq 0$; $95\%CI_{high}$: -2.53).

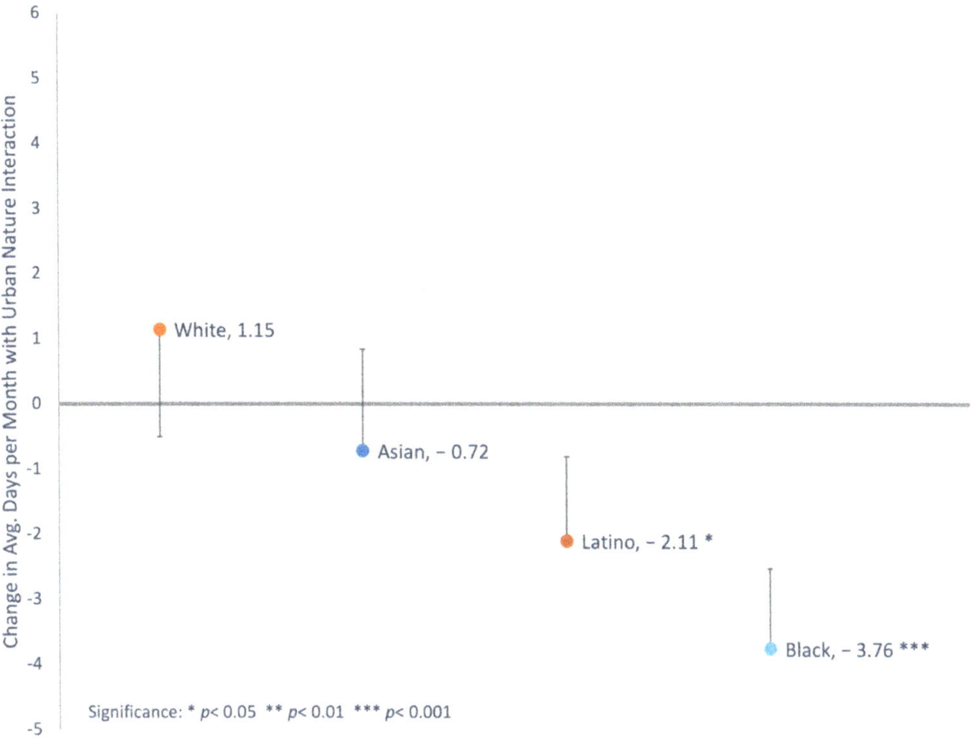

Figure 1. Change to average frequency of urban nature interaction among each racial/ethnic group.

To test whether COVID-19 impacted frequency of urban nature interaction significantly unequally amongst racial/ethnic groups, changes to frequency of urban nature interaction were compared across racial/ethnic groups. The Kruskal–Wallis non-parametric equivalent to ANOVA was used. This test was significant ($p < 0.001$), meaning the COVID-19 pandemic did not impact frequency of urban nature interaction equally across racial/ethnic groups.

Black participants were impacted to a greater degree than White participants (adjusted $p < 0.001$). Latino participants were impacted to a greater degree than White participants (adjusted $p = 0.009$). Black participants' frequency of urban nature interaction was also impacted significantly more than that of Asian participants (adjusted $p = 0.009$). No other pairings of racial/ethnic groups significantly differed in change to frequency of urban nature interaction.

3.3. Independent Variables for the Inequitable Effects of COVID-19

With the observed disparate outcomes in frequency of urban nature interaction during COVID-19, we tested whether perceived coronavirus threat, perceived quality of nearby urban nature, and/or sense of belonging in urban nature at least partially contribute to this disparity. Based on an automated stepwise variable selection, it was determined that the combination of race/ethnicity, age, pre-pandemic frequency of urban nature interaction, and sense of belonging best predict the changes to frequency of urban nature interaction of this study's sample (AIC = 1132.52). To attain the smallest AIC value, perceived coronavirus threat, perceived quality of nearby urban nature, gender, and income variables were removed from the regression formula. This result shows that the effects of COVID-19 on frequency of urban nature interaction are associated with sense of belonging in urban nature, pre-pandemic frequency of urban nature interaction, race/ethnicity, and age. This regression shows that participants with a lower sense of belonging in urban nature lost more time in urban nature during COVID-19. It is important to note here that neither race/ethnicity, age, pre-pandemic frequency, sense of belonging, nor the combination of these variables can fully explain the changes to frequency of urban nature interaction. However, it is of interest to see which of the included variables in this study best fit the outcome data.

With sense of belonging in urban nature significantly contributing to changes in frequency of urban nature interaction from 2019 to 2020, we tested whether sense of belonging was significantly associated with either 2019 frequency of urban nature interaction or 2020 frequency of urban nature interaction. Sense of belonging in urban nature did not significantly predict frequency of urban nature interaction in fall 2019 ($p = 0.211$) but predicted frequency of urban nature interaction in fall 2020 ($p = 0.001$).

3.4. Sense of Belonging in Urban Nature

White participants descriptively responded with the highest sense of belonging in urban nature (M = 25.31, SD = 4.4) followed by Asian (M = 23.24, SD = 4.62), Latino (M = 20.12, SD = 5.21), and Black participants (M = 18.56, SD = 4.75) (see Figure 2, Table 3). A Kruskal–Wallis test shows disparities in sense of belonging in urban nature across racial/ethnic groups ($p < 0.001$). White participants had a significantly higher sense of belonging in urban nature than that of Latino (adjusted $p < 0.001$) and Black participants (adjusted $p < 0.001$). Sense of belonging among Asian participants was significantly higher than that of Latino (adjusted $p < 0.002$) and Black (adjusted $p < 0.001$) participants.

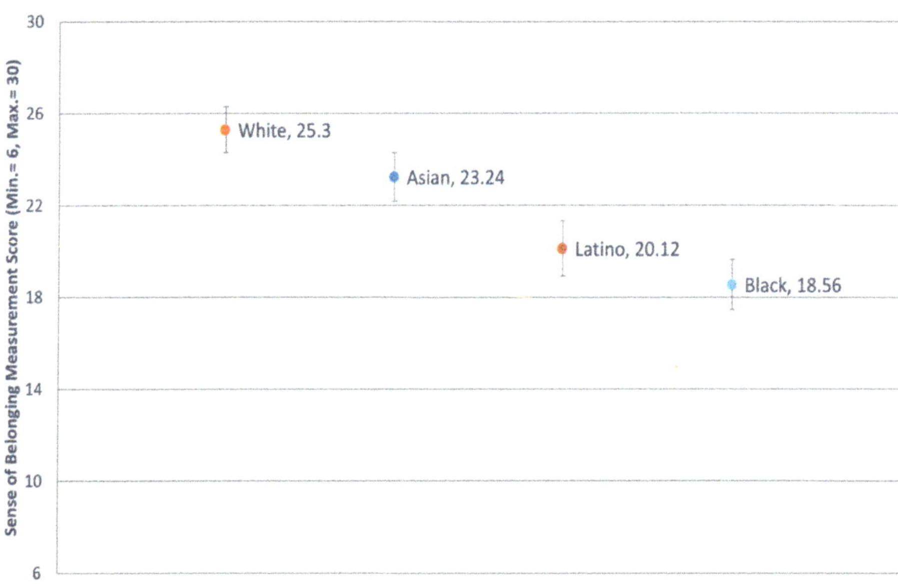

Figure 2. Average sense of belonging in urban nature for each racial/ethnic group.

Using a linear regression with control variables for race/ethnicity, age, gender, and income, sense of belonging in urban nature was found to significantly predict participants' reported importance of urban nature conservation ($p = 0.02$). Participants with a higher sense of belonging in urban nature put a higher level of importance on urban nature conservation.

3.5. Perceived Quality, COVID-19 Threat Level, and Importance of Urban Nature Conservation

A Kruskal–Wallis test revealed that at least one pair of racial/ethnic groups significantly differed in their level of importance of nearby urban nature conservation ($p < 0.001$). Dunn's test shows that the average response to the Perceived Coronavirus Threat Questionnaire among Black residents (M = 4.44, SD = 0.78) was significantly lower than that of Asian residents (M = 4.80, adjusted $p = 0.0034$) and White residents (M = 4.97, adjusted $p < 0.001$) (see Table 3).

We found significant differences between at least one pair of racial/ethnic groups in terms of perceived quality of nearby urban nature ($p < 0.001$). Black participants scored significantly lower on the Perceived Green Space Quality Scale (M = 21.76, SD = 4.78) compared with both Asian (M = 24.07, adjusted $p = 0.007$) and White participants (M = 24.61, adjusted $p < 0.001$). Latino participants also scored significantly lower (M = 22.31, SD = 4.15) than both Asian (adjusted $p = 0.03$) and White participants (adjusted $p < 0.001$) (see Table 3).

No significant differences in perceived coronavirus threat were observed between racial/ethnic groups.

3.6. Types of Urban Nature Interactions

Prior to the pandemic, participants of this sample descriptively took walks with other people more frequently than they took walks alone. Six months into the pandemic, participants more frequently took walks alone than took walks with other people (see Table 4 for descriptive frequencies of each activity before and during the pandemic). This pattern holds true descriptively within each racial group except for Latino and White participants. Latino participants more frequently took walks alone before the pandemic. White participants more frequently took walks with other people during the pandemic.

Table 4. Types of urban nature interaction before and during the COVID-19 pandemic.

| | Frequency (%) | | | | | | | | | |
| | *n* = 300 | | Asian (*n* =75) | | Black (*n* = 75) | | Latino (*n* = 75) | | White (*n* = 75) | |
Urban Nature Interaction [1]	**Before**	**During**	**Before**	**During**	**Before**	**During**	**Before**	**During**	**Before**	**During**
Took a walk with other people	194 (65)	137 (46)	53 (71)	38 (51)	37 (49)	15 (20)	40 (53)	24 (32)	64 (85)	60 (80)
Took a walk alone	179 (60)	164 (55)	51 (68)	47 (63)	26 (35)	24 (32)	44 (44)	36 (48)	58 (77)	57 (76)
Sat in nature	155 (52)	116 (39)	38 (51)	30 (40)	24 (32)	18 (24)	40 (53)	24 (32)	53 (71)	44 (59)
Enjoyed the stillness and quietness of nature	153 (51)	129 (43)	39 (53)	29 (39)	28 (37)	21 (28)	40 (53)	35 (47)	46 (61)	44 (59)
Watched the sunrise or sunset	137 (46)	108 (36)	32 (43)	14 (19)	14 (15)	20 (27)	36 (48)	26 (35)	55 (73)	48 (64)
Looked out at a large view of water	135 (45)	112 (37)	41 (55)	34 (45)	9 (12)	8 (11)	32 (43)	20 (27)	53 (71)	50 (67)
Ran or jogged	129 (43)	72 (24)	34 (45)	19 (25)	30 (40)	12 (16)	37 (49)	17 (23)	28 (37)	24 (32)
Looked out at a large view of the city	97 (32)	77 (26)	31 (41)	24 (32)	4 (5)	1 (1)	19 (25)	13 (17)	43 (57)	39 (52)
Had a picnic	94 (31)	57 (19)	27 (36)	18 (24)	9 (12)	1 (1)	23 (31)	13 (17)	35 (47)	25 (33)
Looked at wildlife	84 (28)	71 (24)	23 (31)	13 (17)	5 (7)	2 (3)	14 (19)	10 (13)	42 (56)	46 (61)
Rode a bike	83 (28)	62 (21)	16 (21)	11 (15)	11 (15)	6 (8)	21 (28)	15 (20)	35 (47)	30 (40)
Walked a dog	65 (22)	67 (23)	13 (17)	14 (19)	11 (15)	11 (15)	19 (25)	18 (24)	22 (29)	24 (32)
Tended to a garden	57 (19)	60 (20)	10 (13)	12 (16)	7 (9)	6 (8)	11 (15)	11 (15)	28 (37)	31 (41)
Played a sport	53 (18)	19 (6)	20 (27)	8 (11)	11 (15)	3 (4)	13 (17)	3 (4)	9 (12)	5 (7)
Watched my children play	49 (16)	36 (12)	13 (17)	9 (12)	17 (23)	11 (15)	10 (13)	9 (12)	9 (12)	7 (9)
Used a water vessel such as a kayak, canoe, paddle board, or sailboat	44 (15)	30 (10)	6 (8)	7 (9)	3 (4)	1 (1)	10 (13)	2 (3)	25 (33)	20 (27)
Collected berries, nuts, mushrooms, greens, or other edible items	44 (15)	29 (10)	11 (15)	8 (11)	4 (5)	1 (1)	8 (11)	3 (4)	21 (28)	17 (23)
Swam or submerged in water	37 (12)	24 (8)	5 (7)	5 (7)	4 (5)	2 (3)	9 (12)	3 (4)	19 (25)	14 (19)
Rode a skateboard or scooter	18 (6)	12 (12)	3 (4)	1 (1)	3 (4)	0 (0)	4 (5)	5 (7)	8 (11)	6 (8)
Volunteered with an organization outside	14 (5)	5 (2)	4 (5)	0 (0)	0 (0)	1 (1)	4 (5)	0 (0)	6 (8)	4 (5)
Other	5 (2)	5 (2)	0 (0)	1 (1)	0 (0)	1 (1)	0 (0)	0 (0)	5 (7)	3 (4)

[1] The above types of interaction were presented to participants in a "choose all that apply" question. They were asked to indicate all that they had enacted at each time point.

4. Discussion

In the last decade, the research has increasingly shown that accessing and interacting with nature is important for people's physical and mental wellbeing (see, e.g., Bratman et al. [69]; Frumkin et al. [70] for comprehensive reviews). Nature experience is linked, for example, to improved immune functioning, reduced diabetes, lower blood pressure, better eyesight, improved postoperative recovery, and reduced mortality; and to increased positive affect, improved manageability of life tasks, and decreases in mental distress. Thus, during the beginnings of a pandemic, during a time of enormous uncertainty and hardship on individuals, communities, and social systems, it seems to us prima facie obvious that it would be good if all people can access nature, and thereby potentially buffer some of the negative physical and mental outcomes caused by the pandemic.

With People of Color in the US being affected more acutely by COVID-19 than the White population [1–6], it is reasonable to believe they stand to gain the most from urban nature interaction. Yet, the COVID-19 pandemic may be disproportionately impacting People of Color's urban nature interaction just as it has disproportionately affected People of Color in many other dimensions.

Within this framing, and with an equity focus, we investigated residents' change in frequency of nature access during the first 6 months of the pandemic across four racial/ethnic populations in Seattle, WA. We found that Black and Latino Seattle residents experienced a significant decrease in their frequency of urban nature interaction while Asian and White residents experienced no change. This is not to say, however, that Asian Seattle residents in other ways were not disproportionately affected by the pandemic during this time; they were. For example, hate crimes against Asian Seattle residents increased 56% from 2019 to 2020 [71].

Among the various studies that have investigated differences across racial/ethnic groups in urban nature interaction before and during the pandemic, the results of this study support People of Color spending less time in urban nature after the onset of the pandemic. Our results specifically align with those of Larson et al. [31], who found that Black and Hispanic urban residents across cities of North Carolina have experienced a decrease in urban park use during the COVID-19 pandemic.

What might have contributed to the inequitable outcomes in urban nature interaction for Black and Latino residents? This study sought to uniquely approach this question through the lens of belongingness. Across the environmental justice literature, there are repeated themes of People of Color feeling excluded from urban nature spaces (see Table 1). This novel study seeks to examine the role of belongingness in urban nature inequities by developing an exploratory Sense of Belonging questionnaire. Results found sense of belonging in urban nature to be significantly associated with participants' changes in frequency of urban nature interaction during COVID-19; specifically, those with a lower sense of belonging (Black and Latino residents) experienced a greater loss of time in urban nature, while those with a higher sense of belonging (Asian and White residents) experienced no change.

While sense of belonging in urban nature was found to significantly contribute to change in frequency of urban nature interaction, perceived coronavirus threat and perceived urban nature quality did not. In terms of peoples' wariness around COVID-19, it may be the case that after 6 months of the pandemic people became less afraid of catching the virus and so were comfortable in situations where they may be in closer proximity to others. Especially when considering the reduced risk of infection in open air, people may have been more willing to take risks to spend time in urban nature. It may also be that since Seattle's urban nature spaces are relatively spacious people felt comfortable spending time in urban nature regardless of their perceived COVID-19 threat level.

It is not clear why perceived urban nature quality was not associated with changes to frequency of urban nature interaction. We may hypothesize that perhaps the pandemic had little effect on residents' perceptions of the quality of nearby urban nature, and so it did not affect whether people increased or decreased their time spent in urban nature.

In addition to sense of belonging in urban nature, age and race/ethnicity were significantly associated with changes to frequency of urban nature interaction. Race/ethnicity being a significant variable is expected as we saw from previous results that racial/ethnic groups experienced different effects on their time spent in urban nature during the first 6 months of the pandemic. It is less clear what the relationship between age and change in frequency of urban nature interaction during the pandemic is. The regression formula resulting from the stepwise variable selection process suggest that those older in age experienced less of a decrease in time spent in urban nature and were more likely to experience no change. Further research is needed to investigate the role of age in changes to urban nature interaction during the COVID-19 pandemic.

Other demographic variables, including gender and income, were not significantly associated with changes to frequency of urban nature interaction during the COVID-19 pandemic. Given the existing literature, one might have expected that income would be a significant independent variable. The fact that both income and gender were not, suggests that the inequitable changes in frequency of urban nature interaction may be more closely tied to racial/ethnic inequities than income or gender inequities.

While sense of belonging in urban nature was significantly associated with frequency of urban nature interaction in fall 2020 (during COVID-19), it was not associated with pre-pandemic frequency of urban nature interaction. This finding, combined with the fact that sense of belonging was significantly associated with change in frequency of urban nature interaction from 2019 to 2020, suggests that there may be some interplay between sense of belonging in urban nature and the COVID-19 pandemic that has affected urban nature interaction. It may be the case that COVID-19 exacerbated the exclusion of Black and Latino residents from public spaces. For example, Hoover and Lim [16] described how in New York City there were more police present in urban parks during COVID-19 to enforce social distancing between visitors. This increased police presence likely heightened the exclusion of Black individuals from those urban nature spaces [16].

In the coming years, as COVID-19 either abates or becomes endemic, an open and important question is whether Black and Latino urban residents experience a rise in urban nature visitation or return to their pre-pandemic frequencies of urban nature interaction. Given the existing racial/ethnic disparities in urban nature, and urban environments in general, as well as the disproportional impacts that COVID-19 has had on Black and Latino communities, it seems possible that Black and Latino urban nature interaction will not fully recover. With results of this study showing that sense of belonging is associated with loss of time in urban nature, sense of belonging in urban nature may have decreased among Black and Latino residents during the pandemic. This may lead to a cycle, wherein a lower sense of belonging leads to less frequent urban nature interaction, which leads to lower sense of belonging. If less frequent urban nature interaction observed during the COVID-19 pandemic becomes a new normal, urban nature conservation values, and nature conservation values as a whole, may decrease over time. These implications are further discussed below.

Some evidence has shown that meaningful experiences in nature may be associated with strong conservation values [72,73]. For those living in dense cities, the most readily available nature that one may experience is urban nature. It is therefore plausible that increasing access to urban nature may increase urban nature conservation values. Conversely, if decreases in urban nature interaction among Black and Latino communities are maintained or continue post-pandemic, urban nature conservation values may decrease among Black and Latino populations. With urban populations expected to continue to increase and the degradation of the natural world persisting, it may become imperative to foster strong nature conservation values among those living in the city. For urban residents who may have limited experiences in more rural nature, meaningful experiences in urban nature can act as a bridge towards becoming interested in the conservation of larger nature areas.

Our results additionally show that participants with a higher sense of belonging in urban nature reported a higher level of importance of urban nature protection. This finding, combined with sense of belonging being associated with changes in frequency of urban nature during the pandemic, provide some support for increased nature interaction being associated with stronger conservation values. Furthermore, these results suggest it is worth continuing investigations of sense of belonging in urban nature to provide evidence of whether efforts to increase sense of belonging in urban nature, in addition to having meaningful experiences in urban nature, would have impacts on conservation values.

Limitations

Achieving equal representation of the included racial/ethnic groups of this study allowed for comparisons between groups to be made with higher confidence. This study excluded several racial/ethnic groups including American Indian, Alaska Native, Native Hawaiian, and Pacific Islander individuals. Those who identified as belonging to more than one racial/ethnic group were also not eligible for participation. Research that includes and appropriately represents these racial/ethnic groups is of importance in future research given that these racial/ethnic groups are frequently underrepresented or not represented at all.

Although recruiting participants for research studies through social media ads has become fairly common, it introduces a bias in that it allows only for those with internet connection, an internet-capable device, and a social media profile to participate.

This study's definition of urban nature did not include natural elements one may experience briefly, such as street trees, or experience from ones' residence, such as nature window views. The intent of excluding these types of nature was to focus on urban nature spaces one may intentionally visit to experience slightly "larger" urban nature. However, we recognize that there is a large body of literature examining the human benefits of urban street trees (e.g., Mullaney et al. [74], Seamans [75], and Taylor et al. [76]) and window nature views (e.g., Kahn et al. [77], Kaplan [78], Taylor et al. [79], and Ulrich [80]), as well as characterizing racial inequities in the accessibility of these types of nature within the US (e.g., Flocks et al. [81], Landry and Chakraborty [82], Li et al. [83]). Inequities in accessibility to urban street trees and window views of nature during the pandemic may be associated with changes in frequency of urban nature interaction. It may be the case that those who benefit from accessibility to urban street trees and nature window views, more often those who are wealthy and White, might not have been as inclined to visit what our study defined urban nature spaces during COVID-19. Additionally, although we observed inequities in changes to urban nature interaction over the course of the first 6 months of the pandemic, this study does not fully capture inequities in the buffering effects of nature during the pandemic, as benefits may have been gained by urban residents from types of nature not included in our definition of urban nature (e.g., street trees and nature window views).

The Perceived Green Space Quality scale and the Sense of Belonging in Urban Nature Questionnaire were provided at a single time point approximately 6 months into the pandemic. The responses to these measurements were used alongside data that pertained to both 6 months into the pandemic and fall 2019, before the COVID-19 pandemic. From fall 2019 to fall 2020 there may have been changes to both perceived quality of nearby urban nature and sense of belonging, and these changes may have been inequitably distributed across racial/ethnic groups. Urban parks in predominantly Black neighborhoods tend to be smaller in area [25]. During the pandemic, this may have meant those smaller parks were more densely crowded. Due to COVID-19 precautions, one may consider more dense parks to be of lower quality. This is one way in which the pandemic may have inequitably affected the quality of urban nature in predominantly non-White neighborhoods. Sense of belonging may have also been impacted to various degrees across racial/ethnic groups. With the rise of anti- Asian American and Pacific Islanders (AAPI) violent attacks during the pandemic, as have occurred in urban parks across the US [66], urban nature spaces may have become less welcoming for Asian residents 6 months into the pandemic.

Two other limitations are worth noting: First, when participants took the survey in winter 2020, they were asked to recall their experiences in fall 2019 and respond to certain questions accordingly. The pre-pandemic data are therefore not as reliable as it would have been had this been a longitudinal study with two data collection periods. Second, the Sense of Belonging in Urban Nature Questionnaire developed for this study, while achieving high internal reliability, remains unvalidated. Further exploratory and confirmatory factor analyses would be needed to validate this questionnaire.

5. Conclusions

Increasing access to urban nature among Black and Latino Seattle residents may narrow inequities in who benefits from urban nature, including during times of major disruption such as a global pandemic. Addressing the inequitable distribution of urban nature benefits is sometimes discussed as one way to increase resiliency of predominantly non-White neighborhoods [84]. However, the goal of increased resiliency, in some ways, places the responsibility of recovering from disruptive events on Communities of Color and does not address the underlying issues that lead to inequities in how Communities of Color are affected by such events in the first place [85]. In the field of ecology, there exists the concept of resilient and resistant plant species. Resilient plants are able to quickly "bounce back" from damage such as being trampled. Resistant species, on the other hand, are more impervious to trauma in the first place [86]. Mapping these terms onto inequities in the effects of disruptive events among Communities of Color, it may be more beneficial to address resilience and resistance among Communities of Color. Increasing urban nature access may be one way of doing that. If a community already benefits from interacting with accessible urban nature, they may be less prone to significant damages when major disruptive events occur.

How may resistance and resilience among urban Communities of Color be strengthened to decrease the inequitable effects of future disruptive events? This study supports that addressing inequitable access to urban nature and sense of belonging in urban nature may be ways of doing so. Everyone in a city should feel equally welcomed in their city's urban nature regardless of race and ethnicity. Thus, the six themes identified across the existing literature that compose the Sense of Belonging in Urban Nature Questionnaire, Ease of Access, Safety, Feeling Out of Place, Unwelcomeness, Institutional Acceptance, and Different Ways of Interacting with Nature Acceptance, may be entry points for city governments to begin to increase sense of belongingness among minoritized groups. Examples of actions that city government agencies can take to target the inequity themes of the Sense of Belonging in Urban Nature Questionnaire include:

- Direct urban parks budget to urban nature spaces predominantly serving People of Color to increase Ease of Access, Safety, and Institutional Acceptance of urban nature spaces near People of Color;
- Organize urban nature programming, specifically for Black or Latino urban residents, to improve representation and increase belongingness in urban nature among Black and Latino communities;
- Present urban nature information on signage and online in multiple languages to increase urban nature accessibility;
- Increase representation of People of Color on urban park signage and websites;
- Improve racial/ethnic diversity of people hired into city government, especially departments which oversee urban nature areas (such as Parks and Recreation departments).

Examining how urban residents interact with nature prior to and during a major disruptive event (such as the COVID-19 pandemic) may also be of aid in assessing the potentially inequitable types of nature that different racial/ethnic groups have access to and reimagining urban centers to be more sustainable, resilient, and resistant in the face of future disruptive events. This study conducted exploratory investigations into the types of interactions participants engaged in prior to the pandemic and during the pandemic (presented in Table 4). In different parks in different locations, there should be equity, for example, in how the parks allow people to sit in nature, have a picnic, engage in sports, run or jog, ride a bike, walk a dog, sit, watch a sunset, watch one's children play, look out on a water view or a city view, tend to a garden, and/or simply enjoy the quietness of nature. This list is part of a larger approach to urban design, Interaction Pattern Design, that seeks to maximize ways for people not only to access nature, but to interact with that they access so that the interactions are engaging, meaningful, and self-reinforcing [46–49].

This list in Table 4 of interactions with nature can provide some insights for how to better design urban nature for future pandemics and for increasing density. For example, prior to the pandemic, Seattle residents took walks with other people more frequently than they took walks alone. Six months into the pandemic, residents took walks alone more frequently than with other people. This is likely due to social distancing mandates and attempts to limit risk of contracting COVID-19. Thus, wide walking trails may be increasingly important to implement. They not only allow for social distancing during times of a pandemic but create the urban nature infrastructure that plans for what, in most urban areas, will be increasing population density.

The results of this study may have larger international relevance in places that experience discrimination based on race, ethnicity, and other dimensions such as religion. Regardless of whether these societal inequities are similar to those of Seattle, WA, or the US, the general conclusions and implications may remain: that minoritized or vulnerable populations may be suffering more in terms of urban nature interaction during the pandemic, and that these inequities may have significant effects on minoritized populations both during the pandemic and moving forward. While the discussions and recommendations included in this paper are directed towards Seattle, WA, the same principles may be applied to other US and international locations where certain populations have lost more time in urban nature during the pandemic than others.

Author Contributions: Conceptualization, A.N., P.H.K.J., J.J.L. and G.N.B.; Data curation, A.N.; Formal analysis, A.N.; Funding acquisition, J.J.L. and P.H.K.J.; Investigation, A.N., P.H.K.J., J.J.L. and G.N.B.; Methodology, A.N., P.H.K.J., J.J.L. and G.N.B.; Project administration, J.J.L.; Supervision, P.H.K.J.; Visualization, A.N.; Writing—original draft, A.N. and P.H.K.J.; Writing—review and editing, P.H.K.J., J.J.L. and G.N.B. All authors have read and agreed to the published version of the manuscript.

Funding: This research was funded by Washington Department of Fish and Wildlife, grant number 20-16075. The APC was waived by MDPI Land.

Institutional Review Board Statement: This study was conducted according to the guidelines of the Declaration of Helsinki and approved by the University of Washington Institutional Review Board (IRB ID: STUDY00011290, approved 30 September 2020).

Informed Consent Statement: Informed consent was obtained from all subjects involved in this study.

Data Availability Statement: The data presented in this study are available on request from the corresponding author. The data are not publicly available due to containing possibly sensitive participant information.

Acknowledgments: Special thanks to Amanda Zhou, Claire Zhang, Thea Weiss, Carly Gray, Sarena Sabine, and Star Berry.

Conflicts of Interest: The authors declare no conflict of interest. The funders had no role in the design of the study; in the collection, analyses, or interpretation of data; in the writing of the manuscript, or in the decision to publish the results.

Notes

1. We recognize that the terms "People of Color" and "Communities of Color" can homogenize the experiences of different racial/ethnic groups. Terms such as Black, Indigenous, and People of Color (BIPOC) are used to highlight the shared experiences of colonization among Black and Indigenous communities. However, this study's sample does not include indigenous people, therefore we chose to use People of Color and Communities of Color throughout this paper. Additionally, we acknowledge that within each racial/ethnic group included in this study, there is great variability in cultural norms and expectations, upbringing, etc.

2. Throughout this paper, 'Black or African American' is shortened to 'Black'.

3. Latinx and Latiné have been used as alternatives to Latino in efforts to be more gender inclusive [7,8]. We recognize that there are people who oppose the use of each of the terms Latino, Latinx, and Latiné within the Hispanic and Latino community [9,10].

We also recognize the issue with researchers, often outside the Hispanic and Latino community, imposing Western norms and altering the way that Hispanic and Latino individuals identify themselves and their community [9,10]. Here, Latino/a/x is used to be inclusive of those within the community who identify as Latino or Latina, and those who wish to use an ungendered term. Latino/a/x is shortened to Latino throughout this paper.

[4]   Urban nature in the case of this study refers to parks, green areas, and places with water, vegetation, and/or animals within a city.

[5]   This White perception of "untouched" nature erases the existence of Indigenous peoples of North America that molded the North American landscape through burning and silvicultural practices for at least 20,000 years before European colonization [15].

References

1. Bathina, K.C.; Thij, M.T.; Valdez, D.; Rutter, L.A.; Bollen, J. Declining Well-Being during the COVID-19 Pandemic Reveals US Social Inequities. *PLoS ONE* **2021**, *16*, e0254114. [CrossRef]
2. Van Dorn, A.; Cooney, R.E.; Sabin, M.L. COVID-19 Exacerbating Inequalities in the US. *Lancet* **2020**, *395*, 1243–1244. [CrossRef]
3. Fortuna, L.R.; Tolou-Shams, M.; Robles-Ramamurthy, B.; Porche, M.V. Inequity and the Disproportionate Impact of COVID-19 on Communities of Color in the United States: The Need for a Trauma-Informed Social Justice Response. *Psychol. Trauma* **2020**, *12*, 443–445. [CrossRef]
4. Gemelas, J.; Davison, J.; Keltner, C.; Ing, S. Inequities in Employment by Race, Ethnicity, and Sector during COVID-19. *J. Racial Ethn. Health Disparities* **2022**, *9*, 350–355. [CrossRef]
5. Hoernke, K. A Socially Just Recovery from the COVID-19 Pandemic: A Call for Action on the Social Determinants of Urban Health Inequalities. *J. R. Soc. Med.* **2020**, *113*, 482–484. [CrossRef]
6. Pareek, M.; Bangash, M.N.; Pareek, N.; Pan, D.; Sze, S.; Minhas, J.S.; Hanif, W.; Khunti, K. Ethnicity and COVID-19: An Urgent Public Health Research Priority. *Lancet* **2020**, *395*, 1421–1422. [CrossRef]
7. Azmitia, M. Latinx Adolescents' Assets, Risks, and Developmental Pathways: A Decade in Review and Looking Ahead. *J. Res. Adolesc.* **2021**, *31*, 989–1005. [CrossRef]
8. Blas, T. "Latinx" Is Growing in Popularity. I Made a Comic to Help You Understand Why. Available online: https://www.vox.com/the-highlight/2019/10/15/20914347/latin-latina-latino-latinx-means (accessed on 26 April 2022).
9. Guerra, G.; Orbea, G. *The Argument against the Use of the Term "Latinx"*; The Phoenix: Swarthmore, PA, USA, 2015.
10. Tlapoyawa, K. Can We Please Stop Using 'Latinx'? Thanx. *Human Parts* 2020. Available online: https://humanparts.medium.com/can-we-please-stop-using-latinx-thanx-423ac92a87dc (accessed on 8 May 2022).
11. Roberts, J.D.; Dickinson, K.L.; Hendricks, M.D.; Jennings, V. "I can't breathe": Examining the legacy of American racism on determinants of health and the ongoing pursuit of environmental justice. *Curr. Environ. Health Rep.* **2022**, *9*, 211–227.
12. Finney, C. *Black Faces, White Spaces: Reimagining the Relationship of African Americans to the Great Outdoors*; UNC Press: Chapel Hill, NC, USA, 2014; ISBN 9781469614496.
13. Byrne, J.; Wolch, J. Nature, Race, and Parks: Past Research and Future Directions for Geographic Research. *Prog. Hum. Geogr.* **2009**, *33*, 743–765. [CrossRef]
14. Johnson, C.Y.; Bowker, J.M. African-American Wildland Memories. *Environ. Ethics* **2004**, *26*, 57–68.
15. Abrams, M.D.; Nowacki, G.J. Native Americans as active and passive promoters of mast and fruit trees in the Eastern USA. *Holocene* **2008**, *18*, 1123–1137.
16. Hoover, F.-A.; Lim, T.C. Examining Privilege and Power in US Urban Parks and Open Space during the Double Crises of Antiblack Racism and COVID-19. *Socio. Ecol. Pract. Res.* **2021**, *3*, 55–70. [CrossRef]
17. Bittel, J. People Called Police on This Black Birdwatcher so Many Times That He Posted Custom Signs to Explain His Hobby. *Washington Post*, 5 June 2020.
18. Newsome, C. Available online: https://twitter.com/hood_naturalist/status/1266387163727486977 (accessed on 26 April 2022).
19. Heynen, N. Green Urban Political Ecologies: Toward a Better Understanding of Inner-City Environmental Change. *Environ. Plan. A* **2006**, *38*, 499–516. [CrossRef]
20. Shinew, K.J.; Stodolska, M.; Floyd, M.; Hibbler, D.; Allison, M.; Johnson, C.; Santos, C. Race and Ethnicity in Leisure Behavior: Where Have We Been and Where Do We Need to Go? *Leis. Sci.* **2006**, *28*, 403–408. [CrossRef]
21. Stodolska, M.; Shinew, K.J.; Acevedo, J.C.; Roman, C.G. "I Was Born in the Hood": Fear of Crime, Outdoor Recreation and Physical Activity among Mexican-American Urban Adolescents. *Leis. Sci.* **2013**, *35*, 1–15. [CrossRef]
22. Wolch, J.R.; Byrne, J.; Newell, J.P. Urban Green Space, Public Health, and Environmental Justice: The Challenge of Making Cities 'Just Green Enough'. *Landsc. Urban Plan.* **2014**, *125*, 234–244. [CrossRef]
23. Weiss, C.C.; Purciel, M.; Bader, M.; Quinn, J.W.; Lovasi, G.; Neckerman, K.M.; Rundle, A.G. Reconsidering Access: Park Facilities and Neighborhood Disamenities in New York City. *J. Urban Health* **2011**, *88*, 297–310. [CrossRef]
24. Muqueeth, S. As COVID Cases Rise, Make Parks A Public Health Priority. Available online: https://www.tpl.org/blog/staff-QA-sadiya-muqueeth (accessed on 26 April 2022).
25. Rigolon, A. A Complex Landscape of Inequity in Access to Urban Parks: A Literature Review. *Landsc. Urban Plan.* **2016**, *153*, 160–169. [CrossRef]
26. Nesbitt, L.; Meitner, M.J.; Sheppard, S.R.J.; Girling, C. The Dimensions of Urban Green Equity: A Framework for Analysis. *Urban For. Urban Green.* **2018**, *34*, 240–248. [CrossRef]
27. Roberts, A.; Zamore, S. Black in Nature. *Milwaukee Courier Weekly Newspaper*, 10 June 2020.

28. Roberts, D.J.D. Central Park: Black Bodies, Green Spaces, White Minds. Medium 2020. Available online: https://medium.com/@ActiveRoberts/central-park-black-bodies-green-spaces-white-minds-3efebde69077 (accessed on 8 May 2022).

29. Grima, N.; Corcoran, W.; Hill-James, C.; Langton, B.; Sommer, H.; Fisher, B. The Importance of Urban Natural Areas and Urban Ecosystem Services during the COVID-19 Pandemic. *PLoS ONE* **2020**, *15*, e0243344. [CrossRef]

30. Plitt, S.; Pregitzer, C.C.; Charlop-Powers, S. Brief Research Report: Case Study on the Early Impacts of COVID-19 on Urban Natural Areas across 12 American Cities. *Front. Sustain. Cities* **2021**, *3*, 725904. [CrossRef]

31. Larson, L.R.; Zhang, Z.; Oh, J.I.; Beam, W.; Ogletree, S.S.; Bocarro, J.N.; Lee, K.J.; Casper, J.; Stevenson, K.T.; Hipp, J.A.; et al. Urban Park Use During the COVID-19 Pandemic: Are Socially Vulnerable Communities Disproportionately Impacted? *Front. Sustain. Cities* **2021**, *3*, 710243.

32. Lopez, B.; Kennedy, C.; Field, C.; McPhearson, T. Who Benefits from Urban Green Spaces during Times of Crisis? Perception and Use of Urban Green Spaces in New York City during the COVID-19 Pandemic. *Urban For. Urban Green.* **2021**, *65*, 127354. [CrossRef]

33. Pipitone, J.M.; Jović, S. Urban Green Equity and COVID-19: Effects on Park Use and Sense of Belonging in New York City. *Urban For. Urban Green.* **2021**, *65*, 127338. [CrossRef]

34. Joassart-Marcelli, P. Leveling the Playing Field? Urban Disparities in Funding for Local Parks and Recreation in the Los Angeles Region. *Environ. Plan. A* **2010**, *42*, 1174–1192. [CrossRef]

35. Wolch, J.; Wilson, J.P.; Fehrenbach, J. Parks and Park Funding in Los Angeles: An Equity-Mapping Analysis. *Urban Geogr.* **2005**, *26*, 4–35. [CrossRef]

36. Byrne, J. When Green Is White: The Cultural Politics of Race, Nature and Social Exclusion in a Los Angeles Urban National Park. *Geoforum* **2012**, *43*, 595–611. [CrossRef]

37. Hagerty, B.M.K.; Lynch-Sauer, J.; Patusky, K.L.; Bouwsema, M.; Collier, P. Sense of Belonging: A Vital Mental Health Concept. *Arch. Psychiatr. Nurs.* **1992**, *6*, 172–177. [CrossRef]

38. Hagerty, B.M.K.; Patusky, K. Developing a Measure of Sense of Belonging. *Nurs. Res.* **1995**, *44*, 9–13. [CrossRef]

39. Shamai, S. Sense of Place: An Empirical Measurement. *Geoforum* **1991**, *22*, 347–358. [CrossRef]

40. Williams, D.R.; Vaske, J.J. The Measurement of Place Attachment: Validity and Generalizability of a Psychometric Approach. *For. Sci.* **2003**, *49*, 830–840. [CrossRef]

41. Peters, K.; Stodolska, M.; Horolets, A. The Role of Natural Environments in Developing a Sense of Belonging: A Comparative Study of Immigrants in the U.S., Poland, the Netherlands and Germany. *Urban For. Urban Green.* **2016**, *17*, 63–70. [CrossRef]

42. Jennings, V.; Reid, C.E.; Fuller, C.H. Green infrastructure can limit but not solve air pollution injustice. *Nat. Commun.* **2021**, *12*, 4681.

43. Rugel, E.J.; Carpiano, R.M.; Henderson, S.B.; Brauer, M. Exposure to Natural Space, Sense of Community Belonging, and Adverse Mental Health Outcomes across an Urban Region. *Environ. Res.* **2019**, *171*, 365–377. [CrossRef]

44. U.S. Census Bureau Quick Facts: Seattle, Washington. Available online: https://www.census.gov/quickfacts/seattlecitywashington (accessed on 8 May 2022).

45. Williams, T.G.; Logan, T.M.; Zuo, C.T.; Liberman, K.D.; Guikema, S.D. Parks and Safety: A Comparative Study of Green Space Access and Inequity in Five US Cities. *Landsc. Urban Plan.* **2020**, *201*, 103841. [CrossRef]

46. O'Neil-Dunne, J. 2016 Seattle Tree Canopy Assessment. Available online: https://www.seattle.gov/documents/Departments/Trees/Mangement/Canopy/Seattle2016CCAFinalReportFINAL.pdf (accessed on 8 May 2022).

47. Silva, C. Racial Restrictive Covenants: Enforcing Neighborhood Segregation in Seattle-Seattle Civil Rights and Labor History Project. Available online: https://depts.washington.edu/civilr/covenants_report.htm (accessed on 26 April 2022).

48. Seattle Parks and Recreation Parks and Open Space Plan. Available online: https://www.seattle.gov/documents/Departments/ParksAndRecreation/PoliciesPlanning/2017Plan/2017ParksandOpenSpacePlanFinal.pdf (accessed on 8 May 2022).

49. Shaver, L.G.; Khawer, A.; Yi, Y.; Aubrey-Bassler, K.; Etchegary, H.; Roebothan, B.; Asghari, S.; Wang, P.P. Using facebook advertising to recruit representative samples: Feasibility Assessment of a cross-sectional survey. *J. Med. Internet Res.* **2019**, *21*, e14021.

50. Reagan, L.; Nowlin, S.Y.; Birdsall, S.B.; Gabbay, J.; Vorderstrasse, A.; Johnson, C.; D'Eramo Melkus, G. Integrative Review of Recruitment of research participants through Facebook. *Nurs. Res.* **2019**, *68*, 423–432.

51. Kahn, P.H., Jr.; Weiss, T.; Harrington, K. Modeling Child–Nature Interaction in a Nature Preschool: A Proof of Concept. *Front. Psychol.* **2018**, *9*, 835. [CrossRef]

52. Kahn, P.H., Jr.; Weiss, T.; Harrington, K. Child-Nature Interaction in a Forest Preschool. In *Research Handbook on Childhood Nature*; Cutter-Mackenzie, A., Malone, K., Hacking, E.B., Eds.; Springer: Cham, Switzerland, 2018; pp. 41–57. ISBN 991012875400502368.

53. Kahn, P.H., Jr.; Lev, E.M.; Perrins, S.P.; Weiss, T.; Ehrlich, T.; Feiinberg, D.S. Human-Nature Interaction Patterns: Constituents of a Nature Language for Environmental Sustainability. *J. Biourbanism* **2018**, *1*, 41–57.

54. Lev, E.; Kahn, P.H., Jr.; Chen, H.; Esperum, G. Relatively Wild Urban Parks Can Promote Human Resilience and Flourishing: A Case Study of Discovery Park, Seattle, Washington. *Front. Sustain. Cities* **2020**, *2*.

55. Conway, L.; Woodard, S.; Zubrod, A. Social Psychological Measurements of COVID-19: Coronavirus Perceived Threat, Government Response, Impacts, and Experiences Questionnaires; PsyArXiv Preprint 2020. Available online: psyarxiv.com/z2x9a (accessed on 8 May 2022).

56. Dzhambov, A.; Hartig, T.; Markevych, I.; Tilov, B.; Dimitrova, D. Urban residential greenspace and mental health in youth: Different approaches to testing multiple pathways yield different conclusions. *Environ. Res.* **2018**, *160*, 47–59.

57. Dai, D. Racial/Ethnic and Socioeconomic Disparities in Urban Green Space Accessibility: Where to Intervene? *Landsc. Urban Plan.* **2011**, *102*, 234–244. [CrossRef]

58. Jennings, V.; Johnson Gaither, C.; Gragg, R.S. Promoting Environmental Justice Through Urban Green Space Access: A Synopsis. *Environ. Justice* **2012**, *5*, 1–7. [CrossRef]

59. McConnachie, M.; Shackleton, C.M. Public Green Space Inequality in Small Towns in South Africa. *Habitat Int.* **2010**, *34*, 244–248. [CrossRef]

60. Powers, S.L.; Lee, K.J.; Pitas, N.A.; Graefe, A.R.; Mowen, A.J. Understanding Access and Use of Municipal Parks and Recreation through an Intersectionality Perspective. *J. Leis. Res.* **2020**, *51*, 377–396. [CrossRef]

61. Shinew, K.J.; Floyd, M.F.; Parry, D. Understanding the Relationship between Race and Leisure Activities and Constraints: Exploring an Alternative Framework. *Leis. Sci.* **2004**, *26*, 181–199. [CrossRef]

62. Zhang, X.; Lu, H.; Holt, J.B. Modeling Spatial Accessibility to Parks: A National Study. *Int. J. Health Geogr.* **2011**, *10*, 31. [CrossRef]

63. Madge, C. Public Parks and the Geography of Fear. *Tijdschr. Voor Econ. En Soc. Geogr.* **1997**, *88*, 237–250. [CrossRef]

64. Roe, J.; Aspinall, P.A.; Ward Thompson, C. Understanding Relationships between Health, Ethnicity, Place and the Role of Urban Green Space in Deprived Urban Communities. *Int. J. Environ. Res. Public Health* **2016**, *13*, 681. [CrossRef]

65. Ho, C.-H.; Sasidharan, V.; Elmendorf, W.; Willits, F.K.; Graefe, A.; Godbey, G. Gender and Ethnic Variations in Urban Park Preferences, Visitation, and Perceived Benefits. *J. Leis. Res.* **2005**, *37*, 281–306. [CrossRef]

66. Floyd, M.F.; Shinew, K.J.; McGuire, F.A.; Noe, F.P. Race, Class, and Leisure Activity Preferences: Marginality and Ethnicity Revisited. *J. Leis. Res.* **1994**, *26*, 158–173. [CrossRef]

67. Hutchison, R. Ethnicity and Urban Recreation: Whites, Blacks, and Hispanics in Chicago's Public Parks. *J. Leis. Res.* **1987**, *19*, 205–222. [CrossRef]

68. Payne, L.L.; Mowen, A.J.; Orsega-Smith, E. An Examination of Park Preferences and Behaviors Among Urban Residents: The Role of Residential Location, Race, and Age. *Leis. Sci.* **2002**, *24*, 181–198. [CrossRef]

69. Bratman, G.N.; Anderson, C.B.; Berman, M.G.; Cochran, B.; de Vries, S.; Flanders, J.; Folke, C.; Frumkin, H.; Gross, J.J.; Hartig, T.; et al. Nature and Mental Health: An Ecosystem Service Perspective. *Sci. Adv.* **2019**, *5*, eaax0903. [CrossRef]

70. Frumkin, H.; Bratman, G.N.; Breslow, S.J.; Cochran, B.; Kahn, P.H., Jr.; Lawler, J.J.; Levin, P.S.; Tandon, P.S.; Varanasi, U.; Wolf, K.L.; et al. Nature Contact and Human Health: A Research Agenda. *Environ. Health Perspect.* **2017**, *125*, 075001. [CrossRef]

71. Levin, B. *Report to the Nation: Anti-Asian Prejudice & Hate Crime*; Center for the Study of Hate and Extremism, California State University: San Bernardino, CA, USA, 2021.

72. DeVille, N.V.; Tomasso, L.P.; Stoddard, O.P.; Wilt, G.E.; Horton, T.H.; Wolf, K.L.; Brymer, E.; Kahn, P.H., Jr.; James, P. Time Spent in Nature Is Associated with Increased Pro-Environmental Attitudes and Behaviors. *Int. J. Environ. Res. Public Health* **2021**, *18*, 7498. [CrossRef]

73. Zelenski, J.M.; Dopko, R.L.; Capaldi, C.A. Cooperation Is in Our Nature: Nature Exposure May Promote Cooperative and Environmentally Sustainable Behavior. *J. Environ. Psychol.* **2015**, *42*, 24–31. [CrossRef]

74. Mullaney, J.; Lucke, T.; Trueman, S.J. A review of benefits and challenges in growing street trees in paved urban environments. *Landsc. Urban Plan.* **2015**, *134*, 157–166.

75. Seamans, G. Mainstreaming the environmental benefits of street trees. *Urban For. Urban Green.* **2013**, *12*, 2–11.

76. Taylor, M.S.; Wheeler, B.W.; White, M.P.; Economou, T.; Osborne, N.J. Research note: Urban street tree density and antidepressant prescription rates—a cross-sectional study in London, UK. *Landsc. Urban Plan.* **2015**, *136*, 174–179.

77. Kahn, P.H., Jr.; Friedman, B.; Gill, B.; Hagman, J.; Severson, R.L.; Freier, N.G.; Feldman, E.N.; Carrère, S.; Stolyar, A. A plasma display window?—The shifting baseline problem in a technologically mediated natural world. *J. Environ. Psychol.* **2008**, *28*, 192–199.

78. Kaplan, R. The nature of the view from home. *Environ. Behav.* **2001**, *33*, 507–542.

79. Taylor, A.F.; Kuo, F.E.; Sullivan, W.C. Views of nature and self-discipline: Evidence from Inner City Children. *J. Environ. Psychol.* **2002**, *22*, 49–63.

80. Ulrich, R.S. View through a window may influence recovery from surgery. *Science* **1984**, *224*, 420–421.

81. Flocks, J.; Escobedo, F.; Wade, J.; Varela, S.; Wald, C. Environmental justice implications of urban tree cover in Miami-Dade County, Florida. *Environ. Justice* **2011**, *4*, 125–134.

82. Landry, S.M.; Chakraborty, J. Street trees and equity: Evaluating the spatial distribution of an urban amenity. *Environ. Plan. A: Econ. Space* **2009**, *41*, 2651–2670.

83. Li, X.; Zhang, C.; Li, W.; Kuzovkina, Y.A. Environmental inequities in terms of different types of urban greenery in Hartford, Connecticut. *Urban For. Urban Green.* **2016**, *18*, 163–172.

84. Colding, J.; Barthel, S. The potential of 'urban green commons' in the Resilience Building of Cities. *Ecol. Econ.* **2013**, *86*, 156–166.

85. Ranganathan, M.; Bratman, E. From urban resilience to abolitionist climate justice in Washington, DC. *Antipode* **2019**, *53*, 115–137.

86. Ma, J.F. Role of silicon in enhancing the resistance of plants to biotic and abiotic stresses. *Soil Sci. Plant Nutr.* **2004**, *50*, 11–18.

 *land*

Article

Changing Perceptions and Uses of "Companion Animal" Public and Pseudo-Public Spaces in Cities during COVID-19 Pandemic: The Case of Beijing

Haoxian Cai and Wei Duan *

School of Landscape Architecture, Beijing Forestry University, Beijing 100083, China
* Correspondence: duanwei@bjfu.edu.cn

Abstract: This paper examines the debate over the place of "companion animal" public space in China's cities. With the COVID-19 outbreak, this debate has entered a new phase, where the social response to the outbreak may have fundamentally changed the public's use and perception of "companion animal" public and pseudo-public space. This paper combines quantitative and qualitative analysis of posts and comments on two of China's largest social media platforms with a big data approach, based on a case study in Beijing, China. There were statistically significant differences in the perception and use of "companion animal" public spaces and pseudo-public spaces before and after the pandemic. We attribute the impact of the pandemic on "companion animal" spaces to three pathways: changes in opportunity, changes in ability, and changes in motivation. We found that the pandemic led to an increase in the amount of time available to some people but a decrease in the amount of "companion animal" public space available due to the pandemic closure. In addition, the use of "companion animal" public spaces in pseudo-public spaces declined, while those located within the open urban green space on the city's outskirts stood out after the outbreak. With the normalisation of the pandemic, there will be new challenges for the development and operation of companion-animal-related public spaces in cities, which will be the next focus of research. In addition, governments and social media should work together to promote and support sustainable animal ethical practices to better respond to the crisis. These findings will help complement the urban services system and guide future planning, design, and evaluation of related spaces.

Keywords: public space; privatisation; companion animals; animal ethics; China; COVID-19

Citation: Cai, H.; Duan, W. Changing Perceptions and Uses of "Companion Animal" Public and Pseudo-Public Spaces in Cities during COVID-19 Pandemic: The Case of Beijing. *Land* **2022**, *11*, 1475. https://doi.org/10.3390/land11091475

Academic Editors: Francesca Ugolini and David Pearlmutter

Received: 14 August 2022
Accepted: 1 September 2022
Published: 3 September 2022

Publisher's Note: MDPI stays neutral with regard to jurisdictional claims in published maps and institutional affiliations.

1. Introduction

Cities, as spaces where high densities of people and goods congregate, provide an important channel for spreading infectious diseases. Urbanisation promotes spatial overlap between hosts within vectors, which facilitates the rapid spread of pathogens [1], as recently shown by COVID-19. At least 60% of newly developing infectious illnesses are thought to be spread from wild to domesticated animals and people [2]. According to recent research investigations, it is likely that COVID-19 originates from zoonotic diseases [3]. In complex urban systems, urban livestock and pet-keeping practices, the mobility of animals in urban spaces, and the direct impact of urbanisation on their physical environment become driving forces that may generate diverse transmission chains at the wildlife–domestic animal–human interface [1,4]. Among these, urban companion animals, the leading domestic animal species in urban spaces, have essential links and critical functions in the pandemic transmission interface. In contrast to rural regions, companion animals are completely included into family life in urban settings, where their living circumstances, such as free range and frequent outdoor activities, may result in intimate encounters between human and urban wildlife [5]. As a result, there is a renewed social debate about urban companion animals [6,7].

The study of companion animals In urban spaces has a long history in academia. Since the development of zoogeography, the question of human–animal interactions in cities and how different urban spaces shape complex, mutually constructed human–companion animal relationships have been important research topics [8,9]. In the literature related to humans and companion animals, in public spaces, in particular, dogs have been a very important area of research [10], with research currently focusing on discussions around the dichotomous relationship between human and companion animal oppositions in urban spaces and the contradiction between better integrating dogs into social life and regulating their rights in urban space [6,11].

At the same time, debates about the end of public space have gradually come to the forefront with the rise of urban privatisation [12]. The most controversial of these is privately owned public open space (POPOS) or private public space (POPS), as opposed to publicly owned open spaces, such as parks and squares, which the government traditionally provides; it is an outdoor or semi-outdoor space on private land or private property, built and managed by private investment in exchange for public use through government incentives [12]. Therefore, some scholars have begun to examine the impact of the privatisation of urban public space on companion animal space, mainly from a politics of rights perspective. For example, Sue Donaldson has repeatedly discussed the negative impact of the privatisation of public space on animal rights, highlighting the significant practical and conceptual challenges facing "companion animal" public space [13]. Marie Carmen argues that the privatisation of public space has given capitalist corporations free reign over urban space to maximise their profits and infringe on the rights of companion animals [14]. "Companion animal" public space is a relatively new but growing area of urban research. In many countries, urban planners are beginning to incorporate this companion animal element into land use decisions, with "companion animal" public space being one of these [6]. In the United States, dog parks are common as "companion animal" public spaces in cities across the country [10,15]. However, most cities in China lack dedicated companion animal parks for humans to interact with their companion dogs, and thus the relevant spaces in China are all rather vague.

The shift from an agrarian to an industrial, post-industrial, digital society has primarily influenced the perception of companion animal spaces as "companion animal" public spaces. However, the COVID-19 outbreak seems to have been a turning point, with the government setting restrictions to limit social gatherings and crowding and to avoid contagion. People have experienced a degree of lockdown, and society has changed significantly in various ways [16,17]. The COVID-19 outbreak is likely to affect the perception and use of "companion animal" public spaces. The possible impact of COVID-19 on public space has been discussed in the literature [18–21]. However, there is still a lack of empirical research on the "companion animal" public space in China. Thus, in the Chinese context that is experiencing a new type of urbanism very different from that of the West [22,23], we based our social media data on Weibo and Xiaohongshu, using content analysis methods in conjunction with natural language (NLP) analysis and GIS spatial analysis to investigate the extent to which the epidemic will change the way people perceive and use "companion animal" public spaces and pseudo-public spaces. At the same time, we conducted semi-structured in-depth interviews to support the validation and correction of the big data findings in order to conduct a more in-depth study. The results of this study will help to complement the functional system of urban services and guide the planning, design, and evaluation of related spaces in the future.

2. Materials and Methods

Beijing is one of China's most crowded and dynamic metropolises, with a relatively well-developed public space infrastructure. Then, as one of the most thriving real-estate markets in China, property values are well above the national average, with shopping centres and commercial complexes ranked in the top tier in China. High property values lead to gentrification, and there is a strong need for social isolation and spatial control of the city's pseudo-public spaces. In addition, as Beijing is the capital, it is also representative of the management of the pandemic. Thus, by choosing this city, we can better investigate the effectiveness of urban governance of public and pseudo-public spaces for "companion animals" during the outbreak.

As the study was conducted on "companion animals" in public and pseudo-public spaces, urban companion animals do not have language skills, so we turned to the owners who are closest to their companion animals and reflect their needs, as well as the public who have a close relationship with them. According to the China Pet Industry White Paper 2021, the number of pets in China is predicted to reach 220 million in 2022, with dogs accounting for the largest share. Therefore, we decided to study the pet dogs that are kept in the largest numbers and most studied in the literature. We first ruled out the questionnaire approach due to its inherent reliance on participant responses, as there may be issues of recall and social desirability bias [24]. In addition, the information received from participants is difficult to verify independently, especially for green space surveys in the parkland category, where there is a large margin of error [25]. There are also significant limitations to direct in situ observations, as samples outside of a given observation time cannot be reliably estimated, thus requiring multiple observations over different days and seasons to ensure reliability [26], and thus direct observation studies that require significant time and often lack longitudinal depth and breadth. At the same time, there are certain shortcomings in big data methods of measurement, which can be well compensated for by in-depth interviews [27]. In addition, Flick et al. mention that respondents' views are more likely to be expressed in a reasonably open design rather than in a standardised questionnaire [28]. Therefore, we selected a method of social media big data combined with semi-structured in-depth interviews to investigate. The method we chose has three advantages.

Firstly, our comprehensive research methodology allows us to collect a wealth of data. Social media data mining allows us to collect much larger volumes of data than traditional field research, and computational profiling of the data helps us to perceive better and identify the changing focus of urban perceptions and uses of public and pseudo-public spaces for "companion animals". Our analysis is based on verbatim and textual big data analysis, which allows us to fully reflect the media landscape on the subject of urban companion animals and urban multi-species interactions and relationships. The flexibility and adaptability of in-depth interviews allows us to obtain more in-depth information and evidence, and to explore more perspectives, layers, and dimensions of the issues uncovered by social media data.

Second, the study data we used are reliable and general. Due to the internet's ongoing development over the past few years, it has integrated into people's lives. The fast growth of social media and the widespread use of smartphones have created more and more well-liked and respected platforms for individuals to openly express their thoughts. New methods for comprehending the traits of social activity are made available by the large volume of data on social media. Textual data from social media create a large database of public impressions, including information that is challenging to obtain through conventional polls. In addition, the atmosphere on the internet is more relaxed and less morally constrained, which better reflects the true psychological state of the interviewees. In-depth interviews, although they may be subject to greater ethical and identity constraints, allow for more in-depth and repeated discussion of certain topics and an understanding of the reasons behind them. Fitting the results from the in-depth interviews to the big data, thus, allows for a high degree of accuracy and generalisability of the data.

Thirdly, social media data combined with a semi-structured in-depth interview approach have allowed us to reflect public perceptions during the outbreak accurately. The Chinese central government encouraged the populace to limit their exposure to public settings after the COVID-19 outbreak. Provincial and local governments have proceeded to create more stringent community access management measures in accordance with the central government's epidemic prevention plans, mandating the populace to remain indoors in order to further stop the spread of the virus. Popular social media has consequently evolved into the fundamental platforms for individuals to learn about the progression of the disease and share their opinions. Therefore, the greatest approach to comprehend public thoughts and attitudes during an outbreak is through textual data from social media. Additionally, because social media data are instantaneous, they may give quick and efficient feedback on shifts in perception as the pandemic develops. At the same time, the semi-structured in-depth interview approach makes up for the shortcomings of big data by providing a great deal of in-depth information compared to the relatively short and fast information conveyed by big data. It also provides a new perspective on public perceptions during the outbreak.

However, it is equally important to recognise the limits of our strategy. Highly diverse social media users occasionally influence and hold the perceptions of the general population, limiting the representativeness of social media data for profiling. In addition, the non-social-media-loving public may introduce more error into the analysis part of our big data study, thus limiting the generalisation of the findings. Moreover, the study is only at a preliminary stage, the scope of the study is limited to Beijing, China, and the number of respondents in the in-depth interviews is limited by the study and, thus, the sample size is small. In addition, the overall perspective of the study is from a macro perspective, and future research is needed to combine more methods for multiple perspectives and further segmentation.

Our research methodology was divided into primary data collection, data processing, data analysis (including natural language (NLP) analysis and spatial and content analysis), and semi-structured in-depth interviews, as shown in Figure 1. To collect data for the study, we used a Python web crawler to search the original text posted by Weibo and Xiaohongshu users from 1 May 2018 to 1 May 2022, restricting the search to Beijing, China, based on the two Chinese keywords "dog walking space" and "good places to walk your dog", with the time interval divided into before (January 2020) and after the outbreak, as shown in Appendix A Figure A1. We save the crawled blog data to a local server as the main data source. Weibo is one of the most important social media platforms in China. Similar to Twitter in its powerful interactive features and timely information updates, it has a significant impact on the organisation of social life and public opinion. In the textual resources of Weibo, key and commonly used words can reflect various public narratives and the extent to which the public pays attention to these narratives [29]. Xiaohongshu (Little Red Book), a lifestyle platform and consumer decision portal, can effectively complement and corroborate the Weibo data with its "place seeding" and how-to reviews. Our next step was to clean up the data by removing duplicate content and ads. We obtained a sample of 26,550 valid blog posts, of which 13,150 were posted before the pandemic and 13,400 during the pandemic, totalling over 4.4 million words.

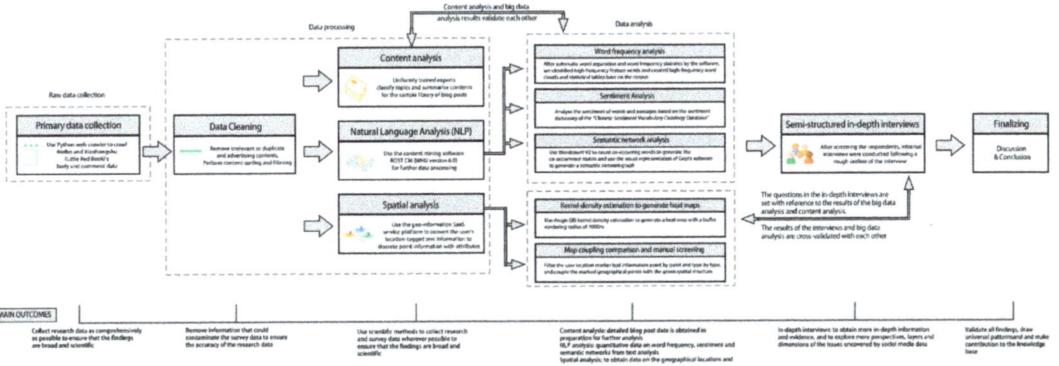

Figure 1. Framework diagram of research ideas.

For the data processing, we used the content mining software ROST CM (WHU version 6.0) to perform a natural language processing (NLP) analysis of the acquired textual material. The starting point for the content analysis was to identify the most frequently occurring semantic units in all the text material to provide an overview of potential research topics. Thus, the first NLP analysis we performed was a word frequency analysis. After building a custom lexicon relevant to the research subject, we filtered and segmented the crawled Weibo and Xiaohongshu text data. We removed auxiliary words (e.g., "due to", "this") and merged duplicate words (e.g., "don't" and "can't", "dog", and "hairy child"). High-frequency feature words regarding public perceptions of urban companion animals were identified through automatic software word separation and word frequency statistics. Eventually, we produced perceptual word cloud maps of the most commonly used words in the corpus, as shown in Figures 2 and 3, word frequency statistics for the top 40 words before and during the pandemic, as shown in Table 1 and Word frequency rose charts, shown in Figures 4 and 5.

Table 1. Frequency statistics for the top 40 words before and during the pandemic.

(Before the Outbreak)			(During the Outbreak)		
Rank	Word	Frequency	Rank	Word	Frequency
1	Address	3562	1	Tickets	3120
2	Dog walking	2819	2	Location	2916
3	Traffic	2787	3	Parking	2752
4	Parking	2680	4	Free	2353
5	Tickets	2235	5	Pandemic	2073
6	Beijing	1388	6	Dog walking	1687
7	Netizen	1212	7	Weekends	1223
8	Navigation	975	8	Camping	1089
9	Kilometres	971	9	BBQ	986
10	Hours	863	10	Picnics	957
11	Support	851	11	Less and less	935
12	Pet friendly	796	12	Pets	926
13	Camping	785	13	Beijing	870
14	Location	754	14	The park is huge	855
15	Place	735	15	Photo shoots	823
16	Photo-taking	653	16	Good places to go	812
17	Minutes	640	17	Cost	801
18	Travel with dogs	628	18	Recommended	763
19	Today	617	19	Friendly	728
20	Less crowded	603	20	Address	682
21	District	579	21	Playability	656
22	Disadvantages	562	22	Dogs	621
23	Car journey	537	23	Blowing wind	607
24	Pet friendly park	439	24	Lawn	583
25	Tents	426	25	Recently	529
26	Overall rating	423	26	Next to	514
27	Kite flying	415	27	Cute pets	472
28	Fees	389	28	Minutes	453
29	Shiba Inu	354	29	Enjoy the flowers	422
30	Free parking	341	30	Scenic spots	395
31	Dogs can be walked	339	31	Enjoy the greenery	354
32	Golden retriever	315	32	Everyone	351
33	Self-drive	286	33	Around the area	315
34	No entrance fee	283	34	Suitable for	278
35	Recommended	274	35	Parking fees	263
36	Route description	271	36	It's all about the dogs	251
37	Parking lot	265	37	Leash	229
38	Approximate	261	38	With hills and water	217
39	Beijing Adoption	247	39	Specific location	214
40	Opening hours	239	40	Tents	208

Figure 2. Perceptive word cloud map (before the outbreak).

Figure 3. Perceptive word cloud map (during the outbreak).

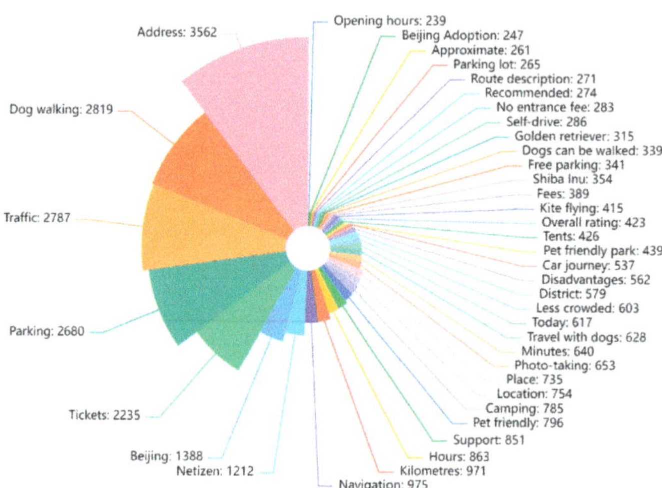

Figure 4. Word frequency rose chart (before the outbreak).

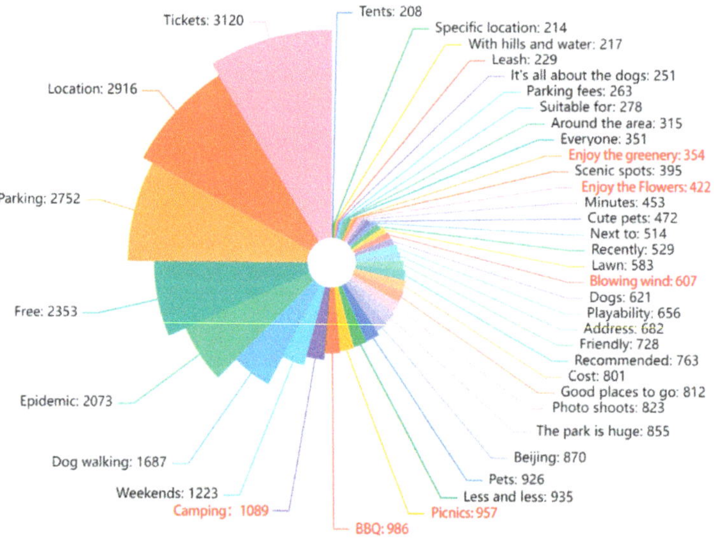

Figure 5. Word frequency rose chart (during the outbreak).

After reading the details of the sample of valid blog posts, we analysed the correlations between high-frequency words and used Wordcount V2 to generate a co-occurrence matrix of co-occurring words. Then, we used the visual representation of the Gephi software to generate semantic network diagrams, as shown in Figures 6 and 7. We then performed the sentiment analysis of text data by using a sentiment lexicon based on the Chinese Sentiment Vocabulary Ontology Library to understand the public's general feelings towards companion animal space, as shown in Figures 8 and 9, and Appendix A Figure A2.

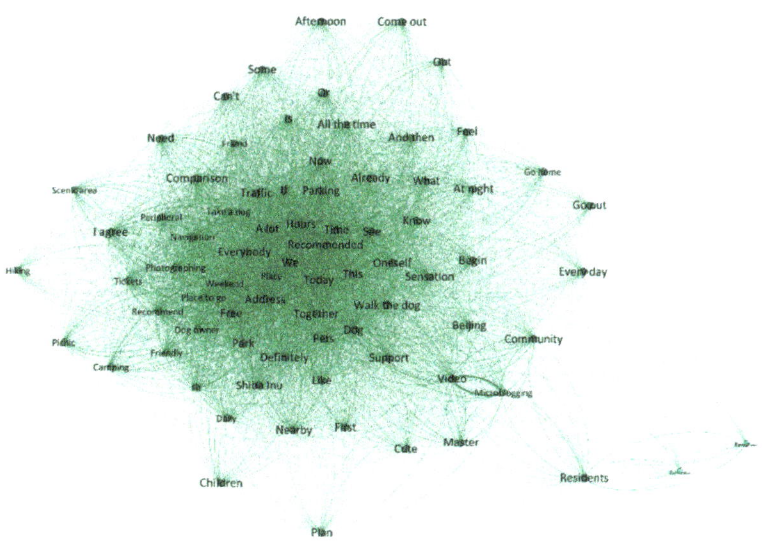

Figure 6. Semantic network diagram (before the outbreak).

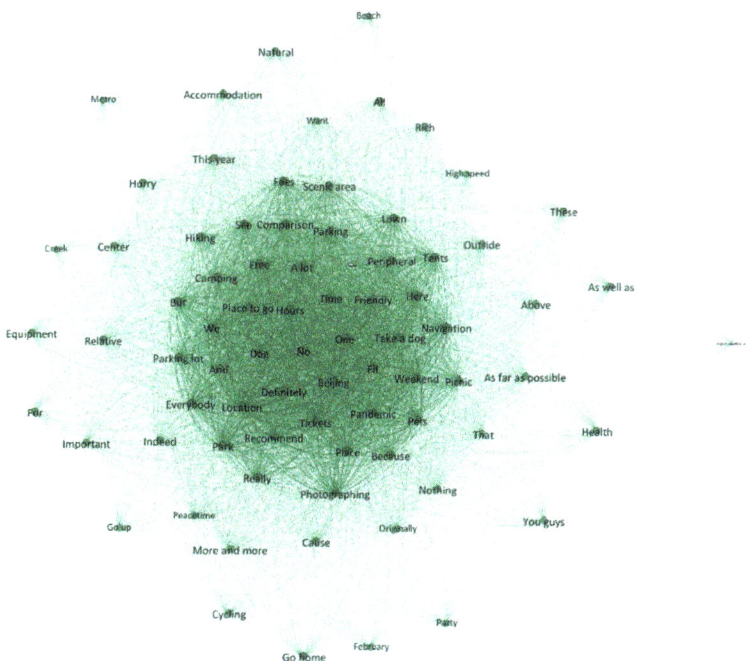

Figure 7. Semantic network diagram (during the outbreak).

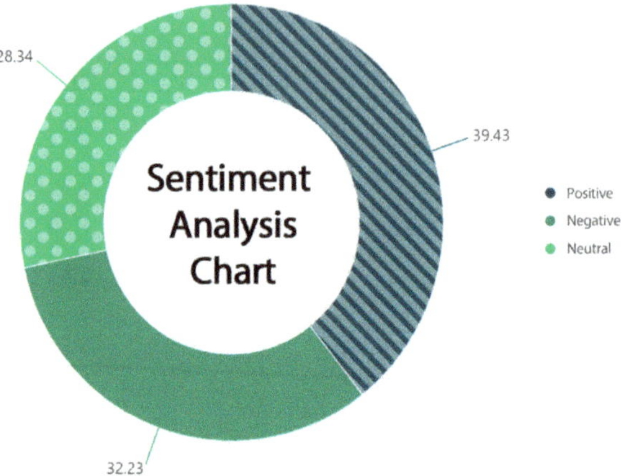

Figure 8. Sentiment analysis chart (before the outbreak).

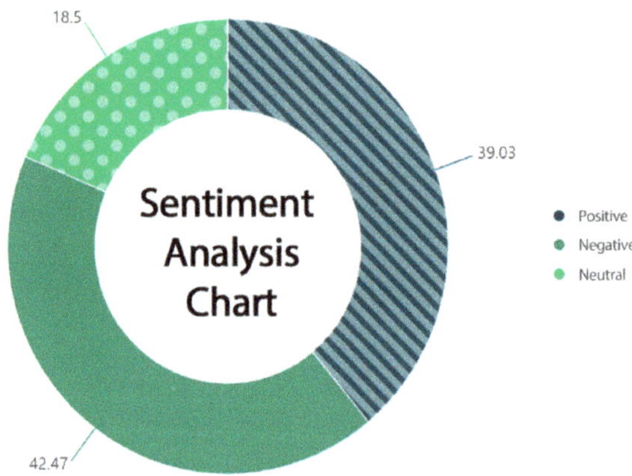

Figure 9. Sentiment analysis chart (during the outbreak).

The next step is spatial analysis. We use the Yijingzhilian Geographic Information SaaS platform to convert the user-tagged location text information into discrete point information with attributes, as shown in Figure 10. In addition, to better present the visualisation results, we generated a heat map by Arch GIS kernel density analysis with a buffer rendering radius of 1000 m, as shown in Figure 11. In addition, we also identified, recorded, and analysed the marked points in the pseudo-public space, generating the heat map shown in Figure 12. Geotagging by users on social media is generally motivated by the idea of marking "places worth remembering", which is highly credible and representative [30]. The location of the user's marker reflects, to some extent, the number and spatial distribution of "companion animals" in public spaces. In addition, to further analyse the types of "companion animal" public spaces and improve the accuracy, we conducted a point-by-point screening to identify the classification and coupled the marked geographical points with the green space structure plan of Beijing, as shown in Figure 13.

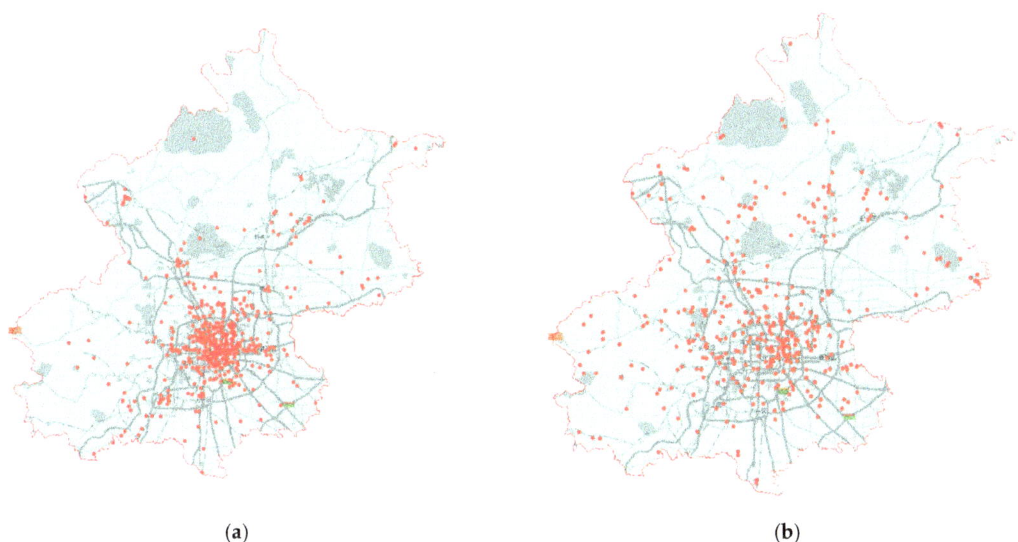

(**a**) (**b**)

Figure 10. (**a**) Geographical distribution of user markers (before the outbreak); (**b**) geographical distribution of user markers (during the outbreak).

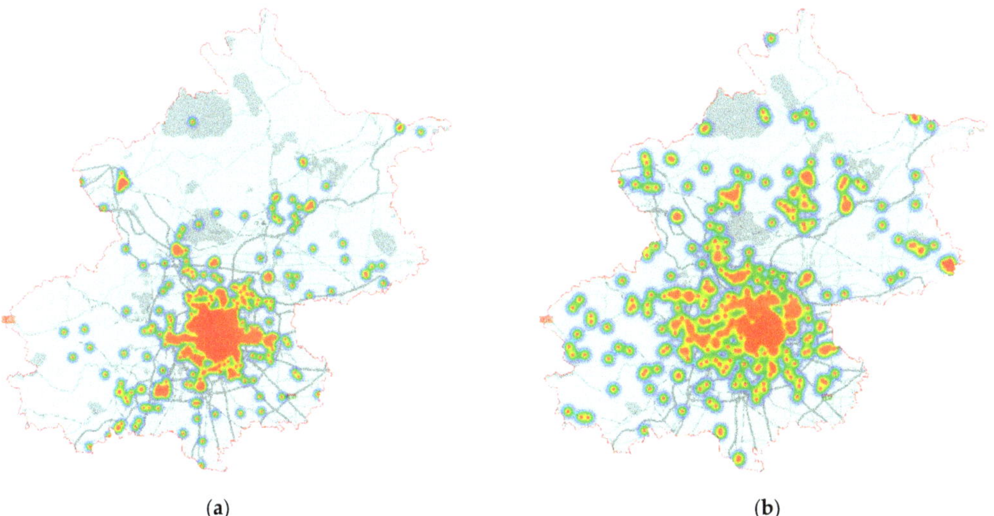

(**a**) (**b**)

Figure 11. (**a**) Heat map of the geographic distribution of user markers (before the outbreak); (**b**) heat map of the geographic distribution of user markers (during the outbreak).

(a)

(b)

Figure 12. (**a**) Heat map of the distribution of pseudo-public space markers (before the outbreak); (**b**) heat map of the distribution of pseudo-public space markers (during the outbreak).

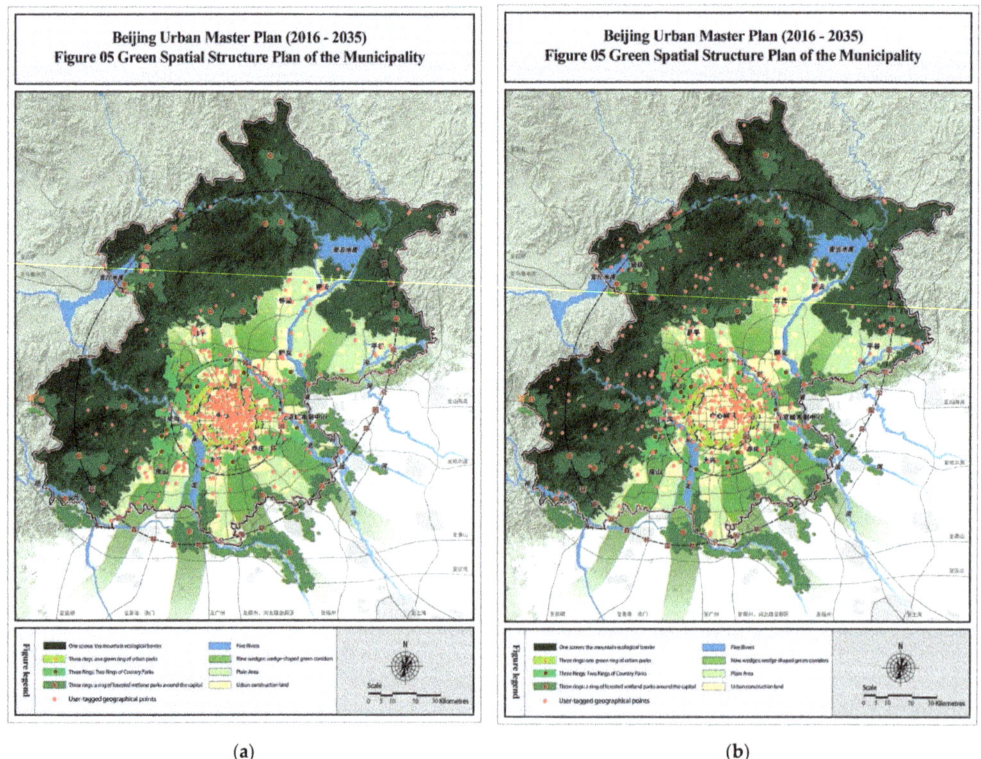

(a)

(b)

Figure 13. (**a**) Coupling of user geographic markers and planning maps (before the outbreak); (**b**) coupling of user geographic markers and planning maps (during the outbreak).

Then, a content analysis and topic summary of the sample database of blog posts were conducted by uniformly trained experts. In order to raise accuracy, the topics in the sample database were individually identified by three experts. Cross-referencing was carried out after the identification was completed. In the case of inconsistent themes, experts proofread and discussed until an agreed outcome was reached. We then used the detailed blog post data combined with the results of the big data analysis to interpret relevant social dynamics that reflect public perceptions of the relationship between public and pseudo-public spaces for urban companion animals and COVID-19.

As a final step, we invited respondents to participate in in-depth interviews, Sargeant defined how to ensure the quality of qualitative research participants [31], and we built on his work to select participants who could best inform our research questions and improve our understanding of changes in public perception and use of "companion animal" public spaces and pseudo-public spaces before and after the outbreak. For example, the inclusion criteria for respondents were Beijing residents, having lived with the dogs for more than 6 months before and during the outbreak, being between 18 and 60 years old, etc. Due to the impact of the epidemic containment, we mainly used Tencent Meetings online face-to-face and telephone interviews. This is considered one of the safest methods compared to the restrictions of movement and risk of infection that would result from more traditional face-to-face public surveys. A total of 12 Beijing residents were interviewed in semi-structured in-depth interviews. Consistent with rooting theory, we used a maximum variation sampling strategy to select residents with different lifestyles and conditions, aiming to provide a variety of perspectives. We did not offer any remuneration or incentives to participants. Interviews with respondents ranged from 19 to 33 min in length and were conducted by two interviewers working together in a semi-structured question-and-answer format, divided into an interviewer who was responsible for the interview and an interviewer who was responsible for recording and asking additional questions. We used digital equipment for recording and further transcription and manual thematic coding. All participants gave verbal and written consent to record their interview sessions and followed their views on whether to anonymise their identities. Similar studies, such as C Mayen Huerta et al. [32] and Charlotte Collins et al. [33], have used in-depth interviews to better understand the association between UGS use and other variables. To enhance the scientific validity of the interview questions, we drew, in part, on the Monash Dog Owner Relationship Scale (MDORS) developed by Fleur Dwyer et al. in designing the questions [34]. The use of this scale allows researchers to increase their understanding of human–companion dog relationships by allowing direct comparisons between groups of participants from different demographic or cultural contexts. We also refer, in part, to the Development and Reliability of the Dogs and Physical Activity (DAPA) Tool developed by Hayley E. Cutt et al., which can be used to retest individual, social context, physical environment, and policy-related factors that influence dog owners' dog walking behaviour confidence assessment [35]. The in-depth interview questions consisted of several basic sections, as shown in Appendix A Figure A3, with the questions being fine-tuned to the respondents. The first part collected information about the interviewees and their families, educational level, employment status, and dog status; the second part asked about the general and material impact of the epidemic on the perception and use of the "companion animal" space by the respondents and their pets. In the third part, the focus of the survey was on the mental and psychological effects.

3. Results

We derived perceptual word cloud maps before and during the pandemic by performing word frequency analysis on the cleaned social media data for NLP analysis, as shown in Figures 2 and 3. In this case, the frequency of words in the text is proportional to their size. During the COVID-19 outbreak, for the public space of "companion animals", the preoutbreak crawl showed that the most frequently used words were "Address", "Traffic", "Dog walking", "Parking", and "Tickets". We can see that accessibility and ease of parking are the top priorities for people choosing public spaces for companion animals. There is also much discussion of dog-related topics, such as "Dog hunting", "Chinese Field Dogs", and "Walking dogs without a leash", to name a few popular and common topics.

Surprisingly, the results of the post-pandemic word frequency survey show that words such as "Tickets", "Location", "Parking", and "Free" continue to top the list of words that express dog owners' strong need for "companion animal" public space, while the frequency of activities such as "Camping", "BBQ", "Picnic", and "Blowing wind" and so on has also increased significantly, as shown in Figures 4 and 5. In contrast to public opinion during the pandemic, where public perceptions of urban companion animals tended to be mainly negative experiences [6], for dog owners in Beijing, the word "epidemic" is also at the top of the list, but it is clearly not the main focus of attention. Perhaps as a result of the normalisation of the pandemic, many dog owners tend to see the "epidemic" simply as a social backdrop, focusing mainly on helping their dogs to find new outdoor spaces or as an after-dinner talker, such as *"I can't get out of Beijing because of the epidemic, but I went to Shentang Yu at the weekend, which is one of the few pet-friendly places, and I highly recommend it." "It's rare to see a dog wearing a breathing mask, lol." "Taking Old Man Bear (dog's name) for a walk after lunch, I met Tammy's mother in full armor walking her dog in the yard too!".*

To further explore issues related to public perception, we conducted a semantic network analysis of high-frequency words before and after the pandemic, as shown in Figures 6 and 7. We can easily see that, before and during the pandemic, the semantic network diagram presents a distinct core, basically with several core words as a group expanding outwards diffusely. The hierarchical relationships between the core words are relatively similar. After careful analysis, we identified that "Address", "Traffic", "Park", "Navigation", and "Parking lot" were strongly correlated with each other before the pandemic. Meanwhile, during the pandemic, "Tickets", "Friendly", "Location", "Pandemic", "Parking", "Dog", and "Lawn" have a strong correlation between the words.

In addition, we conducted sentiment analysis on the blog posts before and after the pandemic, mainly analysing the sentiment polarity (i.e., positive, neutral, and negative sentiment) and sentiment intensity of the words with sentiment components within each utterance, and then calculated the total value of each utterance to determine its sentiment category. In order to determine the overall attitude and sentiment tendency of the total opinion data sample, we statistically integrated all statements, as shown in Figures 8 and 11, and found that, before the pandemic, positive sentiment accounted for 39.43% (5283), negative sentiment accounted for 32.23% (4318), and neutral sentiment accounted for 28.34% (3799). During the pandemic, the proportion of positive sentiments was 39.03% (5132), negative sentiments 42.47% (5584), and neutral sentiments 18.5% (2434). We can see that, especially during the pandemic, most bloggers expressed more personal sentiment on the topic of "companion animals" in public spaces, with negative sentiment increasing by 10.24%.

Subsequently, in order to analyse the number and spatial distribution of "companion animal" public spaces, we extracted the geographical locations marked by users from social media data, with a total of 1586 valid points, including 749 before and 837 during the pandemic, as shown in Figure 10. Heat maps were generated to better present the results, as shown in Figure 11. It is clear that the prominent distribution locations were in the urban areas of Beijing, both before and during the pandemic. However, when it comes to identifying the types of "companion animal" public spaces before and after the pandemic, we find that the largest proportion of points is still in scenic areas and parks, with 83% (621) before the pandemic and 91% (761) during the pandemic. In contrast, the remaining points are mainly in pseudo-public spaces, such as shopping plazas and public spaces attached to buildings and certain streets. Comparing the heat maps before and during the pandemic, as shown in Figure 11, it is clear that the density of geographic markers in the built-up area decreases and the density of markers in the surrounding areas increases. After coupling the markers with the Green Spatial Structure Plan in Beijing Urban Master Plan (2016–2035), as shown in Figure 13, we can find that the markers in the peripheral areas are mainly located in parks and green areas, as well as major scenic spots, which coincides with our differentiation of marker types. Comparing the heat map of pseudo-public space markers, we can also find that the proportion of pseudo-public space has a clear tendency to decrease during the pandemic compared to the pre-pandemic period, as shown in Figure 12.

Afterwards, the content of the blogs was evaluated by uniformly trained experts and the posts were categorised into two themes: 1. the need for and feelings about public spaces for "companion animals" and 2. the experience of owning a companion animal. Finally, following semi-structured in-depth interviews, three central themes were identified: 1. changes in time and space for dog walking, 2. changes in motivations and attitudes towards dog walking, and 3. crisis management of pets during epidemics. Based on these three themes and the results of the analysis above, we continue in the next section of the discussion with a further analysis of the changing perceptions and use of public and pseudo-public spaces for "companion animals" in the city.

4. Discussion

The pandemic has profoundly changed public perceptions and usage of public and pseudo-public spaces for "companion animals". After reading all the comments, combined with the results of the big data analysis and the results of the in-depth interviews, referring to the COM-B behavioural model developed by Michie et al. [36], we conclude that the impact of the pandemic on the dynamics of direct interaction between humans and their pets with "companion animal" public and pseudo-public spaces can be summarised in three distinct but not mutually exclusive pathways: changes in opportunity, changes in capacity, and changes in motivation, as in Figure 14. Of course, the intensity and direction of these three pathways may vary considerably depending on demographic, regional, and national socioeconomic, political, cultural, and social factors and environmental factors, and potentially relevant changes in severity and response to the pandemic.

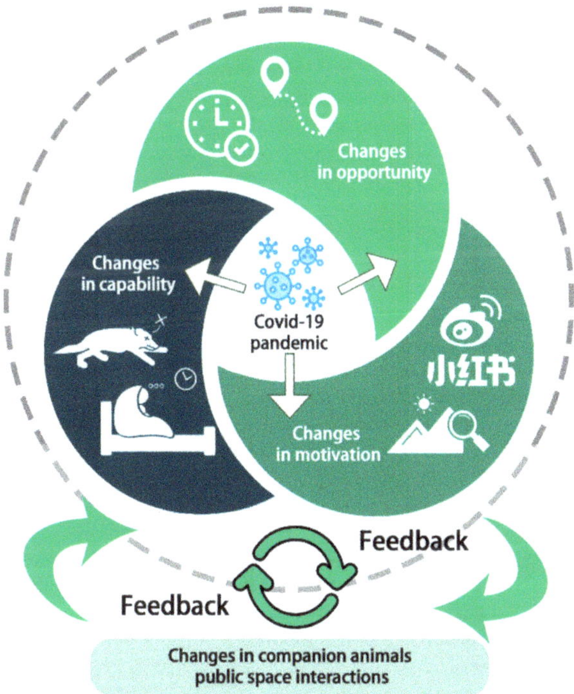

Figure 14. A conceptual framework of pandemic influences on the direct interaction of humans and their pets with "companion animals" public and pseudo-public spaces.

(1) Changes in opportunities related to the factors that facilitate or make possible spatial interactions with "companion animal" space, including the amount of interaction, the duration of contact, and how they interact. Firstly, regarding the number of interactions and the timing of contact, the pandemic had both positive and negative effects on humans and their pets with the "companion animal" space. On the positive side, the use of "companion animal" spaces increased for some residents during the pandemic. Survey data showed over 2000 blog posts expressing more time and similar sentiments during the pandemic, perhaps due to the use of teleworking during the pandemic increasing the time available for other activities for some residents [37]. As one user, "buzz Lightyear's wife", posted: *"I've spent a lot more time with my pals and had more opportunities to walk my dog this year because of the pandemic. It has probably been the most life-affirming year I have had."* However, there is also a negative side to this, with less access to "companion animal" space for the wider public. During the pandemic, many places, such as shopping centres, community spaces, and major urban parks, were closed to reduce infection rates [38,39]. For low-risk and above-containment areas, working from home for long periods of time under the epidemic, people become distant from each other [40–42] and dogs' physical and mental health and well-being are seriously affected by reduced range of motion, less exercise, and less dog-to-dog communication [43]. As a result, the survey data show that, compared to the pre-pandemic period, the word frequency statistics show that words such as "Tickets", "Location", "Parking", and "Free", which express the strong demand of dog owners for "companion animal" public spaces, are still at the top of the list, while the frequency of activities such as "Camping", "BBQ", "Picnic", and "Blowing wind" and so on has also increased significantly, as shown in Figures 4 and 5. The reduced access to "companion animal" spaces may, therefore, be one of the main drivers of the increased demand for such green spaces during the pandemic. For example, one user named "Ollie with long eyelashes" excitedly posted about her desire: *"The closure was lifted at 12 pm, so I quickly asked my friends to take my pups out for a run, and I was very happy to have just a shallow 3-min run"*.

During our in-depth interviews, interviewee 01 told us that, although some parks and green spaces were off-limits to dogs before the epidemic, in practice, the relevant no-dog rule existed in name only and, in some cases, it was possible to slip in and walk dogs when park guards were not looking. There was only a sign indicating the ban, and the associated penalties were almost non-existent. However, during the epidemic control period, almost all respondents mentioned that they found that even city parks near their residential areas were no longer allowed to walk their dogs, and that regulations were much stricter than before the epidemic. At the same time, the monitoring of dog walking in street green areas and green areas attached to buildings has also become very strict. Overall, respondents generally felt that, while, before the epidemic, it was mainly a subjective desire to take their pets to "companion animal" public spaces, after the epidemic, it was forced to become an immediate necessity, and the frequency and time spent walking dogs was significantly reduced. As a result of the epidemic, they rarely met other dog walkers, and thus many respondents reported that their and their dogs' social needs were not being met and that they were more depressed than before the epidemic. At the same time, respondent 04 mentioned that her dog walking behaviour and the way she used the public space for "companion animals" was very much dependent on the "dog walking culture" in the small area she visited, for example, whether dogs are walked on leashes, whether owners can follow their pets onto lawns, etc, which the big data cannot measure. After the epidemic, however, this dynamic use of "companion animal" public spaces has, to some extent, disappeared, as people are not allowed to gather in large numbers. In conclusion, the pandemic has increased the amount of time available to some people, but the amount of public space available for "companion animal" has decreased due to the closure and, with the normalisation of the pandemic, the construction and operation of urban public space will face new challenges. The next focus of research will be the balance between planning and design, management and operation, and policy implementation.

In addition, the pandemic may not only change the amount of "companion animal" space a person has access to, but also the way they interact with "companion animal" space. For urban green spaces, the closure of many major urban parks may have increased people's use of nearby natural environments and reduced their access to natural environments away from where they live. However, this is not absolute. For example, Figure 11 show that the general use of "companion animal" spaces by Beijing residents for several years has shifted from a concentration in built-up areas and city parks in the main urban areas before the pandemic to areas such as country parks outside of the city. We can see from this that, under the normal state of the pandemic, people are more inclined to go to suburban green spaces than before. For example, "Zhao Erdog loves chillies" posted: *"Under the cherished sun, I took my dog for a walk in this small wood. The city is becoming more and more green, but there are fewer and fewer places for the dogs to roam. Even without the towering walls, the walls built by the human heart are unbreakable. Do we have to take our dogs into the mountains to have a place in the valley? Alas! For now, let's just walk our dogs and cherish them."* For the built-up areas of the city, as parks and other green spaces are not the first places visited by pets and their owners, they are most often used in built-up areas [44]; usually, pet owners walk their dogs underneath their neighbourhoods or on the streets without a clear destination and, naturally, they do not need to make a point to record it. Thus, the main types of sites in our marker data are scenic and public green spaces, while built-up areas are mainly pseudo-public spaces, as shown in Figure 12. Pseudo-public spaces were more heavily regulated during the pandemic, so the number of geographic markers recommended was low.

In the in-depth interviews, the majority of respondents said that they went to pseudo-public spaces, such as shopping centres and plazas to walk their dogs prior to the epidemic in order to meet their own related needs and to take their dogs for a walk. However, there were some exceptions; for example, interviewee 12 mentioned going to private pet parks specifically for their dogs to give them an amusement-park-like experience, while some respondents said they went for grooming and maintenance and to show off their dogs and give their dogs a taste of "busy city life". However, during the epidemic, there were more risks to be avoided due to the ever-changing policies, and some respondents even said they did not walk their dogs in pseudo-public spaces at all after the outbreak. Unsurprisingly, the big data analysis points to a clear downward trend in the proportion of pseudo-public space. Although we cannot confirm whether this downward trend is due to the pandemic or privatisation, or a combination of both, it also reveals another problem: the use of "companion animal" public space in pseudo-public spaces decline in status, and those located within the open urban green space on the outskirts of the city stand out during the pandemic outbreak. How Beijing's economy and urban planning will respond to this change in spatial usage and its consequences will be an essential consideration for future urbanisation.

(2) Competence is the psychological and physical ability of pets and their owners to interact with "companion animal" public spaces. Firstly, on a physical level, it is clear that, if a person is infected with COVID-19, the pet will also be placed in temporary quarantine by the pet facility, which will stop their use of the "companion animal" space. However, for pets in pandemic areas, in addition to being taken away for isolation, there are also many one-size-fits-all policies, such as the one in Beijing's Daxing district, where a week-long dog-catching spree and forced "innocuity treatment" has caused discontent among dog owners. A netizen named "Christa-meng" commented on the news of the pet quarantine: *"Can we pay attention to the recent dog arrests in Daxing District, with police cars squatting in front of the district in the early hours of the morning? Why not go straight to the breeding source if they are not allowed to breed? It is not fair to euthanize a fur baby raised for seven or eight years outside the Fifth Ring Road because it was taken away with a single word!"* "A small milk dumpling said, *"This may cause some people who don't really love dogs to abandon them and turn them into stray dogs. It adds to the burden on society."* These comments quickly sparked outrage as the public began to expose various extreme COVID-19 prevention measures. Statistics show that negative sentiment increased by 10.24%. A careful reading of the blog post reveals three

main sources of negative comments; the first is dissatisfaction with the local community's one-size-fits-all policy of compulsory confiscation and "innocuity treatment" disposal of pets on health grounds, the second is dissatisfaction with the increasing difficulty in finding places to walk pets, and the third is dissatisfaction with the extreme negativity against pets in society, such as the extreme comments about directly linking pets to COVID-19.

In addition, in the in-depth interview, interviewee 06 told us that he had never travelled with his dog again afterwards because he had a need to transport his pets but, also, because the epidemic quarantine delayed a lot of time during the transport process and his own pets died of starvation and thirst as a result. At the same time, many interviewees also mentioned the complete lack of measures taken by the government during the epidemic, which was basically people-oriented, with dogs being treated as accessories to people. Some respondents said that they were forced to separate from their dogs because of the quarantine and could leave them with neighbours who had dogs or pet shops, while others said that there were no measures in place to deal with the situation and that their pets were forced to starve to death. It is also worth noting that the government has been lax in the management of dogs, as the epidemic is more loosely controlled in rural Beijing. As a result, survey respondents in rural areas reported that they were largely unaffected or minimally affected. A few respondents said that dogs were never taken out of their compounds, so, naturally, the impact was minimal. Overall, however, the cruel treatment of companion animals and the abandonment and even killing of animals to prevent humans from being infected in the above incidents reflect a strong anthropocentrism. Animals are reduced to resources for human growth when anthropocentrism is strong, and there is no ethical analysis of whether human demands and aspirations are appropriate. [45]. Emerging human-to-human infectious illnesses are seen as public health emergencies that solely endanger human health due to species barriers; hence, most matching emergency plans only offer treatment and refuge for people, while excluding the companion animals that share their homes. Although the central government and local authorities in China have begun to actively correct excessive pandemic prevention measures by issuing circulars to educate the public on animal protection, there are still no contingency plans proposed for the arrangement of relevant spaces.

On a psychological level, our sentiment analysis survey of the public opinion data sample showed a slight increase in overall negative sentiment, as the uncertainty and fear associated with the outbreak, as well as the massive blockade and economic recession, may have led to increased symptoms of anxiety, depression, post-traumatic stress disorder, and other forms of psychological disorders in the general population, even if people were not infected with COVID-19. At the same time, pets may experience behavioural changes due to changes in owners and the outside environment [46,47], such as frequent barking or fear of noise and the inability to be left alone in the home [48]. Such changes may facilitate the possibility of spatial interaction with "companion animals". As a result, the frequency of activity words has increased significantly in the survey data. For example, "Kafka Chou Chou" says: *"Every time the city was closed during the pandemic, I tried to take my dog Bao out to play, hiding her in the trunk every time, and then letting her sit in the car after the pandemic checkpoint, because she was not used to seeing people outside. Then depression usually stayed by itself, hiding under the table for half a day, and it was only afterwards that she was taken out for walks every day for about 3 months that she slowly got better"*. Thus, we can also see from this that changes in ability sometimes lead to changes in motivation and, ultimately, to changes in behaviour and that there is a complex inter-relationship between the three drivers of the behavioural model.

(3) Motivation is a process in the human brain that motivates and guides behaviour, and the spread of the COVID-19 disease may have significantly altered people's motivation to interact with "companion animals" public space. In addition to the changes mentioned above in motivation due to changes in the available time, opportunities and accessibility, and physiological and psychological changes caused by the pandemic, during the pandemic, many users recommend "companion animal" spaces through geolocation marking and

the creation of various tags, such as "#Dangers of having a dog without a walk during a pandemic", created an information ecosystem defined by an unprecedented amount of data that profoundly influenced other users' motivations and behaviours [49,50]. For example, "small bok choy" said: *"I've always seen online that it's good to walk your dog here, so I stopped by today to take a look. I don't know, it's pretty big, love it, love it, and my boyfriend played in it all afternoon, tomorrow I'll bring Fu (dog's name) to experience it"*. In our in-depth interviews, our respondents said that local residents in Beijing chose places to walk their dogs before the epidemic mainly out of local and lived experience, either by experiencing them first or by seeing people walking their dogs on the road and then recording them before taking them for a walk, or, in some cases, by recommendations from friends and family or by looking at tips on the internet. However, after the epidemic and, therefore, the ready change in policy, there was a lot of reliance on recommendations from the internet. However, this is not without its exceptions, as some respondents said that, in order to save time, they tended to go to familiar places rather than those recommended on the internet when there was a suitable space to walk their dog, and that, if they really wanted to walk their dog for a break, they basically chose to go to the more remote suburban green areas.

In addition, on the one hand, we can see from the pattern of changes in high-frequency words in Table 1, Figures 4 and 5, such as the significant increase in the frequency of activity words, that the positive attitudes of pet owners toward the public space of "companion animal" increased during the outbreak. On the other hand, however, the overall public's negative attitudes toward companion animals also increased significantly, even to pathological fear [7], which partly contributed to the statistical increase in the proportion of negative sentiments posted by pet owners during the pandemic. Although the two groups, pet owners and the wider public, partially overlap, the conflict between them exacerbates the social perception of multi-species conflict, leading to a variety of vicious conflicts [7] that are detrimental to the sustainability of the "companion animal" public space. We argue that social media plays a vital role during a pandemic, influencing the motivations and behaviours of users. Government interventions in the "companion animal" public space based on social media theories of critical knowledge, attitudes, intentions, and behaviour change could play a significant role. A shift from self-regulation of social media platforms to government intervention may be a positive step.

5. Conclusions

This paper examines the changing perceptions and use of "companion animals" in public and pseudo-public spaces in the city during the COVID-19 pandemic, using Beijing, China, as an example. The study was based on a Python web crawler that collected relevant text and comment data from Chinese social media Weibo and Xiaohongshu, followed by natural language (NLP) analysis, spatial analysis, and content analysis of the pre- and post-pandemic social media data, and, finally, semi-structured in-depth interviews were conducted. The big data revealed statistically significant differences in the perception and use of "companion animal" public spaces and pseudo-public spaces before and after the outbreak. Referring to the COM-B behavioural model developed by Michie et al., we conclude that the impact of the pandemic on the dynamics of direct interaction between humans and their pets with "companion animal" public and pseudo-public spaces can be attributed to three pathways: changes in opportunity, changes in ability, and changes in motivation, with complex inter-relationships between the three drivers. We found that the pandemic has acted as a mirror and catalyst to expose the multi-species coexistence of humans and animals in China's cities. The pandemic has increased the amount of time available to some people but reduced the amount of public space available for "companion animals" due to the pandemic's closure. As pandemics become normalised, the operation of public spaces related to "companion animals" in cities will face new challenges, and the balance between planning, design, management, operation, and policy implementation will be the next focus of research. In addition, due to species barriers, emerging human-to-human infectious diseases are perceived as public health crises that threaten only human health. Therefore, most corresponding emergency plans only provide treatment and shelter for humans while neglecting the companion animals that live with them, especially in "companion animal" public spaces. There is still a lack of contingency planning by the Chinese central government and local authorities to arrange these spaces. In addition, the survey showed that the use of "companion animal" public spaces in pseudo-public spaces declined, while those in open urban green spaces on the outskirts of the city stood out after the outbreak, creating new requirements for the future development of Beijing's urban green space system and related policies. At the same time, pet owners' perceptions and behaviours regarding "companion animal" public spaces are heavily influenced by social media during pandemics, and the potential for abandonment and cross-species infections resulting from the acceptance of negative information about them can make companion animals a new public safety hazard. However, discussions about "companion animal" public spaces on social media are not actively regulated and appropriately intervened in, in some cases, leaving a very conflicted and confusing impression and stirring up group antagonism. We argue that there could be a shift from self-regulation on social media platforms to government intervention and joint efforts between government and social media to advocacy and support for sustainable animal ethics practices to better respond to the crisis.

Humans and companion animals merge, constitute, and permeate each other in everyday urban spaces. More research is needed to determine how to more thoroughly analyse the more complex relationships between "companion animal" public spaces and urban spaces and how these relationships are expressed in social media. The future of public space is uncertain due to the normalisation of pandemics, and more research is needed from different parts of the world.

Author Contributions: Conceptualisation, H.C. and W.D.; methodology, W.D.; software, H.C.; validation, W.D. and H.C.; formal analysis, H.C.; investigation, H.C.; resources, W.D.; data curation, H.C.; writing—original draft preparation, H.C.; writing—review and editing, W.D.; visualisation, H.C.; supervision, W.D.; project administration, W.D.; funding acquisition, W.D. All authors have read and agreed to the published version of the manuscript.

Funding: This research was funded by fundamental research funds for central universities (2018ZY10) and National Natural Science Foundation of China, project No. 51508024.

Institutional Review Board Statement: Ethical review and approval were waived for this study, due to involving no more than minimal risk.

Informed Consent Statement: Informed consent was obtained from all subjects involved in the study.

Data Availability Statement: Data supporting reported results can be found by contacting the authors.

Acknowledgments: I would like to thank Associate Professor Duan Wei for his strong support for this research work. Also, I would like to thank Haoning Liu and Shaojie Wang for their great contributions for this study.

Conflicts of Interest: The authors declare no conflict of interest.

Appendix A

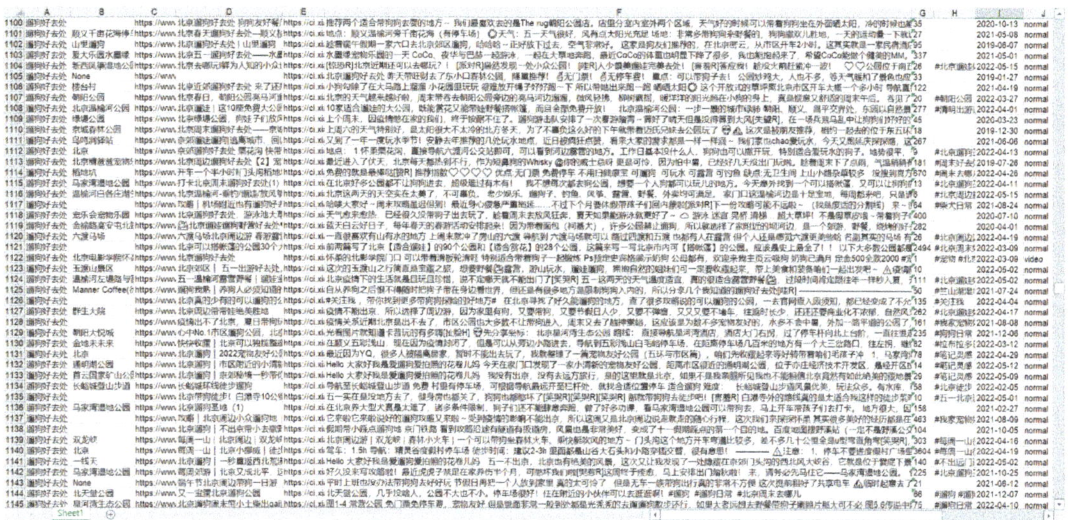

Figure A1. Some of the original text posted by Xiaohongshu users crawled using a web crawler.

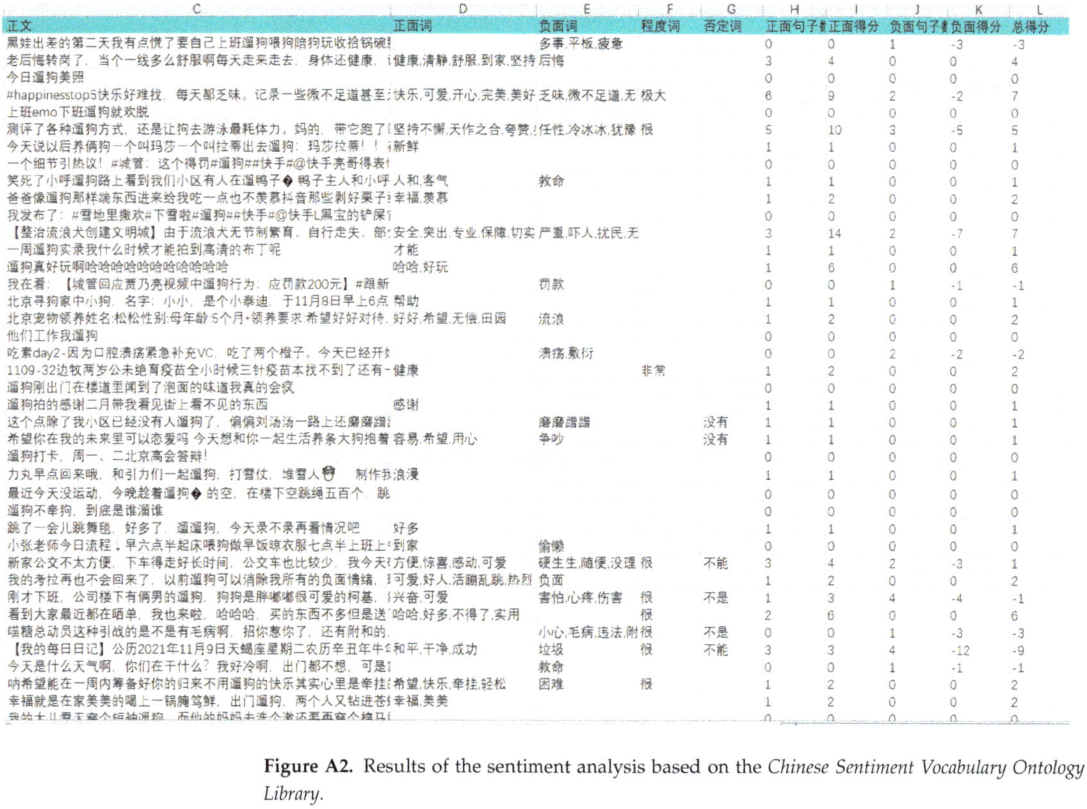

正文	正面词	负面词	程度词	否定词	正面句子数	正面得分	负面句子数	负面得分	总得分
黑娃出差的第二天我有点慌了要自己上班遛狗喂狗陪狗玩收拾锅碗		多事,平板,疲惫			0	0	1	-3	-3
老后悔转岗了，当个一线多么舒服啊每天走来走去，身体还健康，	健康,清静,舒服,到家,坚持	后悔			3	4	0	0	4
今日遛狗美照					0	0	0	0	0
#happinesstop5快乐好难找，每天都乏味。记录一些微不足道甚至	快乐,可爱,开心,完美,美好	乏味,微不足道,无	极大		6	9	2	-2	7
上班emo下班遛狗就欢脱					0	0	0	0	0
测评了各种遛狗方式，还是让狗去游泳最耗体力，妈的，带它跑了	坚持不懈,天作之合,夸赞	任性,冷冰冰,犹豫	很		5	10	3	-5	5
今天说以后养俩狗一个叫玛莎一个叫拉蒂出去遛狗：玛莎拉蒂！！	新鲜				1	1	0	0	1
一个细节引热议！#城管：这个得罚#遛狗##快手#@快手亮哥得表					0	0	0	0	0
笑死了小呼遛狗路上看到我们小区有人在遛鸭子� 鸭子主人和小呼	人和,客气	救命			1	1	0	0	1
爸爸像遛狗那样端东西进来给我吃一点也不羡慕抖音那些剥好栗子	幸福,羡慕				1	2	0	0	2
我发布了：#雪地里撒欢#下雪啦#遛狗##快手#@快手L黑宝的铲屎					0	0	0	0	0
【整治流浪犬创建文明城】由于流浪犬无节制繁育、自行走失、部	安全,突出,专业,保障,切实	严重,吓人,扰民,无			3	14	2	-7	7
一周遛狗实录我什么时候才能拍到高清的布丁呢	才能				1	1	0	0	1
遛狗真好玩啊哈哈哈哈哈哈哈哈哈哈	哈哈,好玩				1	6	0	0	6
我在看：【城管回应贾乃亮视频中遛狗行为：应罚款200元】#跟新		罚款			0	0	1	-1	-1
北京寻狗家中小狗，名字：小小，是个小泰迪，于11月8日早上6点 帮助					1	1	0	0	1
北京宠物领养姓名:松松性别:母年龄:5个月·领养要求:希望好好对待，	好好,希望,无偿,田园	流浪			1	2	0	0	2
他们工作我遛狗					0	0	0	0	0
吃素day2-因为口腔溃疡紧急补充VC，吃了两个橙子。今天已经开		溃疡,敷衍			0	0	2	-2	-2
1109-32边牧两岁公未绝育疫苗全小时候三针疫苗本找不到了还有-	健康		非常		1	2	0	0	2
遛狗刚出门在楼道里闻到了泡面的味道我真的会馋					0	0	0	0	0
遛狗拍的感谢二月带我看见街上看不见的东西	感谢				1	1	0	0	1
这个点除了我小区已经没有人遛狗了，偏偏刘汤汤一路上还磨磨蹭		磨磨蹭蹭		没有	1	1	0	0	1
希望你在我的未来里可以恋爱吗 今天想和你一起生活养条大狗抱着	容易,希望,用心	争吵		没有	1	1	0	0	1
遛狗打卡，周一、二北京高会答辩！					0	0	0	0	0
力丸早点回来哦，和引力们一起遛狗，打雪仗，堆雪人 制作我浪漫					1	1	0	0	1
最近今天没运动，今晚趁着遛狗� 的空，在楼下空跳绳五百个，跳					0	0	0	0	0
遛狗不牵狗，到底是谁溜谁					0	0	0	0	0
跳了一会儿跳舞毯，好多了，遛遛狗，今天录不录再看情况吧	好多				1	1	0	0	1
小张老师今日流程↓早六点半起床喂狗做早饭晾衣服七点半上班上‘到家		偷懒			0	0	0	0	0
新家公交不太方便，下车得走好长时间，公交车也比较少，我今天	方便,惊喜,感动,可爱	硬生生,随便,没理	很	不能	3	4	2	-3	1
我的考拉再也不会回来了，以前遛狗可以消除我所有的负面情绪，	可爱,好人,活蹦乱跳,热烈	负面			1	2	0	0	2
刚才下班，公司楼下有俩男的遛狗，狗狗是胖嘟嘟很可爱的柯基，	兴奋,可爱	害怕,心疼,伤害	很	不是	1	3	4	-4	-1
看到大家最近都在晒单，我也来啦，哈哈哈，买的东西不多但是送	哈哈,好多,不得了,实用		很		2	6	0	0	6
喵糖总动员这种引战的是不是有毛病啊，招你惹你了，还有附和的，		小心,毛病,违法,附	很	不是	0	0	1	-3	-3
【我的每日日记】公历2021年11月9日天蝎座星期二农历辛丑年牛	和平,干净,成功	垃圾	很	不能	3	3	4	-12	-9
今天是什么天气啊，你们在干什么? 我好冷啊，出门都不想，可是		救命			0	0	1	-1	-1
呐希望能在一周内筹备好你的归来不用遛狗的快乐其实心里是牵挂	希望,快乐,牵挂,轻松	因难	很		1	2	0	0	2
幸福就是在家美美的喝上一锅腌笃鲜，出门遛狗，两个人又钻进被	幸福,美美				1	2	0	0	2
我的女儿要天穿个短袖遛狗，而他的妈妈非穿个裙还要再穿个棉马					0	0	0	0	0

Figure A2. Results of the sentiment analysis based on the *Chinese Sentiment Vocabulary Ontology Library*.

Broad question set for the interview:

Part 1 - Basic information

- What is your age? Gender? Level of education (tertiary or higher, below tertiary)? Employment status (employed, unemployed) and occupation? Household income (A. Below RMB 5,000 B. RMB 5,000-10,000 C. Above RMB 10,000)? Location of residence (urban, suburban, rural)? Self-rated physical condition (healthy, unhealthy)?

- How many years have you had the dog (how old is the dog)? Number? What breed? Size of dog?

Part2—General and material impact

- What were the main places where dogs were usually walked before the outbreak? What were the main types? What was the frequency and duration? What would you say is particularly convenient or inconvenient about the place where you walk your dog?

- What are the main places where dogs are usually walked after the epidemic? What were the main types? What was the frequency and duration? What would you say is particularly convenient or inconvenient about the place where you walk your dog?

- Do you think time spent with dogs and walking them has increased or decreased since the outbreak and why?

- How were your dogs handled during the closure? Did the government give any support or issue any policy on this?

- How do you usually find a suitable place to walk your dog? By what means? Do you have any online recommendations for places to walk your dog?

- Have you tried private pet parks and taking your dog for grooming and maintenance? Has the frequency of visits to relevant places changed since the outbreak? Why?

- Have you tried walking your dog in shopping centres and plazas, such as Sanlitun and Outlets, and has the frequency of visits to these places changed since the outbreak? Why?

Part3—Psychological and spiritual influences

- Has there been any significant change in your dog's mood or behaviour since the outbreak? Has there been any change in the number of interactions with you?

- Has the mindset and reasons for walking dogs changed or are there any concerns about walking dogs after the outbreak compared to before the outbreak? Specifically?

- Do you think the quality of life for the dogs went up or down during the epidemic especially during the closure? What were the overall benefits and drawbacks?

Figure A3. Broad question set for the semi-structured in-depth interviews.

References

1. Hassell, J.M.; Begon, M.; Ward, M.J.; Fèvre, E.M. Urbanization and disease emergence: Dynamics at the wildlife–livestock–human interface. *Trends Ecol. Evol.* **2017**, *32*, 55–67. [PubMed]
2. Karesh, W.B.; Dobson, A.; Lloyd-Smith, J.O.; Lubroth, J.; Dixon, M.A.; Bennett, M.; Aldrich, S.; Harrington, T.; Formenty, P.; Loh, E.H. Ecology of zoonoses: Natural and unnatural histories. *Lancet* **2012**, *380*, 1936–1945. [PubMed]
3. Nguyen, T.T.; Abdelrazek, M.; Nguyen, D.T.; Aryal, S.; Nguyen, D.T.; Reddy, S.; Nguyen, Q.V.H.; Khatami, A.; Nguyen, T.T.; Hsu, E.B. Origin of novel coronavirus causing COVID-19: A computational biology study using artificial intelligence. *Mach. Learn. Appl.* **2022**, *9*, 100328. [PubMed]
4. Soulsbury, C.D.; White, P.C. Human–wildlife interactions in urban areas: A review of conflicts, benefits and opportunities. *Wildl. Res.* **2015**, *42*, 541–553.
5. Braun, B. Environmental issues: Writing a more-than-human urban geography. *Prog. Hum. Geogr.* **2005**, *29*, 635–650.
6. Middle, I. Between a dog and a green space: Applying ecosystem services theory to explore the human benefits of off-the-leash dog parks. *Landsc. Res.* **2019**, *45*, 81–94.
7. Yin, D.; Gao, Q.; Zhu, H.; Li, J. Public perception of urban companion animals during the COVID-19 outbreak in China. *Health Place* **2020**, *65*, 102399.
8. Emel, J.; Urbanik, J. *Animal Geographies: Exploring the Spaces and Places of Human-Animal Encounters*; Lantern Books: New York, NY, USA, 2010.
9. Barua, M. Volatile ecologies: Towards a material politics of human—Animal relations. *Environ. Plan. A* **2014**, *46*, 1462–1478.
10. Urbanik, J.; Morgan, M. A tale of tails: The place of dog parks in the urban imaginary. *Geoforum* **2013**, *44*, 292–302.
11. Schneider, T.J.; Maguire, G.S.; Whisson, D.A.; Weston, M.A. Regulations fail to constrain dog space use in threatened species beach habitats. *J. Environ. Plan. Manag.* **2020**, *63*, 1022–1036.
12. Carmona, M. The "public-isation" of private space–towards a charter of public space rights and responsibilities. *J. Urban. Int. Res. Placemak. Urban Sustain.* **2022**, *15*, 133–164.
13. Donaldson, S. Animal agora: Animal citizens and the democratic challenge. *Soc. Theory Pract.* **2020**, *46*, 709–735.
14. Shingne, M.C. The more-than-human right to the city: A multispecies reevaluation. In *Animals in the City*; Routledge: Abingdon, UK, 2021; pp. 30–55.
15. Gómez, E.; Baur, J.W.; Malega, R. Dog park users: An examination of perceived social capital and perceived neighborhood social cohesion. *J. Urban Aff.* **2018**, *40*, 349–369.
16. Paköz, M.Z.; Sözer, C.; Doğan, A. Changing perceptions and usage of public and pseudo-public spaces in the post-pandemic city: The case of Istanbul. *Urban Des. Int.* **2022**, *27*, 64–79.
17. Appiah, E.; Enyetornye, B.; Ofori, V.; Enyetornye, J.; Kwamena Abbiw, R. Health and economic consequences: How COVID-19 affected households with pet and their pets. A systematic review. *Cogent Soc. Sci.* **2022**, *8*, 2060542.
18. Wang, S.; Li, A. Demographic Groups' Differences in Restorative Perception of Urban Public Spaces in COVID-19. *Buildings* **2022**, *12*, 869.
19. Noszczyk, T.; Gorzelany, J.; Kukulska-Kozieł, A.; Hernik, J. The impact of the COVID-19 pandemic on the importance of urban green spaces to the public. *Land Use Policy* **2022**, *113*, 105925.
20. Bustamante, G.; Guzman, V.; Kobayashi, L.C.; Finlay, J. Mental health and well-being in times of COVID-19: A mixed-methods study of the role of neighborhood parks, outdoor spaces, and nature among US older adults. *Health Place* **2022**, *76*, 102813.
21. Feng, X.; Astell-Burt, T. Perceived Qualities, Visitation and Felt Benefits of Preferred Nature Spaces during the COVID-19 Pandemic in Australia: A Nationally-Representative Cross-Sectional Study of 2940 Adults. *Land* **2022**, *11*, 904.
22. Jiang Haiyan, H.F.; Liu, W.; Ma, Y. Research Progress on Privately Owned Open Space and Enlightenment to Publicization of Affiliated Green Space. *Urban Stud.* **2020**, *27*, 7–11.
23. Wang Yiming, C.J. Dimensions of the Publicness of Urban Space and Defining Features of "Public" and "Private" in Western Research. *Urban Plan. Int.* **2017**, *32*, 59–67.
24. Evenson, K.R.; Wen, F.; Hillier, A.; Cohen, D.A. Assessing the contribution of parks to physical activity using GPS and accelerometry. *Med. Sci. Sports Exerc.* **2013**, *45*, 1981. [PubMed]
25. Evenson, K.R.; Wen, F.; Golinelli, D.; Rodríguez, D.A.; Cohen, D.A. Measurement properties of a park use questionnaire. *Environ. Behav.* **2013**, *45*, 526–547.
26. Cohen, D.A.; Sehgal, A.; Williamson, S.; Marsh, T.; Golinelli, D.; McKenzie, T.L. New recreational facilities for the young and the old in Los Angeles: Policy and programming implications. *J. Public Health Policy* **2009**, *30*, S248–S263. [PubMed]
27. Ruslin, R.; Mashuri, S.; Rasak, M.S.A.; Alhabsyi, F.; Syam, H. Semi-structured Interview: A Methodological Reflection on the Development of a Qualitative Research Instrument in Educational Studies. *IOSR J. Res. Method Educ. IOSR-JRME* **2022**, *12*, 22–29.
28. Flick, U. *An Introduction to Qualitative Research*; SAGE: Newbury Park, CA, USA, 2018.
29. Lu, Y.; Qiu, J.; Jin, C.; Gu, W.; Dou, S. The influence of psychological language words contained in microblogs on dissemination behaviour in emergency situations–mediating effects of emotional responses. *Behav. Inf. Technol.* **2022**, *41*, 1337–1356.
30. Sharag-Eldin, A.; Ye, X.; Spitzberg, B.; Tsou, M.-H. The role of space and place in social media communication: Two case studies of policy perspectives. *J. Comput. Soc. Sci.* **2019**, *2*, 221–244.
31. Sargeant, J. Qualitative research part II: Participants, analysis, and quality assurance. *J. Grad. Med. Educ.* **2012**, *4*, 1–3.

32. Mayen Huerta, C.; Cafagna, G. Snapshot of the use of urban green spaces in Mexico City during the COVID-19 pandemic: A qualitative study. *Int. J. Environ. Res. Public Health* **2021**, *18*, 4304.

33. Collins, C.; Haase, D.; Heiland, S.; Kabisch, N. Urban green space interaction and wellbeing–investigating the experience of international students in Berlin during the first COVID-19 lockdown. *Urban For. Urban Green.* **2022**, *70*, 127543.

34. Dwyer, F.; Bennett, P.C.; Coleman, G.J. Development of the Monash dog owner relationship scale (MDORS). *Anthrozoös* **2006**, *19*, 243–256. [CrossRef]

35. Cutt, H.E.; Giles-Corti, B.; Knuiman, M.W.; Pikora, T.J. Physical activity behavior of dog owners: Development and reliability of the Dogs and Physical Activity (DAPA) tool. *J. Phys. Act. Health* **2008**, *5*, S73–S89.

36. Michie, S.; Van Stralen, M.M.; West, R. The behaviour change wheel: A new method for characterising and designing behaviour change interventions. *Implement. Sci.* **2011**, *6*, 42. [CrossRef] [PubMed]

37. Day, B.H. The value of greenspace under pandemic lockdown. *Environ. Resour. Econ.* **2020**, *76*, 1161–1185. [CrossRef] [PubMed]

38. Derks, J.; Giessen, L.; Winkel, G. COVID-19-induced visitor boom reveals the importance of forests as critical infrastructure. *For. Policy Econ.* **2020**, *118*, 102253. [CrossRef] [PubMed]

39. Grima, N.; Corcoran, W.; Hill-James, C.; Langton, B.; Sommer, H.; Fisher, B. The importance of urban natural areas and urban ecosystem services during the COVID-19 pandemic. *PLoS ONE* **2020**, *15*, e0243344.

40. Kleinschroth, F.; Kowarik, I. COVID-19 crisis demonstrates the urgent need for urban greenspaces. *Front. Ecol. Environ.* **2020**, *18*, 318. [CrossRef] [PubMed]

41. Venter, Z.S.; Aunan, K.; Chowdhury, S.; Lelieveld, J. COVID-19 lockdowns cause global air pollution declines. *Proc. Natl. Acad. Sci. USA* **2020**, *117*, 18984–18990. [CrossRef]

42. Venter, Z.S.; Barton, D.N.; Gundersen, V.; Figari, H.; Nowell, M. Urban nature in a time of crisis: Recreational use of green space increases during the COVID-19 outbreak in Oslo, Norway. *Environ. Res. Lett.* **2020**, *15*, 104075.

43. Hoffman, C.L. The experience of teleworking with dogs and cats in the United States during COVID-19. *Animals* **2021**, *11*, 268. [CrossRef]

44. Gaunet, F.; Pari-Perrin, E.; Bernardin, G. Description of dogs and owners in outdoor built-up areas and their more-than-human issues. *Environ. Manag.* **2014**, *54*, 383–401.

45. Norton, B.G. Environmental ethics and weak anthropocentrism. *Environ. Ethics* **1984**, *6*, 131–148.

46. Oliva, J.L.; Johnston, K.L. Puppy love in the time of Corona: Dog ownership protects against loneliness for those living alone during the COVID-19 lockdown. *Int. J. Soc. Psychiatry* **2021**, *67*, 232–242. [PubMed]

47. Parente, G.; Gargano, T.; Di Mitri, M.; Cravano, S.; Thomas, E.; Vastano, M.; Maffi, M.; Libri, M.; Lima, M. Consequences of COVID-19 lockdown on children and their pets: Dangerous increase of dog bites among the paediatric population. *Children* **2021**, *8*, 620.

48. Bowen, J.; García, E.; Darder, P.; Argüelles, J.; Fatjó, J. The effects of the Spanish COVID-19 lockdown on people, their pets, and the human-animal bond. *J. Vet. Behav.* **2020**, *40*, 75–91.

49. Van Stipriaan, B.; Kearns, R.A. Bitching about a billboard: Advertising, gender and canine (re) presentations. *N. Z. Geogr.* **2009**, *65*, 126–138. [CrossRef]

50. Limaye, R.J.; Holroyd, T.A.; Blunt, M.; Jamison, A.F.; Sauer, M.; Weeks, R.; Wahl, B.; Christenson, K.; Smith, C.; Minchin, J. Social media strategies to affect vaccine acceptance: A systematic literature review. *Expert Rev. Vaccines* **2021**, *20*, 959–973.

MDPI

St. Alban-Anlage 66

4052 Basel

Switzerland

www.mdpi.com

Land Editorial Office

E-mail: land@mdpi.com

www.mdpi.com/journal/land